Weathering as a Predisposing Factor to Slope Movements

Geological Society books refereeing procedures

The Society makes every effort to ensure that the scientific and production quality of its books matches that of its journals. Since 1997, all book proposals have been refereed by specialist reviewers as well as by the Society's Books Editorial Committee. If the referees identify weaknesses in the proposal, these must be addressed before the proposal is accepted.

Once the book is accepted, the Society Book Editors ensure that the volume editors follow strict guidelines on refereeing and quality control. We insist that individual papers can only be accepted after satisfactory review by two independent referees. The questions on the review forms are similar to those for *Journal of the Geological Society*. The referees' forms and comments must be available to the Society's Book Editors on request.

Although many of the books result from meetings, the editors are expected to commission papers that were not presented at the meeting to ensure that the book provides a balanced coverage of the subject. Being accepted for presentation at the meeting does not guarantee inclusion in the book.

More information about submitting a proposal and producing a book for the Society can be found on its web site: www.geolsoc.org.uk.

It is recommended that reference to all or part of this book should be made in one of the following ways:

Calcaterra, D. & Parise, M. (eds) 2010. *Weathering as a Predisposing Factor to Slope Movements.* Geological Society, London, Engineering Geology Special Publications, **23**.

Yamakoshi, T., Mathys, N. & Klotz, S. Visual observation of erosion processes on the Black Marl badlands in the southern Alps, France. *In*: Calcaterra, D. & Parise, M. (eds) *Weathering as a Predisposing Factor to Slope Movements.* Geological Society, London, Engineering Geology Special Publications, **23**, 201–212.

GEOLOGICAL SOCIETY ENGINEERING GEOLOGY SPECIAL PUBLICATION NO. 23

Weathering as a Predisposing Factor to Slope Movements

EDITED BY

D. CALCATERRA
Federico II University of Naples, Italy

and

M. PARISE
National Research Council of Italy

2010
Published by
The Geological Society
London

THE GEOLOGICAL SOCIETY

The Geological Society of London (GSL) was founded in 1807. It is the oldest national geological society in the world and the largest in Europe. It was incorporated under Royal Charter in 1825 and is Registered Charity 210161.

The Society is the UK national learned and professional society for geology with a worldwide Fellowship (FGS) of over 9000. The Society has the power to confer Chartered status on suitably qualified Fellows, and about 2000 of the Fellowship carry the title (CGeol). Chartered Geologists may also obtain the equivalent European title, European Geologist (EurGeol). One fifth of the Society's fellowship resides outside the UK. To find out more about the Society, log on to www.geolsoc.org.uk.

The Geological Society Publishing House (Bath, UK) produces the Society's international journals and books, and acts as European distributor for selected publications of the American Association of Petroleum Geologists (AAPG), the Indonesian Petroleum Association (IPA), the Geological Society of America (GSA), the Society for Sedimentary Geology (SEPM) and the Geologists' Association (GA). Joint marketing agreements ensure that GSL Fellows may purchase these societies' publications at a discount. The Society's online bookshop (accessible from www.geolsoc.org.uk) offers secure book purchasing with your credit or debit card.

To find out about joining the Society and benefiting from substantial discounts on publications of GSL and other societies worldwide, consult www.geolsoc.org.uk, or contact the Fellowship Department at: The Geological Society, Burlington House, Piccadilly, London W1J 0BG: Tel. +44 (0)20 7434 9944; Fax +44 (0)20 7439 8975; E-mail: enquiries@geolsoc.org.uk.

For information about the Society's meetings, consult *Events* on www.geolsoc.org.uk. To find out more about the Society's Corporate Affiliates Scheme, write to enquiries@geolsoc.org.uk.

Published by The Geological Society from:
The Geological Society Publishing House, Unit 7, Brassmill Enterprise Centre, Brassmill Lane, Bath BA1 3JN, UK

(*Orders*: Tel. +44 (0)1225 445046, Fax +44 (0)1225 442836)
Online bookshop: www.geolsoc.org.uk/bookshop

British Library Cataloguing in Publication Data

A catalogue record for this book is available from the British Library.
ISBN 978-1-86239-297-7

Typeset by Techset Composition Ltd, Salisbury, UK
Printed by CPI Antony Rowe, Chippenham, UK

Distributors

North America
For trade and institutional orders:
The Geological Society, c/o AIDC, 82 Winter Sport Lane, Williston, VT 05495, USA
Orders: Tel. +1 800-972-9892
Fax +1 802-864-7626
E-mail: gsl.orders@aidcvt.com

For individual and corporate orders:
AAPG Bookstore, PO Box 979, Tulsa, OK 74101-0979, USA
Orders: Tel. +1 918-584-2555
Fax +1 918-560-2652
E-mail: bookstore@aapg.org
Website: http://bookstore.aapg.org

India
Affiliated East-West Press Private Ltd, Marketing Division, G-1/16 Ansari Road, Darya Ganj, New Delhi 110 002, India
Orders: Tel. +91 11 2327-9113/2326-4180
Fax +91 11 2326-0538
E-mail: affiliat@vsnl.com

Contents

Weathering as a predisposing factor to slope movements: an introduction

Domenico Calcaterra[1]* & Mario Parise[2]

[1] Department of Hydraulic, Geotechnical and Environmental Engineering, Section of Engineering Geology, Federico II University of Naples, Naples, Italy
[2] National Research Council of Italy, IRPI, Bari, Italy
**Corresponding author (e-mail: domenico.calcaterra@unina.it)*

Recognition of weathering as a primary process in the development of landforms dates back at least to the late 19th and early 20th century, as indicated by several contributions from various countries (e.g. Cortese 1895; Branner 1896; Falconer 1911; Jutson 1914; Walther 1915); nevertheless, it was only in the second half of the last century that systematic studies on weathering and related topics such as geomorphology, engineering geology and petrography were carried out. In a recent review paper Ehlen (2005) found through an on-line search more than 9000 hits where the term 'weathering' was used, dating from the mid-1950s to the beginning of the 21st century. Such a huge number of citations clearly demonstrates the attention that the scientific community has dedicated to the in-place breakdown of rocks by chemical, physical and biological processes. However, despite the high frequency of landslides and erosional phenomena in weathered materials, and the damage and casualties they repeatedly cause, not very much is known about the direct and indirect relationships between weathering and slope movements. The matter is further complicated by the high variability of landslide features (Working Party on World Landslide Inventory 1993). According to the local conditions, a variety of slope movements may take place in the weathered rock masses. Shallow soil slips evolving to rapid and catastrophic debris flows are probably the most common type in steep residual soil slopes, whereas thicker failures, also including deep-seated complex movements, can affect large volumes of weathered rocks.

Following the first, simplified attempts to classify slides in natural slopes of residual soils (e.g. Morgenstern & de Matos 1975; Vargas & Pichler 1975), Durgin (1977) provided a comprehensive scheme of the relationships existing between landslides and weathering (Table 1). In general terms, studies on weathering-related landslides showed that in the less weathered rock volumes joints preferentially control the occurrence of instability phenomena such as rock falls and toppling failures, whereas in soil-like terrains failures are generally controlled by the interactions between the constituent elements (e.g. grains, clasts, blocks).

Although the fundamental geological and geomorphological features of weathering-related slope movements have been recognized with respect to both soil-like and rock-like materials, little attention has been paid to the mechanics and kinematics of the most important instability typologies in weathered materials. In this respect, in recent times, matric suction has been usefully considered in some studies dealing with the weathered cover, following the basic assumptions of the unsaturated soil theory (Fredlund & Morgenstern 1977). Matric suction, or negative pore pressure, increases the apparent strength of a soil; its role in controlling the stability of steep residual soil slopes after heavy and prolonged rainfall has been analysed in regional settings worldwide (e.g. Rahardjo *et al.* 2004; Cascini *et al.* 2006; and papers in this volume by **Alonso *et al.*** and **Picarelli & Di Maio**), thus representing a promising direction toward a better comprehension of the behaviour of weathered slopes and the occurrence and development of the related instabilities.

This Special Publication brings together papers presented at the Fourth General Assembly of the European Geosciences Union in Vienna, 15–20 April 2007, together with invited contributions. In this volume, following previous work (see Parise *et al.* 2004; Ollier *et al.* 2007), geomorphologists, engineering geologists and geotechnical engineers contribute to an up-to-date overview of approaches, methods and techniques devoted to better understanding the weathering conditions of rock masses on slopes.

The book includes worldwide case-studies, in which a variety of geological and geomorphological settings are treated, and is divided into three main sections, dealing with broad aspects of the weathering–landslides issue, and the analysis of slope movements in igneous or metamorphic and sedimentary weathered rocks.

The intention of the editors was to consider the subject from a different point of view than the usual climatic perspective in which the response to weathering agents is related to the specific environmental conditions typical of low, intermediate and high latitudes, respectively; such a

From: CALCATERRA, D. & PARISE, M. (eds) *Weathering as a Predisposing Factor to Slope Movements.*
Geological Society, London, Engineering Geology Special Publications, **23**, 1–4.
DOI: 10.1144/EGSP23.1 0267-9914/10/$15.00

Table 1. *Weathering–mass movement relationships according to Durgin (1977) as reported by* ***Migoń***

Stage of weathering	*Mass movements*
Saprolite (thoroughly decomposed rock)	Rotational slides, slumps
Decomposed granite (<15% of fresh rock left)	Debris flows, debris avalanches, debris slides
Corestone stage (15–85% of fresh rock left)	Rock fall avalanches, rolling rocks
Fresh rock (<15% of weathered material along joints)	Rock falls, rock slides, block glides; debris avalanches and slides over sheeting surfaces

scheme is based on studies such as that by Peltier (1950), where, starting from nine morphogenetic regions, eight weathering regions were set up, obtained by adding the intensity of both chemical and frost weathering in relation to rainfall and temperature.

At the same time, the nature of bedrock is commonly recognized as one of the main factors controlling the formation of weathered mantles and related landforms. However, although the main weathering-affected regions of the world have been adequately studied from the morphogenetic standpoint, rock weathering research has been mainly focused on granites and limestones, neglecting other rock types (e.g. marls, sandstones, low-grade metamorphic rocks), which are only occasionally dealt with from a weathering perspective.

In this Special Publication, the first, introductory section is opened by **Ollier** with a paper concerned with the principles of deep weathering. The author, claiming that 'perhaps the greatest hazard in deep weathering is the failure to recognize it', highlights that most geologists and engineers do not appreciate the great depths that weathering can attain. Further attention is paid to the current terminology, which, after more than 50 years of studies on weathering, is far from being unanimously shared; in this respect, the terminological confusion between soil, saprolite and regolith, discussed, among others, by Ehlen (2005) and Ollier *et al.* (2007), is typical.

The famous Po Shan landslide (which caused 67 fatalities) that occurred in 1972 in Hong Kong is used by **Ollier** as an example of the relationships between deep weathering and mass movements. This landslide induced the Hong Kong Government to systematically tackle the landslide hazard through a specific department, previously known as the Geotechnical Control Office, at present the Geotechnical Engineering Office (GEO). Hong Kong is also the focus of the paper by **Hencher & Lee**, who review the nature and mechanics of landslides in the weathered terrain there. In this paper, the authors present a relationship that links intensity of landsliding to the 24 h rainfall. The gradual deterioration and internal erosion of slopes prior to detachment is discussed, with emphasis on the growth of natural piping systems and infilling of dilated fracture networks, which are recommended as important indicators of landslide development. The conclusions are supported by case examples of slope failures, the study of some of which has been taken to a forensic level.

A wide review dealing with the contribution of mass movement to the development of landforms in weathered granite and gneiss terrains is offered by **Migoń**, using examples from SE Brazil, southern Mexico, SE China and elsewhere. A summary relationship is discussed, related to the weathering classification scheme by Ruxton & Berry (1957) and the Geological Society Engineering Group Working Party (1990). As described by **Migoń**, the style of slope failures varies depending on the local relief and the grade of weathering (Fig. 1). Shallow debris slides and boulder falls are common in less weathered rock and steep terrains, whereas rotational slides appear to dominate in highly weathered rock and in more subdued topography. In certain topographic circumstances, slides may turn into debris flows or earthflows.

Two contributions in the first section have the theme of the behaviour of fissured, weathered clayey rocks.

Alonso *et al.* analyse the strength degradation of two marl formations from the Iberian peninsula, which show similar responses as a consequence of physical weathering processes (suction changes and stress changes). According to **Alonso *et al.***, unloading changes effective stresses and leads to swelling of rocks if active minerals are present, whereas water content changes (suction changes) induce plastic deformations producing fissuring of the rock. Through an extensive laboratory test campaign, partly related to a foundation–rock interface of a gravity dam and to a tunnel, the evolution in behaviour of the studied formations is demonstrated along with implications for weathering-related slope instability. The paper by **Picarelli & Di Maio**, based upon laboratory testing, field observations and 'pure speculation' (the authors' own words), examines some processes of deterioration of stiff overconsolidated clays and clay shales, which, in turn, are considered among the main causes of landslides in fine-grained soils. In particular, deterioration localized within soil bands or involving large soil masses is described, with reference to its mechanical or physicochemical causes. Again, confusion in terminology is highlighted: softening and strain-softening are frequently mentioned as the main causes of deterioration, as well as other phenomena, such as slaking, destructuring or fatigue.

In the second section, devoted to landslides in igneous and/or metamorphic rocks, four papers are dedicated to weathered magmatic rocks and related slope movements, treating case studies from Italy, Ethiopia, the USA and Brazil.

Calcaterra & Parise review about 40 years of studies focused on landslides in weathered plutonic and metamorphic rocks of the Calabria region in southern Italy. There, some researchers have recognized an ancient

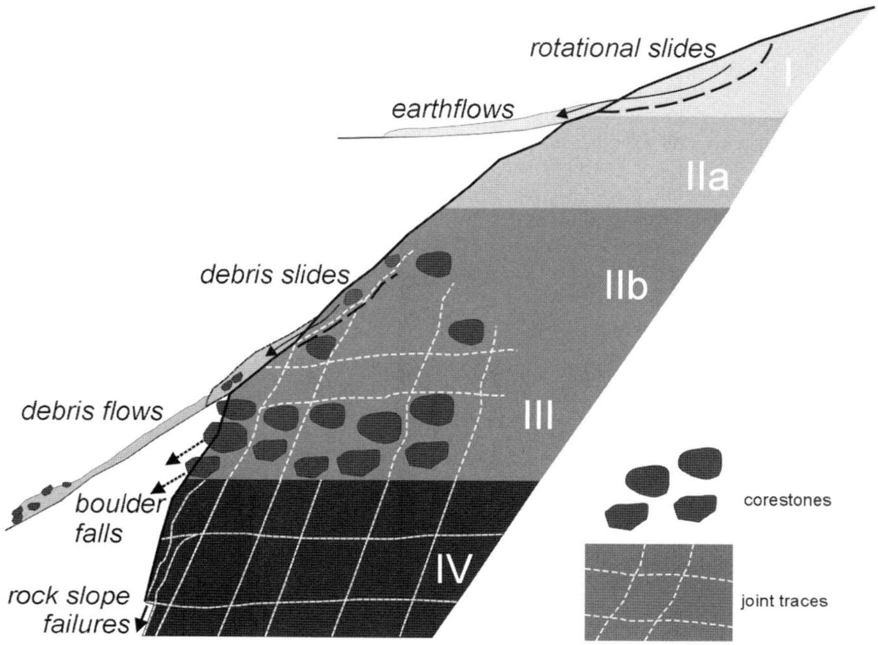

Fig. 1. Weathering grade-mass movement relationships in basement rocks, as described by **Migoń**.

(Tertiary) weathering, which could be regarded as one of the main reasons for the depth of the weathered mantle: in some districts, the depth to unweathered rocks is as great as 150–200 m, displaying a complex profile. On the Calabrian slopes almost all the main typologies of mass movements have been recognized, ranging from shallow soil slips to extremely rapid debris flows and to slow-moving deep-seated gravitational slope deformations. In many cases, development of landslides has to be related to the availability of weathered materials on the slopes, and landsides are triggered by the main factors active in that region, from seismic shocks to intense and clustered rainfall events and, last but not least, anthropogenic actions.

Delmonaco *et al.* consider the weathering processes that affect the stability of the rock-hewn Orthodox churches of Lalibela, in Ethiopia, cut out of weathered basalts and volcanic tuffs some 800 years ago. Field surveys, *in situ* geotechnical analyses and laboratory tests have revealed that water represents the main causative factor of damage, in terms of direct rainfall, soil infiltration, capillarity and diffuse humidity; an important role is played by the presence of montmorillonite in the weathered volcanic materials. Major consequences are the alveolar weathering of the churches' façade, the degradation of the roofs, a reduction in the rock strength and sliding of façades along bedding joints.

Latham *et al.* discuss a case study from Haywood County, North Carolina, USA. The Hunters Crossing landslide, a slow-moving, weathered rock slide, involves a small community of condominiums in the town of Waynesville, where a complex suite of metamorphic rocks (migmatitic metagreywacke, schist, gneiss) crops out. After a detailed site investigation campaign, it appeared that the failure surface was potentially located at about 11 m below the ground surface, possibly at the contact between saprolite or completely decomposed rock and partially decomposed rock. However, inclinometers installed at two locations on the slope have not detected enough movement to corroborate that assessment.

Based upon some decades of specific research, **Lacerda** reviews the main properties of the Brazilian saprolitic and lateritic soils, with special attention paid to shear strength parameters, a basic topic to better comprehend the mechanics of landslides. **Lacerda** concludes his paper by stating that many slides in residual soils are directly related to relict structures inherited from the parent rock, and the residual strength is relevant in these cases. Moreover, the existence of 'true' cohesion in lateritic soils is particularly important in the initiation of shallow slides.

Three case studies dealing with sedimentary weathered formations are presented in the third section of the book.

The interrelationship between slope deformation and fault-induced weathering as a preparatory factor for the development of sliding is analysed by **Pánek *et al.*** through several case studies from the Western Carpathians in the Czech Republic. The study area comprises flysch nappes with alternating sandstone and shale of different permeability. By means of geomorphological analyses, the authors demonstrate how weathered fault zones influenced the development of some of the largest and most active slope failures in the study area, among which some deep-seated lateral spreading and complex movements in massive bedrocks are described.

To clarify the erosion processes on a marly bare slope in the southern French Alps, the erosion processes on a steep and erodible slope composed of Black Marls formations were observed by **Yamakoshi *et al.*** using a time-lapse video camera. The observations revealed that miniature debris flows occurred at the time of the rainfall–runoff event during which the most severe erosion took place over the 3 month observation period. With the term 'miniature debris flows', the authors define small-scale flows inside the slope rills, where sometimes intermittent surges form, identical to debris flow in natural torrents.

Meisina closes the volume with a contribution on a new correlation between the residual shear strength and the methylene blue value for weathered clay soils on argillaceous bedrock and on alluvial soils in the Oltrepo Pavese area of Northern Italy; shallow landslides occur periodically in these soils as a result of high-intensity rainfall. The method, although not an alternative to the classical laboratory tests, is considered helpful for a rapid assessment of residual shear strength for design purposes and hazard studies.

Following four years (2004–2007) of organization of a specific symposium on 'Slope movements in weathered materials: recognition, analysis, and hazard assessment', within the framework of the General Assembly of the European Geosciences Union, our hope is that the present volume might provide useful hints on this issue, and encourage further efforts at both the local and the international level to move in the direction of gaining a more thorough knowledge about the way weathering acts in predisposing natural slopes to the occurrence of mass movements, thus contributing to the process of mitigating the risks related to natural hazards.

References

Branner, J. C. 1896. Decomposition of rocks in Brazil. *Geological Society of America Bulletin*, **7**, 255–314.

Cascini, L., Gullà, G. & Sorbino, G. 2006. Grounwater modelling of a weathered gneissic cover. *Canadian Geotechnical Journal*, **43**, 1153–1166.

Cortese, E. 1895. *Geologic description of Calabria*. Regio Ufficio Geologico d'Italia, Rome [in Italian].

Durgin, P. B. 1977. Landslides and the weathering of granitic rocks. *Reviews in Engineering Geology*, **3**, 127–131.

Ehlen, J. 2005. Above the weathering front: contrasting approaches to the study and classification of weathered mantle. *Geomorphology*, **67**, 7–21.

Falconer, J. D. 1911. *The Geology and Geography of Northern Nigeria*. Macmillan, London.

Fredlund, D. G. & Morgenstern, N. R. 1977. Stress state variables for unsaturated soils. *Journal of the Geotechnical Engineering Division, ASCE*, **103**, 466.

Geological Society Engineering Group Working Party 1990. Tropical residual soils; Geological Society Engineering Group Working Party Report. *Quarterly Journal of Engineering Geology*, **23**, 1–101.

Jutson, J. T. 1914. *An outline of the physiographic geology (physiography) of Western Australia*. Geological Survey of Western Australia, Bulletin, **61**.

Morgenstern, N. R. & De Matos, M. 1975. Stability of slopes in residual soils. *In*: *Proceedings of the 5th Panamerican Conference on Soil Mechanics and Foundation Engineering, Buenos Aires*, Vol. 3, 369–384.

Ollier, C. D., Calcaterra, D. & Parise, M. 2007. Studies in weathering and slope movements—an introduction. *Geomorphology*, **87**, 101–103.

Parise, M., Calcaterra, D. & Larsen, M. C. 2004. Introduction to Weathering and Slope Movement papers. *Quarterly Journal of Engineering Geology and Hydrogeology*, **37**, 76.

Peltier, L. 1950. The geographic cycle in periglacial regions as it is related to climatic geomorphology. *Annals of the Association of American Geographers*, **40**, 214–236.

Rahardjo, H., Aung, K. K., Leong, E. C. & Rezaur, R. B. 2004. Characteristics of residual soils in Singapore as formed by weathering. *Engineering Geology*, **73**, 157–169.

Ruxton, B. P. & Berry, L. 1957. Weathering of granite and associated erosional features in Hong Kong, *Geological Society of America Bulletin*, **68**, 1263–1282.

Vargas, M. & Pichler, E. 1975. Residual soil and rock slides in Santos, Brazil. *In*: *Proceedings of the 4th International Conference on Soil Mechanics and Foundation Engineering, London*, Vol. 2, 394–398.

Walther, J. 1915. Laterites in western Australia. *Zeitschrift für Geologische Gesellschaft*, **67B**, 113–140 [in German].

Working Party On World Landslide Inventory 1993. A suggested method for describing the activity of a landslide. *Bulletin of the International Association of Engineering Geology*, **47**, 53–57.

Very deep weathering and related landslides

C. D. Ollier

School of Earth and Environment, The University of Western Australia,
Crawley, WA 6009, Australia (e-mail: cliffol@cyllene.uwa.edu.au)

Abstract: Some huge landslides have occurred in deeply weathered rock, but most geologists and engineers do not appreciate the great depths that weathering can attain. Rocks can be weathered to depths of hundreds of metres, often in a very irregular manner. Fresh rock is converted to weathered rock called saprolite by isometric chemical alteration. Some saprolite is later eroded, and fresh rock appears in the landscape in distinctive landforms. This erosion of an irregularly weathered landscape is called stripping (meaning the stripping of saprolite from fresh rock). The age of the weathering may be measured in geological periods, as can the age of stripping, and may be related to past climate different from that of today. Deep weathering occurs basically by hydrolysis, which requires groundwater. Slope failure in saprolite depends on the engineering property of the material, and its relation to bedrock and corestones, and relationship to degree of weathering of adjacent saprolite. This paper emphasizes the deep weathering phenomenon itself, but a few examples are given to illustrate the variety of landslides in deeply weathered material.

Landslides can occur even on fresh rocks, and where weathering is only a few metres deep, but in this paper I wish to concentrate on very deep weathering, measured in tens or hundreds of metres. Many engineering reports describe deep weathering and very often the investigators are surprised and think they have come upon something exceptional.

Here I describe the phenomenon of deep weathering, speculate on the causes and history of deep weathering, and show that deep weathering is far from unusual. Deep weathering is not well treated in much geomorphological literature, and is largely absent in engineering literature. A recent book on *Geomorphology for Engineers* (Fookes *et al.* 2005) mentions in only one sentence that there are some deep weathering profiles in the tropics, and the words 'regolith' and 'saprolite' appear in only one figure and are never mentioned in the text. The general background can be applied to both landslides in natural situations and ground failure in engineering situations.

The global distribution of deep weathering

Deep weathering is particularly common in the tropics and there is a common idea that 'tropics' and 'deep weathering' go together. However, deep weathering profiles are also found in many areas outside the tropics. The list includes Sweden, Finland, Scotland, England, the USA, India, Australia and many more. Even in Inner Mongolia the Central China Goldfields announced in an unpublished report of 2007 that deep weathering had oxidized mineralized zones to a depth of 80 m, making gold more accessible.

It should also be noted that a lot of deep weathering is in what are now arid and semi-arid areas of the world. Stierman & Healy (1984) described granite weathered to a depth of 70 m in the Mojave Desert of southern California. An extremely unusual deep weathering profile (9.5 m) from the Dry Valleys of Antarctica has been described by Dickinson (2006).

An important point is that most deep weathering profiles were not formed under the present climate, but under different climates in the past. How far in the past? When European deep weathering remnants were first encountered researchers thought they might relate to interglacial periods. Then they were attributed to 'Tertiary' times, and eventually some were definitely dated to Mesozoic time.

Rock types

In eastern Australia deeply weathered Palaeozoic shales are the dominant source of clay for brickmaking. In SW England weathered granites are the source of commercial kaolin. In fact, almost all rocks except quartzite and limestone are susceptible to deep weathering, and granite is probably over-represented in the literature. Many more examples of deep weathering of granite have been described by Migoń (2006).

In Western Australia deep weathering affects all rocks, but the greenstone (generally metamorphosed basalt) has significantly different profiles from those on granite and gneiss, different surface soils, and carbonate nodules. Whereas granites normally weather mainly to kaolinite, greenstones can produce smectites, which with their higher water-holding capacity and swelling properties are more likely to be

From: CALCATERRA, D. & PARISE, M. (eds) *Weathering as a Predisposing Factor to Slope Movements*.
Geological Society, London, Engineering Geology Special Publications, **23**, 5–14.
DOI: 10.1144/EGSP23.2 0267-9914/10/$15.00

landslide-prone. From Scotland, weathered gabbro possibly up to 14 m thick has been described by Basham (1974).

An unusual engineering and weathering association was reported from investigations for the Channel Tunnel between England and France, where Chalk Marl was the rock of interest. Weathering was 35 m deep on the English coast, 40 m on the French coast, and 15 m in the central Channel. It was necessary to keep the tunnel at least 10 m below the base of weathering, but how was the weathering defined? Harris *et al.* (1996) have provided a detailed account of this engineering problem.

Weathering processes and products

Weathering profiles can be hundreds of metres deep, well beyond the influence of the weather at the ground surface, but the alteration process is still called weathering.

The products of weathering depend on the composition of the original rock. Granitic rocks consist essentially of quartz, feldspar and mica. The quartz remains unaltered whereas both the feldspar and mica turn to clay, so the weathering product (sometimes called grus) consists of sand and clay. Basalt consists of feldspar and a little olivine. All these minerals weather to clay, so there will be no sand in the weathering product. Fundamentally, chemical weathering is the reaction of minerals with air and water, so the presence of pores or cracks in the rock is important.

Physical weathering processes are generally unimportant in deep weathering: chemical alteration is paramount. The dominant process is hydrolysis, or reaction with water. More fundamentally, the process consists of replacing cations in the silicate minerals with hydrogen, which converts them to clay. Oxidation is important in the upper part of weathering profiles, especially in reactions with iron producing the red–brown colours of weathered rocks.

Carbonation is described in elementary textbooks as an important factor, but really it is important only in the soil zone. Carbon dioxide has no significant effect at lower levels and the product of deep weathering of silicate minerals is clay, not carbonate.

The Russian soil scientist Polynov (1937) regarded weathering as the pathology of rocks, and there is increasing evidence that he was right. The subject of biogeochemistry is burgeoning, and the role of micro-organisms is increasingly seen to be important. Inorganic chemistry alone cannot explain many observed features. Even minerals normally thought of as highly resistant are found to be etched by bacteria. Iron bacteria are plentiful, and aluminium may be largely mobilized by organisms. A useful geological summary has been provided by Ehrlich (1998), who pointed out that viable bacteria, algae, fungi and protozoa may be detected in subsurface samples between 10 and 50 m deep, but between depths of 250 m and several thousand metres only viable bacteria are found.

The final product is weathered rock, in place and with no volume change, called saprolite.

Weathering and hydrothermal alteration

For practical engineering purposes, the nature of the altered material is more important than the mechanism that created it, but scientists want to know whether alteration is by hydrothermal or weathering processes. In the Snowy Mountains Hydroelectric scheme in Australia, many kilometres of tunnels were drilled. Some of the tunnelling was in fresh rock, some in very weathered rock. A chief engineer at the time, Danny Moye, told me that in those days the geologists did not believe that weathering could be so deep or extensive, so the soft rock was attributed to hydrothermal alteration. This is still a popular supposition amongst geologists not familiar with deep weathering.

Another example is the formation of china clay in the granites of SW England. These had long been assumed to be of hydrothermal origin (e.g. Exley 1958), but Sheppard (1977) used isotope studies to show they were formed by weathering.

A more recent example comes from tunnel work in eastern Germany. Thuro & Scholz (2003) considered that the evidence shows two phases of hydrothermal alteration before weathering.

Hydrothermal alteration can cause some alteration of silicate minerals to clay, but should also be accompanied by mineralization. Increases in metals such as Au, Sn, Ag, Cu and Zn and cations such as S, Cl and F may be indicators. In some studies the presence of chlorite has been taken as evidence of hydrothermal alteration, but it can also be formed during weathering (Ollier 1983). Of course, I am not disputing that hydrothermal alteration occurs and may form clay bodies in some situations.

The weathering profile

Weathered rock can accumulate to thicknesses of many tens of metres and can be divided into various zones. The succession of zones from ground surface down to fresh rock is called the weathering profile. A typical profile on granite is shown in Figure 1 and reveals several interesting features. The soil zone and the saprolite together are called regolith, which may be defined as everything between fresh rock and fresh air. It is important to avoid terminological confusion between soil, saprolite and regolith. Pedologists, geomorphologists and farmers use the term 'soil' to denote the material at the top of the profile, usually containing soil fauna and plant roots. Engineers use 'soil' to mean all that is not hard rock, including saprolite. Hillside creep may produce a mobile zone in the upper metre or so, but the true saprolite is in place. The usage has been discussed further, together with different views on 'profiles', by Ollier *et al.* (2007).

Weathering penetrates down joints in the granite, isolating the blocks between the joints, which then weather to smaller and smaller blocks called corestones. Each corestone is bounded by its own weathering front, and the corestones

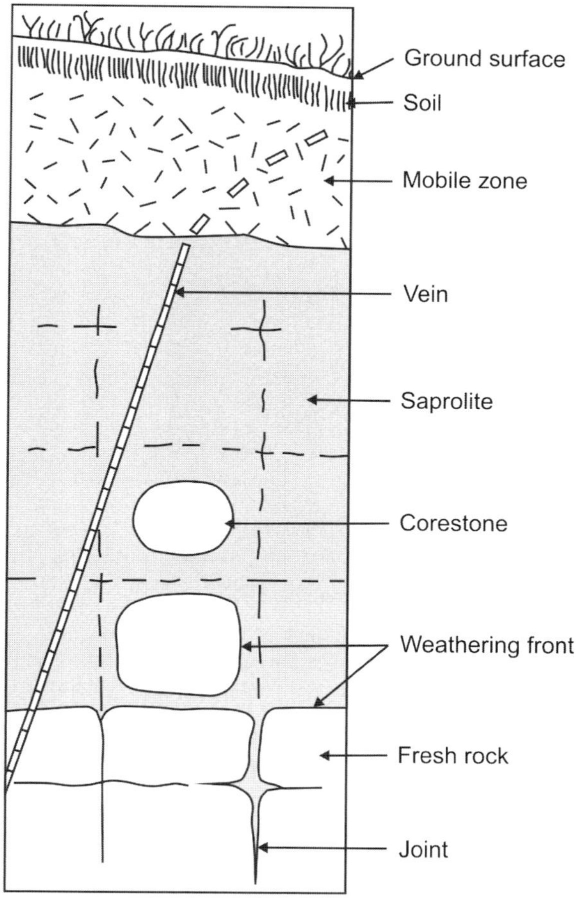

Fig. 1. Diagrammatic profile through deeply weathered granite.

(the cores of original joint blocks) are separated by saprolite (Fig. 2). They start as angular corestones, especially common near the base of the profile, and become more rounded as weathering progresses (Fig. 3). Some researchers have described the profile as if the corestones systematically decrease in size and number up the profile, such as Ruxton & Berry (1957), who described deep weathering in Hong Kong. My own observations in the same area were that the distribution of corestones can be very irregular, and it is possible to find large corestones in isolation high in a mass of saprolite.

Two features are often surprising to observers not familiar with deep weathering profiles. The first is that the weathering does not usually grade slowly into unweathered rock, but instead there is a knife-sharp contact between fresh and weathered rock called the weathering front. The second is that the joints, quartz veins, xenoliths and other structures retain their original position (Fig. 4), showing that there has been no change in volume during weathering, so some materials have been lost in solution. Secondary minerals are often pseudomorphs of the primary minerals they replaced. Weathered rock with no volume change is called saprolite, and apart from soil in the top metre or so, which is disturbed by roots, animals, hillside creep and similar processes, the profile consists of saprolite and perhaps corestones of unaltered rock.

Groundwater fills voids in rocks and weathered rocks, and under gravity it naturally accumulates at the bottom of the weathered profile. Water pressure in saprolite can be an important feature in landsliding. If there is permanent groundwater, the lower saprolite is saturated and anaerobic conditions prevail. The saprolite tends to be grey or green in contrast to the red and brown of the oxidized zone above. In such a profile the base of the groundwater corresponds to the deepest weathering front, and the water table corresponds to the change from green to red. In supergene economic deposits (enriched by weathering) oxidized ores are formed above the water table, and reduced ores (usually sulphides) below the water table.

Occasionally the terms groundwater base and water table are confused in the literature, so some caution is needed. For example, Browne (1964) described the origin of tors by saprolite stripping like this. 'It is therefore most probable that . . . tor topography originated through prolonged differential chemical weathering extending from the surface down to the *water-table where weathering ceases to operate*' (my italics). Clearly he meant groundwater base.

Geomorphologists and engineers have devised many schemes to describe deep weathering profiles, using factors such as corestones, colour, mineralogy, hardness and a host of engineering properties. These will not be reviewed here, but guides to some have been provided by Ollier *et al.* (2007).

Depth of weathering

About a year ago we had a visitor from the USA who gave us a talk on deep weathering. He talked of a weathering profile 2 m deep. We Australians were not impressed, as we are used to profiles measured in hundreds of metres. Clearly, people in different parts of the world have different concepts of what deep weathering is, and they have different problems. Examples of very deep weathering are reported in other parts of the present paper, and here I stress the unequivocal evidence for hundreds of metres of weathering provided by economic deposits.

The people who know most about deep weathering are not engineers or geomorphologists, but economic geologists. There is a group of ore bodies formed by supergene enrichment, and the study of these, with ground truth confirmed by actual mining, reveals more about the depth of weathering than any other study. A supergene ore body may have the following zones: the upper part is the oxidized zone, which in turn may be divided into a leached zone at the top and a zone of oxide ores below, usually becoming richer towards

Fig. 2. A deep weathering profile in Hong Kong, photographed in the 1960s when such profiles were not protected by shotcrete. Isolated corestones, some angular, some rounded, are separated by saprolite, and the joint pattern is preserved in the weathered rock. (Photograph by C. D. Ollier.)

the base. The lower part is the enriched sulphide zone, where the richest ore is usually found. This has been derived from weathering of the material above, and precipitated in the reducing conditions below the water table. The rich sulphide zone may merge into protore below or, more rarely, have a sharp lower boundary.

Fig. 3. The upper part of a deep weathering profile in Sweden, showing a large and well-rounded boulder in saprolite. (Photograph by C. D. Ollier.)

Fig. 4. A deeply weathered profile in granite near Mount Kosciusko, Australia. A large corestone is preserved in almost featureless saprolite, but the aplite vein on the right shows that the material is all in place except for the top metre, and results from isovolumetric alteration. Such deep weathering would not occur under the present climate and topography, and is inherited, in this case from the Mesozoic. (Photograph by C. D. Ollier.)

Thicknesses of oxidized and sulphide zones are measures of the thicknesses of the oxidized and reduced zones of weathering, and the total depth to protore can be regarded as the thickness of the weathering profile. The oxidized zone normally ranges from a few metres to a hundred metres or so in thickness. It is greater still in a number of places including Kennecott, Alaska (600 m), Tintic, Utah (600 m), Zambia (600+ m), Zimbabwe (900 m) and Tsumeb, Namibia (1000 m). Thick sulphide zones include those at Ely, Nevada (150 m), Butte, Montana (300 m), Morenci, Arizona (300 m) and Bingham, Utah (420+ m) (Lindgren 1933).

The age of deep weathering and stripping

After rocks have been deeply weathered, erosion may remove the weathered rock, bringing corestones to the ground surface and eventually even exposing the fresh rock at the base of the weathering profile. The process of partial or complete removal of the weathered rock is called stripping, and it may produce distinctive landforms and landscapes. Some workers, such as Thomas (1994), believe that weathering and stripping may continue together simultaneously, and the local landscape depends on the relative effects of the two opposed processes. Others think there was sometimes a distinct period of ancient weathering, and stripping set in later, perhaps associated with tectonic uplift, creation of new continental margins, or climatic change. This is perhaps an oversimplification, because weathering never stops, but there are whole landscapes where an ancient phase of weathering provides the base on which subsequent landscapes evolve.

One illustrative example comes from Victoria, Australia, where the old gold mines revealed many ancient river sediments (deep leads), which can be dated by overlying basalt formed when lava flowed down the valleys and buried the alluvium. Modern rivers in the area have a mixed pebble load of whatever is in the catchment: basalt, granite, sandstone and so forth. However, such rocks become rarer as the sediments become older, and oldest leads contain almost nothing but quartz, and gold. It seems that when the oldest gravels were laid down the whole landscape was so deeply weathered that only quartz veins were providing pebble-sized material to streams (Ollier 1992*a*). As time went on more and more saprolite was removed, and more fresh rock exposed to erosion. Further examples of the separation of weathering and stripping are given in the next section on planation surfaces.

Another example comes from Uganda (Ollier 1992*b*). Very deep weathering occurs on the African Planation Surface, but it thins to the north and the Acholi Surface, topographically continuous with the African Surface, is stripped down to fresh bedrock. Clearly, the regolith preserved on the African Surface was not produced under present conditions.

A third example comes from the Northern Territory of Australia (Wright 1963). Here there are three erosion surfaces. The highest and oldest is the Bradshaw Surface, below which rocks are deeply weathered to about 100 m, with a lower pallid zone that is silicified near the base. The Maranboy is an intermediate surface produced by stripping of the upper, less silicified parts of the Bradshaw weathering profile. The Tipperary Surface is the lowest surface, produced by removal of part of the Maranboy Surface, and is cut across silicified lower Bradshaw weathering profile and relatively fresh rock. The Bradshaw deep weathering was clearly a distinct event, and the saprolite of the Bradshaw profile was the geological basis that was later eroded to produce the present landscape.

Rocks have been deeply weathered at many different times. Evidence for the age is most positively obtained from stratigraphic relationships. Migoń & Lidmar-Bergström (2001) provided many examples from Europe. In the French Massif Central weathered gneiss 50 m thick is overlain by early Jurassic sediments. A 60 m thick saprolite on granite and gneiss in the Bohemian Massif is covered by Oligocene sediments. Lidmar-Bergström (1988) provided a famous example from Sweden. The granite bedrock was deeply weathered, to form a mass of kaolinite with occasional corestones. A Cretaceous sea then invaded the area, and Cretaceous oysters are found on the corestones. in what is today an economic kaolin pit.

About half of the state of Minnesota is covered by an old saprolite, and one of the classic studies of weathering was of the Morton Gneiss in Minnesota (Goldich 1938). Setterholm & Morey (1989) described an example in the same State where a weathering profile up to 90 m thick is developed on Archaean gneiss, granite and granodiorite, and covered in places by late Cretaceous marine deposits. Ager (1993) came across a similar exposure, and wrote that the profile indicated 'subaerial tropical times in earlier Mesozoic'. Again, we have the assumption of 'tropical climate'. Most ancient weathering was indeed probably tropical, but not necessarily, and all that is needed for deep weathering is groundwater and plenty of time (Ollier 1988).

In the Norseman area of Western Australia, now semi-arid, chains of salt lakes mark old river courses, and cross-sections are afforded by deep open-cut mines such as Central Norseman. The valleys contain terrestrial sediment with leaves, fruit and pollen of Eocene age, and this is overlain by a marine deposit with sponge spicules and sharks' teeth, also Eocene. Evidently, the Eocene sea invaded from the south and extended up the valleys for hundreds of kilometres, apparently without flowing onto the surrounding plain. So by 40 Ma ago the rivers were incised into a very flat plain and had very low gradients. Below the alluvial deposits the rocks are very weathered, and red. If this weathering occurred after the alluvium was deposited, the intense oxidation would have destroyed the organic remains in the alluvium, so the deep weathering is pre-Eocene, and probably very much older.

Non-stratigraphic methods may be used to date weathering profiles. One limit to the age of weathering is the age of the rock itself. Pain & Ollier (1981) described deep weathering (probably 10 m) on a granite in Papua New Guinea dated at only 2.3 Ma. The topography is rugged, and the slope–soil relationships are typical of much of Papua New Guinea, and larger areas of the tropics, with fresh rock exposed only in the actively eroding river channels, and the hillsides and ridge-tops covered in soil and vegetation, with no rock outcrops. Outside the tropics it is more usual to find bare rock cropping out on ridge tops, and the valley bottoms obscured by alluvium.

In Australia palaeomagnetism has been used to date weathering profiles. Schmidt & Ollier (1988) recorded Cretaceous to early Tertiary weathering in New England. Palaeomagnetic dating of ferricretes in southwestern Queensland (Idnurm & Senior 1978) and northern South Australia (Pillans 2005) gave ages of 55–70 Ma (Maastrichtian to Paleocene).

Spier *et al.* (2006) applied $^{40}Ar/^{39}Ar$ constraints on the evolution of a deep weathering profile in Minas Gerais, Brazil, up to 400 m thick. Samples down to 150 m gave ages of 62–14 Ma, but many samples at the weathering–bedrock interface gave ages between 51 and 41 Ma, suggesting that the weathering profiles had already reached their present depth in the Palaeogene, and there was no further advance in the weathering during the Neogene. They wrote: 'The results suggest that weathering in some of these ancient landscapes is not controlled by steady-state advances of the weathering front through time, but may reflect climatic and geomorphological conditions prevailing in a remote past'. Earlier, Carmo & Vasconcelos (2004) examined three weathering profiles in the same large region of Minas Gerais, Brazil. Their $^{40}Ar/^{39}Ar$ analyses showed that the formation of weathering profiles was contemporaneous in three regions up to 400 km apart, suggesting a strong weathering event in the Middle to Late Miocene (10–8 Ma).

Application of a similar technique in west Africa by Hénocque *et al.* (1998) dated manganese nodules around 45–50 Ma (Eocene). In contrast, a study in Western Australia by Dammer *et al.* (1999) found age peaks at 52–49, 44–43, 40–39, 35, 30–29 and 7 Ma, reflecting formation of weathering products over a remarkably extended period, in an episodic manner possibly reflecting past climatic changes.

Planation surfaces and weathering

To make deep weathering profiles by hydrolysis of silicate minerals takes a very long time, during which there must be little erosion of saprolite if the profile is to grow deeper. If erosion exceeds weathering the weathering profile will be reduced; if weathering exceeds erosion then the saprolite profile becomes thicker. This is why it is generally assumed that very thick weathering profiles must be formed on plains where only limited erosion is possible. From Argentina Carignano *et al.* (1999) described landscape antiquity of the Sierras Pampeanas, and wrote that a long period of tectonic quiescence and humid tropical climate in the Jurassic allowed the progressive development of a broad planation surface. Deep weathering profiles, corestones, bornhardts and tors are interpreted as residuals of that surface.

From Brazil it is claimed: 'we may safely conclude that giant supergene iron orebodies are likely to be found only in the high elevation plateaus (Gondwana Surface) that were exposed to the more intense and prolonged weathering history' (Spier *et al.* 2006).

Partridge (1998, p. 167) wrote of the southern Africa situation: 'significant for an appreciation of the present macro-geomorphology of southern Africa is the continent-wide planation surface known as the African Surface—generated by the multi-phase cycle of Cretaceous erosion. This surface forms a readily identifiable datum across the high plains because of the widespread preservation of deep weathering and massive cappings of laterite and silcrete on remnants which have survived later dissection.'

Great depths of weathering are found in the Snowy Mountains region of Australia, where relief was created by erosion of a palaeoplain. Completely weathered rock has been found at depths of 274 m below an undulating ground surface. If a former plain is cast over the area, the highest in Australia, then the depth of postulated weathering becomes staggering, possibly as much as 1500 m. The alternative explanation is that weathering created saprolite up to 300 m deep on some of the most rugged mountain slopes in the country. In either case, deep weathering occurs on a grand scale. Either weathering can go on to depths of over 1000 m, or deep weathering can occur and its products be preserved on very steep slopes.

A similar Australian situation is found at Kiewa, Victoria, intensively studied for a hydroelectric dam. Features include schist weathered to 200 m, gneiss weathered to 320 m, mylonite weathered to 120 m and an oxidized copper lode found at 350 m, and in one large area of granodiorite all bores to 120 m failed to reach bedrock.

On the other hand, in his book on tropical geomorphology Thomas (1994), who stressed the importance of Quaternary events in the tropics, remarked that 'any suggestion that they [weathering mantles] underlie only the ancient planation surfaces of continental interiors is misleading'.

Deep weathering and landslides

This paper is mainly concerned with principles of deep weathering, and detailed descriptions of landslides are presented in the rest of the volume. However, by way of summary, a few examples are discussed here.

One potential hazard is caused by weathering along joints producing clay (Fig. 5). If exposed on a valley side these could lead to sliding of large angular corestones over a lubricated plane. However, much more important is the movement of large masses of saprolite.

Dry saprolite is not especially hazardous, but when water saturated it can be dangerous. This was well illustrated by a tunnel through saprolite and fresh rock at Terrago in Victoria (Ollier 1984). The dry saprolite could be blasted and sealed without difficulty, if the work was completed quickly. However, if left for a few days it became wet and started to flow, even if in a narrow seam. The contractors advanced too far into one extensive body of saturated saprolite and left it for 3 days, a long weekend break from work. The saprolite flowed into the tunnel and filled it for 40 m. If such a seam of saprolite should crop out on a valley side, flow would be inevitable after rain, undermining the overlying rock.

Lacerda (2007) has described some of the mechanisms of landslides in more detail, concentrating on weathering of granite and gneiss in SE Brazil where weathering reaches more than 100 m. The fluctuation of groundwater induces cyclical variations in pore water pressure. This in turn causes expansion and contraction of the saprolite structure, weakening it. This is a fatigue phenomenon. Another mechanism is that of the advance of a saturation front beyond the root zone. The loss of cohesion without the beneficial contribution of the roots to the shearing strength causes a sudden slide.

The combination of topography, climate and deep weathering gives rise to hundreds of landslides in Hong Kong, especially in exceptionally wet weather. Lumb (1975) found that in Hong Kong landslips were generally associated with the total 24 h rainfall on the day of the event, together with the cumulative rainfall over the previous 15 days.

Two large landslides will be described as examples.

The devastating Po Shan Road landslide occurred in 1972 (Geotechnical Engineering Office 1992). The slip started above Po Shan Road, swept on to Conduit Road, and then to Kotewall Road, where it destroyed a four-storey house, obliterated the 12-storey Kotewall Court and partly destroyed a block of unoccupied flats and another construction site. About 40 000 m^3 of debris travelled 270 m down slope and resulted in 67 fatalities. The area is underlain by deeply weathered volcanic rocks. The gradient above Po Shan Road was 35°, and gentler below. Exceptional rain was associated with the event. About 1400 mm was recorded in May and June 1972, and 650 mm from 16 to 18 June 1972.

The second biggest landslide in Hong Kong is the Shum Wan Road landslide, which involved a mass of about 26 000 m^3 of transported material (Geotechnical Engineering Office 1996). The original hillside slope was about 27°. The slide started above Nam Long Shan Road, about 70 m above Sum Wan Road, crossed both roads and demolished structures at Aberdeen Harbour. The volcanic rock here is steeply dipping tuff (70–90°), and weathering ranges from complete to slightly decomposition. A thin cover of colluvium does not appear to have contributed to the failure. A set of joints makes it easy to distinguish saprolite from colluvium at this site, and the backscar is defined by

Fig. 5. The lower part of a deeply weathered profile in Sweden, where weathering has penetrated along joints in granite. In some circumstances the angular joint blocks may slide over the weathered joints. (Photograph by C. D. Ollier.)

steep joints. An important observation was the recognition that the debris that crossed Shum Wan Road was a relatively intact slab of rock about 2–3 m in thickness. It consisted of weathered tuff, which had joints infilled by kaolinitic clay. Careful mapping of the slab showed that there was a continuity of geological structure within it, indicating that it had moved from the hillside to the waterfront in essentially one piece. The main landslide is now in two parts, which

Fig. 6. The Thistle landslide, Utah, summer 1983. The slide is advancing straight towards the camera, and its toe is the brown slope shaped by earth scrapers to create a temporarily more stable dam profile, while overflow water descends an engineered spillway on the left. The end of the lake created by the landslide is just visible beyond the bluff. (Photograph by T. Waltham.)

probably moved together until the final stages of failure. The upper part of the landslide took place as a rotational (spoon-shaped) slip and the lower part was a planar (translational) slip. The slide moved slowly initially, but gathered speed and the lower part detached itself as the 'slab' travelled down the slope. There was a continuous sheet of debris resting on top of the original ground surface (the 'planar scar'). The Shum Wan Road landslide was caused by the effect of elevated groundwater conditions following an exceptional wet period, and during an intense rainstorm. The rainfall in Hong Kong in August 1995 was the highest August rain ever recorded.

The next example is of a slide on deeply weathered mudstone (Waltham 1983). The Thistle landslide developed in April 1983 in a tributary of the Spanish Fork River in Utah, USA. After a wet 1982, an unusual input of groundwater from snow melt activated a landslide. It was soon moving at about a metre per hour, and its toe advanced until it dammed the Spanish Fork valley (Fig. 6). A lake formed rapidly on the upstream side, prompting evacuation of the town of Thistle, the re-routing of a trans-continental highway and railway, and later draining of the lake when the landslide dam was deemed to be unstable. It is the most costly single landslide in the USA. The main landslide mass is 50–100 m thick and consists mainly of deeply weathered mudstone of the Cretaceous North Horn Formation. This rock is well known for its substantial weakening where it is weathered and saturated.

General reviews of landslides in weathered granite have been given by Durgin (1977) and Migoń (2006), who have given references to many others. Thomas (1994, chapter 6) provided a review of landsliding in the tropics, which of course includes examples of deep weathering. Gidigasu (1988) stressed that the products of deep weathering place severe limitations upon the application of classical rock and soil classifications and rating systems for foundation engineering, and he noted that numerous earth failures had resulted from the misapplication of these principles.

Perhaps the greatest hazard in deep weathering is the failure to recognize it. When there are lots of rock outcrops and boulders at the ground surface it is hard to imagine the large volume of thoroughly weathered rock beneath, and roads and other engineering works can run into unforeseen difficulties.

References

AGER, D. V. 1993. *The New Catastrophism*. Cambridge University Press, Cambridge.

BASHAM, I. R. 1974. Mineralogical changes associated with deep weathering of gabbro in Aberdeenshire. *Clay Minerals*, **10**, 189–202.

BROWNE, W. R. 1964. Grey billy and the age of tor topography in Monaro, N. S. W. *Proceedings of the Linnean Society of New South Wales*, **89**, 322–325.

CARIGNANO, C., CIOCCALE, M. & RABASSA, J. 1999. Landscape antiquity of the central–eastern Sierras Pampeanas (Argentina): Geomorphological evolution since Gondwanic times. *Zeitschrift für Geomorphologie, Supplement band*, **118**, 245–268.

CARMO, I. & VASCONCELOS, P. 2004. Geochronological evidence for pervasive Miocene weathering, Minas Gerais, Brazil. *Earth Surface Processes and Landforms*, **29**, 1303–1320.

DAMMER, D., MCDOUGALL, I. & CHIVAS, A. R. 1999. Timing of weathering-induced alteration of manganese deposits in Western Australia: evidence from K/Ar and ^{40}Ar/^{39}Ar dating. *Economic Geology*, **94**, 87–108.

DICKINSON, W. A. 2006. Deep weathering and mineralization in the Sirius Group: a Martian analogue in the Dry Valleys, Antarctica. *Geochimica et Cosmochmica Acta*, **70**, Supplement 1, A141.

DURGIN, P. B. 1977. Landslides and the weathering of granitic rock. *In*: COATES, D. R. (ed.) *Landslides*. Geological Society of America, Reviews in Engineering Geology, **3**, 127–131.

EHRLICH, H. L. 1998. Geomicrobiology: its significance for geology. *Earth-Science Reviews*, **45**, 45–60.

EXLEY, C. S. 1958. Magmatic differentiation and alteration in the St Austell granite. *Quarterly Journal of the Geological Society of London*, **114**, 197–227.

FOOKES, P. G., LEE, E. M. & MILLIGAN, G. (eds) 2005. *Geomorphology for Engineers*. Whittles, Caithness.

GEOTECHNICAL ENGINEERING OFFICE 1992. *Reassessment of the Po Shan Landslide of 18 June 1972*. SPR **16/92**.

GEOTECHNICAL ENGINEERING OFFICE 1996. *Report on the Shum Wan Road Landslide of 13 August 1995. Volume 2. Findings of the Landslide Investigation*. Geotechnical Engineering Office, Hong Kong.

GIDIGASU, M. D. 1988. Potential application of engineering pedology in shallow foundation engineering on tropical residual soils. *In*: *Geomechanics in Tropical Soils, Proceedings of the Second International Conference on Geomechanics in Tropical Soils, Singapore, 1988*, Vol. 1. Balkema, Rotterdam, 17–24.

GOLDICH, S. S. 1938. A study in rock weathering. *Journal of Geology*, **46**, 17–58.

HARRIS, C. S., HART, M. B., VARLEY, P. M. & WARREN, C. D. (eds) 1996. *Engineering Geology of the Channel Tunnel*. Telford, London.

HÉNOCQUE, O., RUFFET, G., COLIN, F. & FERAUD, G. 1998. ^{40}Ar–^{39}Ar dating of west African lateritic cryptomelanes. *Geochimica et Cosmochimica Acta*, **62**, 2739–2756.

IDNURM, M. & SENIOR, B. R. 1978. Palaeomagnetic ages of Late Cretaceous and Tertiary weathered profiles in the Eromanga Basin, Queensland. *Palaeogeography, Palaeoclimatology, Palaeoecology*, **24**, 263–277.

LACERDA, W. A. 2007. Landslide initiation in saprolite and colluvium in southern Brazil: field and laboratory observations. *Geomorphology*, **87**, 104–119.

LIDMAR-BERGSTRÖM, K. L. 1988. Exhumed Cretaceous landforms in south Sweden. *Zeitschrift für Geomorphologie, Supplement band*, **72**, 21–40.

LINDGREN, W. 1933. *Mineral Deposits*. McGraw–Hill, New York.

LUMB, P. 1975. Slope failure in Hong Kong. *Quarterly Journal of Engineering Geology*, **8**, 31–65.

MIGOŃ, P. 2006. *Granite Landscapes of the World*. Oxford University Press, Oxford.

MIGOŃ, P. & LIDMAR-BERGSTRÖM, K. L. 2001. Weathering mantles and their significance for geomorphological evolution of central and northern Europe since the Mesozoic. *Earth-Science Reviews*, **56**, 285–324.

OLLIER, C. D. 1983. Weathering or hydrothermal alteration? *Catena*, **10**, 57–59.

OLLIER, C. D. 1984. *Weathering*. Longman, Harlow.

OLLIER, C. D. 1988. Deep weathering, groundwater and climate. *Geografiska Annaler*, **70A**, 285–290.

OLLIER, C. D. 1992*a*. Global change and long-term geomorphology. *Terra Nova*, **4**, 312–319.

OLLIER, C. D. 1992*b*. Age of soils and landforms in Uganda. *Israel Journal of Earth Science*, **41**, 227–231.

OLLIER, C. D., CALCATERRA, D. & PARISE, M. 2007. Studies in weathering and slope movements – an introduction. *Geomorphology*, **87**, 101–103.

PAIN, C. F. & OLLIER, C. D. 1981. Geomorphology of a Pliocene granite in Papua New Guinea, *Zeitschrift für Geomorphologie, Neue Folge*, **25**, 249–258.

PARTRIDGE, T. C. 1998. Of diamonds, dinosaurs and diastrophism; 150 million years of landscape evolution in southern Africa. *South African Journal of Geology*, **101**, 167–184.

PILLANS, B. 2005. Geochronology of the Australian regolith. *In*: ANAND, R. R. & DE BROEKERT, P. (eds) *Regolith–Landscape Evolution across Australia*. CRC LEME, Perth, 41–61.

POLYNOV, B. B. 1937. *Cycle of Weathering* (translated by A. Muir). Murby, London.

RUXTON, B. P. & BERRY, L. 1957. The weathering of granite and associated erosional features in Hong Kong. *Geological Society of America Bulletin*, **68**, 1263–1292.

SCHMIDT, P. W. & OLLIER, C. D. 1988. Palaeomagnetic dating of Late Cretaceous to early Tertiary weathering in New England. *Earth-Science Reviews*, **25**, 363–372.

SETTERHOLM, D. R. & MOREY, G. B. 1989. An extensive pre-Cretaceous weathering profile in east–central and southwestern Minnesota. *US Geological Survey Bulletin*, **1989-H**, H1–H29.

SHEPPARD, S. M. F. 1977. The Cornubian batholith, SW England. D/H and $^{18}O/^{16}O$ studies of kaolinite and other alteration minerals. *Quarterly Journal of the Geological Society of London*, **133**, 573–592.

SPIER, C. A., VASCONCELOS, P. M. & OLIVIERA, S. M. B. 2006. $^{40}Ar/^{39}Ar$ geochronological constraints on the evolution of lateritic iron deposits in the Quadrilátero Ferrífero, Minas Gerais, Brazil. *Chemical Geology*, **234**, 79–104.

STIERMAN, D. J. & HEALY, J. H. 1984. A study of the depth of weathering and its relationship to the mechanical properties of near-surface rocks in the Mojave Desert. *Pure and Applied Geophysics*, **122**, 425–439.

THOMAS, M. F. 1994. *Geomorphology in the Tropics: a Study of Weathering and Denudation in Low Latitudes*. Wiley, Gloucester.

THURO, K. & SCHOLZ, M. 2003. Deep weathering and alteration in granites – a product of coupled processes. *In*: *Proceedings of the International Congress on Coupled T–H–M–C Processes in Geosystems: Fundamentals, Modelling, Experiments and Applications, GeoProc 2003*. Royal Institute of Technology (KTH) Stockholm, 781–786.

WALTHAM, T. 1983. Utah reroutes round mud mound. *New Civil Engineer*, **557**, 32–37.

WRIGHT, R. L. 1963. Deep weathering and erosion surfaces in the Daly river basin, Northern Territory. *Journal of the Geological Society of Australia*, **10**, 151–164.

Deterioration processes of hard clays and clay shales

L. Picarelli[1]* & C. Di Maio[2]

[1] Department of Civil Engineering, Seconda Università di Napoli, Aversa, Italy
[2] Department of Structures, Geotechnics and Engineering Geology, Università della Basilicata, Potenza, Italy

**Corresponding author (e-mail: luciano.picarelli@unina2.it)*

Abstract: Soil deterioration is one of the main causes of slope failure in hard clays and clay shales, but so far a clear general framework of the causes of deterioration is still missing. Based on laboratory data, field observations and speculation, this paper examines some processes of deterioration of stiff overconsolidated clays and clay shales with particular regard to their influence on the stability of slopes. In particular, deterioration localized within soil bands or involving large soil masses is described with reference to its mechanical or physico-chemical causes. Coupling between mechanical and physico-chemical processes is considered.

The geotechnical literature reports numerous well-documented cases of first-time slides in overconsolidated (OC) clays characterized by an operative shear strength (i.e. the strength that is mobilized at failure) well below the peak strength envelope measured on high-quality large representative samples (bulk strength). The difference between the bulk strength measured in the laboratory and the resistance mobilized at failure is usually ascribed to unpredictable processes of soil deterioration. Softening and strain-softening are frequently mentioned as the main causes of deterioration, but also other phenomena, such as slaking, destructuring or fatigue are sometimes mentioned. In fact, different terms are often used to indicate similar deterioration phenomena, and even the terminology is not clear; as a first consequence, communication among experts is not easy. In addition, often the nature of the processes concerned is not properly clarified, and inconsistencies emerge from data or from their interpretation.

For these reasons, a greater consideration of the deterioration mechanisms seems necessary. A contribution to this is the main goal of this paper, which refers to fine-grained materials such as stiff clays and clay shales, and is based on elements that emerge from laboratory testing, field observations and pure speculation. Weathering will not be considered as such, even though it may include one or more processes discussed in the paper.

The interplay between soil deterioration and slope failure

Human activity, rainfall and earthquakes are the main causes of landslides, but slope failure, especially in hard clays and clay shales, may occur suddenly, without an apparent cause. Because the majority of landslides occur during the wet season, the increase in pore pressure (or the decrease in suction) caused by rainfall is the most important trigger.

Picarelli *et al.* (2004) remarked that in a particular climatic period, the annual rainfall regime generally displays recurrent features and its effects on pore pressures follow a seasonal trend that roughly replicates every year. In particular, pore pressure continuously fluctuates within a range whose extreme values, in general, are lower and higher than, respectively, the maximum and the minimum value attained in the past under different environmental conditions (Fig. 1a). As a consequence of pore pressure fluctuation, the mean effective stress also fluctuates following essentially the same path, which does not reach the initial failure envelope (represented by a continuous line in Fig. 1b). In such cases the only process that can explain the slope failure is soil deterioration, that is a time-dependent decrease in strength (dotted lines in Fig. 1b), which can eventually lead to failure in conjunction with pore pressure increase. Often, the initiation of this process occurred in the past. In some cases, it can have already caused somewhere undetected damage to soil structure as a consequence of local failure (point A in Fig. 1c, d); a general slope failure can be the result of further decrease in the shear strength parameters and contemporaneous increase in pore pressure. Rainfall is hence the triggering factor, but deterioration is the prominent preparatory factor, thus slope failure can be the result of a combination of pore pressure increase and soil deterioration (Fig. 1b, c). Obviously, in other cases deterioration may work in conjunction with different triggering factors.

If such a process involves the entire volume of soil above an assigned depth, the mechanical parameters that are

From: Calcaterra, D. & Parise, M. (eds) *Weathering as a Predisposing Factor to Slope Movements*.
Geological Society, London, Engineering Geology Special Publications, **23**, 15–32.
DOI: 10.1144/EGSP23.3 0267-9914/10/$15.00

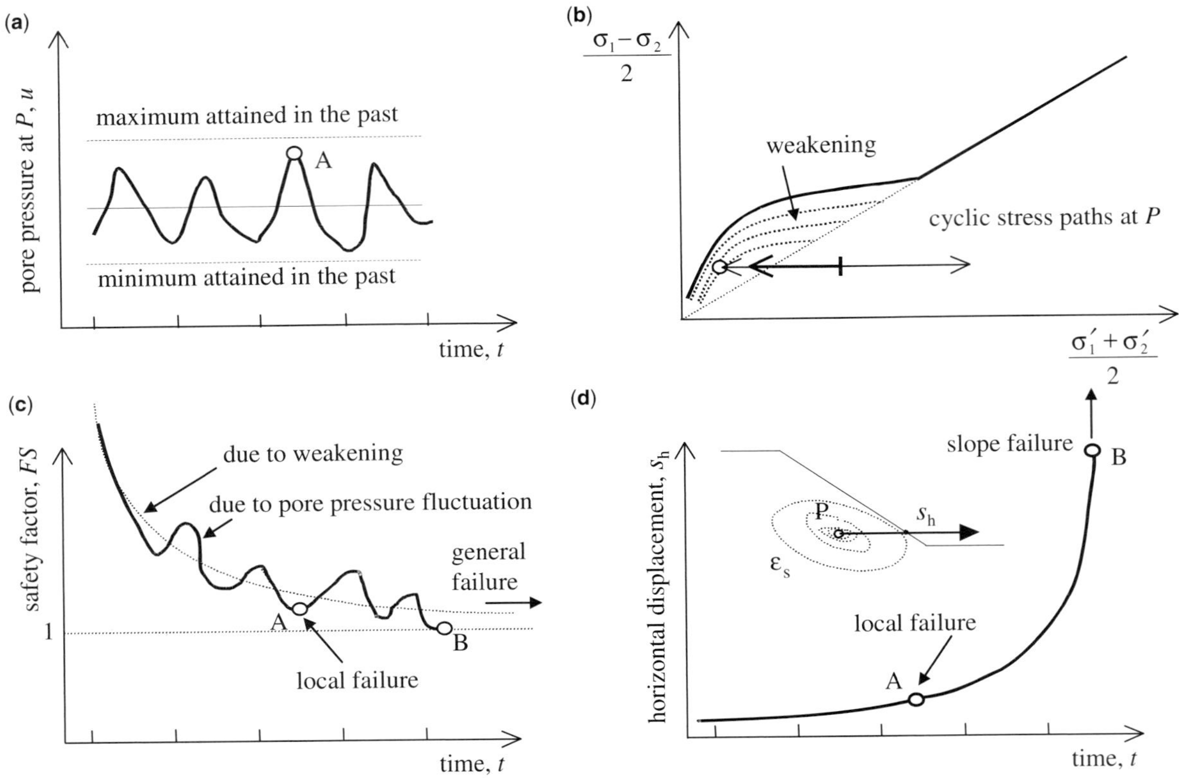

Fig. 1. (**a**) Fluctuating pore pressure at a representative point of a slope; (**b**) local stress paths; (**c**) decrease in the safety factor with time; (**d**) hypothetical horizontal displacement at a point of the slope (Picarelli *et al.* 2004).

provided by relatively recent tests can be consistent with the actual safety factor of the slope. On the other hand, if the process is fast enough, data collected even recently can provide misleading information. Finally, if the deteriorated zone is only a limited part of the slope, such as a thin soil band, the risk that *in situ* and laboratory tests do not provide useful data for slope analysis is much higher. Well-documented back-analyses of slides in hard clays and clay shales often yield significant differences between the operative strength and the bulk strength, which can lead to an overestimation of the safety factor.

Terzaghi (1936) observed that the mobilized shear strength in fissured OC clays and clay shales may be much lower than the bulk strength. He first adopted the term 'softening' to indicate a process of drained shear strength decrease that can justify slope failure. In the following decades similar remarks about the operative strength of stiff clays have been made by other researchers, such as Skempton (1948, 1964, 1970), Bjerrum (1967), Morgenstern (1977) and many others.

Based on well-documented data, today the concept that the mobilized strength of highly OC clays at failure can be less than the peak bulk strength is definitely accepted. In particular, English researchers have carried out systematic investigations on landslides in the London Clay and in other similar formations comparing the operative strength obtained from back-analyses and the bulk strength measured on high-quality large samples. A synthesis of data on the London Clay and the Upper Lias Clay is reported in Figure 2. The difference between the operative and the bulk strength seems to concern essentially the cohesion, which is always very low in spite of its appreciable value provided by tests.

Macroscopic features of deteriorated stiff clays and clay shales

Deterioration is a very general term that is here adopted to indicate a mechanical and/or physico-chemical process causing time-dependent reduction of both stiffness and shear strength of the soil through modifications of its structure. This implies a variation of the mechanical soil parameters (E', c', φ'); hence, a simple change in the state of stress is not necessarily a cause of deterioration.

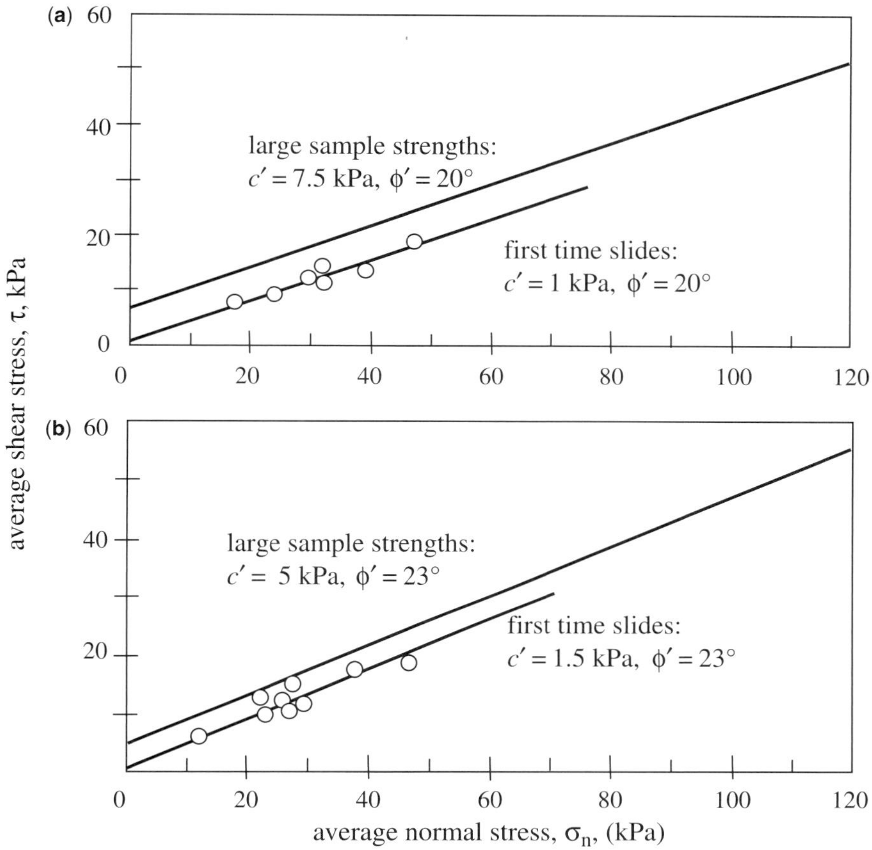

Fig. 2. Failure envelope of the London Clay (**a**) and Upper Lias Clay (**b**) from laboratory tests and back-analysis of first-time slope failures (Chandler 1984).

Deterioration is of little importance in granular soils, whereas it can play a dramatic role in the case of fine-grained soils such as hard clays and clay shales and in soft argillaceous rocks. Sometimes, the effects of deterioration are revealed by macroscopic changes in soil features, such as colour or density, or through evident signs of structural damage, such as fissures, cracks, formation of thin shear bands and so on. Generally, the effects of deterioration are not uniform in the deposit, depending on soil composition and physical condition as well as on depth.

In fissured soils a typical result of deterioration is the formation of very thin soft veneers between intact pieces of soil; this phenomenon typically starts from the faces of fissures. As deterioration advances, the size of intact pieces decreases and the thickness of the veneers increases. Highly fissured deteriorated fine-grained soils consist of a relatively soft matrix including small intact pieces (which are sometimes called lumps or lithorelicts); such a feature progressively disappears with depth (Chandler 1969).

Landslide bodies in stiff clays and clay shales often consist of highly deteriorated soils. An extreme case is represented by mudslides. Hutchinson (1988) described the features of mudslides in stiff clays and clay shales as a mélange of intact elements (the lithorelicts) and soft clay (clay matrix). Vallejo (1989) proposed a classification of the 'structure' of soils involved in mudflows (Fig. 3), including a grain-supported structure (a), a grain–matrix-supported structure (b) and a matrix-supported structure (c). This naturally has a great influence on the mudflow behaviour.

As a result, the local water content is highly variable (Hutchinson 1988; Picarelli 1993). The value that is measured in the laboratory on specimens of ordinary size is only an average value, and cannot be taken as representative of the overall soil state. Picarelli (1993) noted that only the water content of the clay matrix can explain the behaviour of some rapid mudslides during the phases characterized by the highest velocity and by a flow-like style of movement.

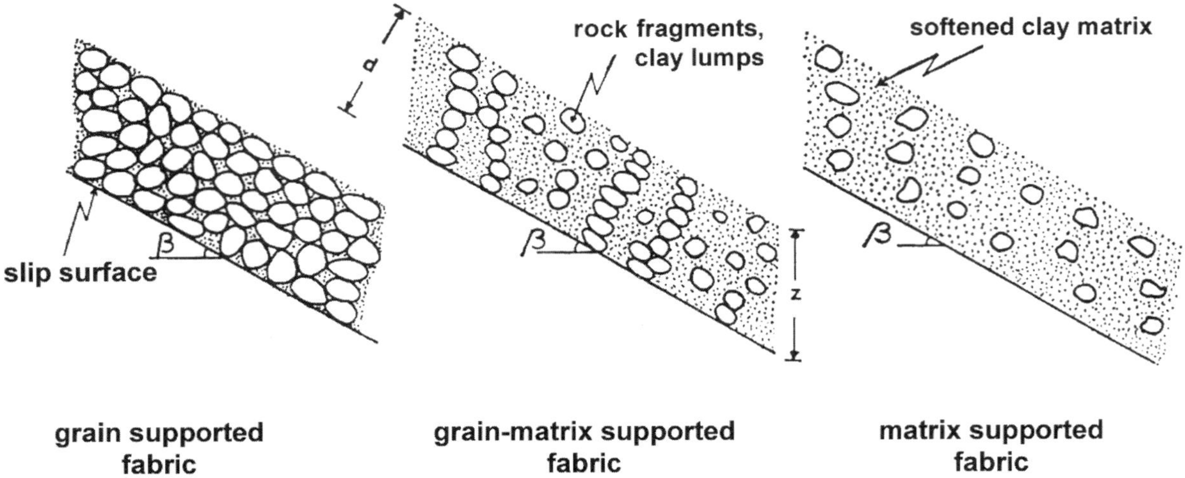

Fig. 3. Fabric of soils involved in mudflows (Vallejo 1989).

Effects, causes and mechanisms of deterioration

Based on experience and back-analysis of landslides in fine-grained soils, deterioration is usually thought to be responsible for partial or complete vanishing of the cohesion (Fig. 2) while the friction angle decreases to the critical or even lower values. The decrease in cohesion may imply breaking of interparticle bonds, a process that is also called destructuring (Leroueil *et al.* 1979), or decrease in the dilation angle. As a consequence of the decrease in the dilation angle, the curved failure envelope tends to become linear. Some deterioration processes can bring about a decrease in the friction angle to a value that is even less than the critical friction angle, because of a change in soil fabric or in the pore fluid composition. Therefore, the most important consequence of deterioration is the loss of the extra strength related to bonding and overconsolidation, with the soil behaviour turning from brittle to ductile, but a further effect may be the loss of a part of the basic frictional resistance.

Some processes of deterioration are described below. For simplicity, they are classified into two basic categories: (1) mechanical; (2) physico-chemical processes.

Mechanical processes of deterioration

Mechanical deterioration is caused by stress changes and associated shear and/or volumetric plastic strains (or both), which cause breaking of bonds (destructuring), increase in the water content and, possibly, modification in the arrangement of soil particles. Mechanical deterioration may also include formation of fissures and cracks. The effects of plastic shear or volumetric strains are examined separately.

Deterioration associated with prevailing plastic shear strains

Plastic shear strains can be responsible for a radical change in soil structure and fabric. The so-called shear zones (Skempton 1967) (i.e. thin soil bands generated by shear strain localization) represent weak zones in overconsolidated clay deposits (Fig. 4). Typically, shear zones present a set of short fissures gently inclined to the direction of movement (minor shears) and may include a slip surface (principal shear), which, in contrast, is consistent with the direction of movement (Riedel 1929; Skempton & Petley 1967). Minor shears form in an advanced stage of the deformation process, revealing local soil failure. The slip surface is a result of the final stage of failure, when the rotation of the principal stresses makes the direction of movement consistent with the direction of shear (Urciuoli 2002). After development of the slip surface, further movement along it favours a progressive orientation in the same direction of clay particles located in a very thin zone around it (Fig. 5).

The mechanism of deterioration induced by shear is called strain-softening. It first causes destructuring, and thus the vanishing of the true cohesion. Successively it causes the decrease in the dilation angle and the formation of minor shears, and finally the formation of the principal shear and the alignment of clay particles. This final process causes the decrease in the basic friction angle from the critical to the residual value.

Shear zones may be the result of any natural phenomenon (such as erosion) or human activity that causes strain localization. The process of plastic adaptation that occurs during the formation of the shear zone often leads to general slope failure (Urciuoli 2002; Picarelli *et al.* 2004), but can also be the effect of a mechanism of soil deformation that does not imply a general soil collapse; typical examples are the plastic ring that forms around tunnels as a consequence of

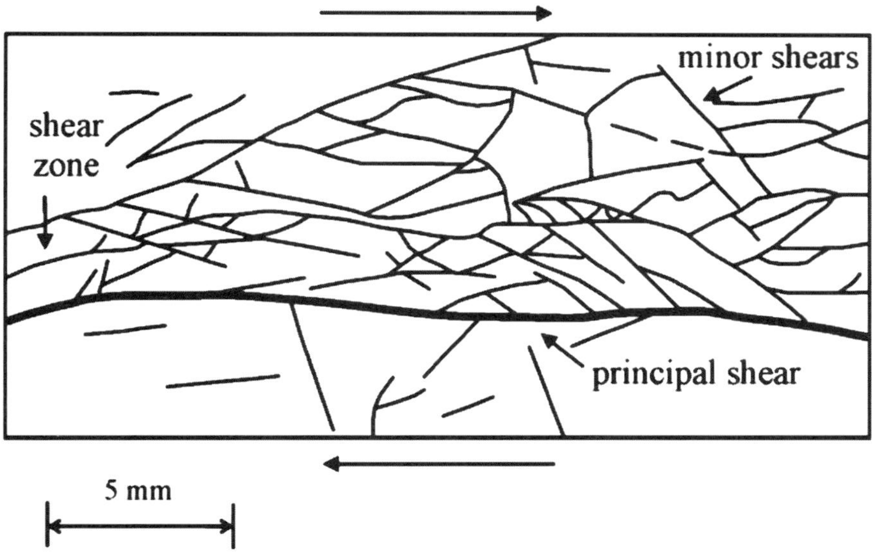

Fig. 4. Shear zone (Skempton & Petley 1967).

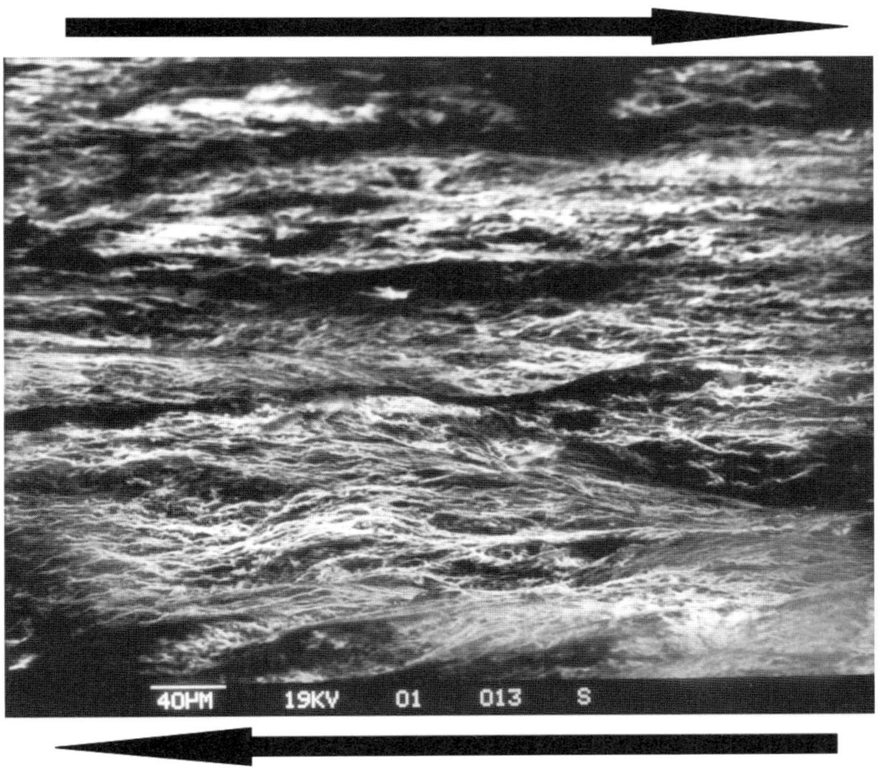

Fig. 5. Micro-fabric of soil in a shear zone (Picarelli *et al.* 2006).

soil excavation (Picarelli & Urciuoli 1993*a*) and the horizontal shear zone that develops at the base of buttes subjected to lateral spreading consequent to valley formation (Picarelli *et al.* 1997). In many cases shear zones involve only small volumes of soil because of the premature arrest of the process of failure. In these cases, which may even show deteriorated zones, the soil can still resist significant stress changes.

Formation of shear zones is often a result of monotonic stress changes, but can be also an effect of cyclic phenomena, such as pore pressure fluctuations and consequent plastic shear strains induced by the associated change in the stress level (Fig. 1b). This has been shown by Eigenbrod *et al.* (1992), who carried out an experimental programme consisting of triaxial tests on undisturbed specimens subjected to constant confining and deviatoric stresses under pulsating back-pressure. As a consequence of accumulated plastic strains, the specimens can fail, provided that the stress level and the number of cycles are high enough. This phenomenon, which is sometimes called fatigue, has similar features to tertiary creep, another mechanical process able to cause soil deterioration through a time-dependent development of plastic shear strains.

Field surveys and laboratory investigations show that the thickness of the shear zone is rather variable. At the base of landslides it has been found to range between a few millimetres (Skempton & Petley 1967) and 1 m (Comegna 2005), but in faulted areas it can be much higher (Tchalenko 1970).

There is evidence suggesting that, after formation, shear zones can experience further deterioration. Tsui *et al.* (1988) described the shear zone found at the base of a ice-thrust clay deposit in Canada: the soil was completely remoulded and included lumps of the parent formation. The shear zones found at the base of mudslides in tectonized clay shales in Italy show the same features. In such a case, deterioration can be so advanced that the lithorelicts are only isolated features in a highly softened clay matrix, as in Figure 3c. Such intense and continuous processes of deterioration may be a consequence of the extremely intense soil deformation that is induced by movement.

On this point, it is worth reporting some considerations by Skempton (1970), who suggested that shear zones could be involved in mechanisms of volumetric softening, as described in next section, caused by opening of minor shears induced by deviatoric stresses, swelling and reconsolidation of soil under its own weight. He believed that such a complex interplay between shear and volumetric strains might cause a decrease in the shear strength to a value very close to the critical strength (the so-called 'fully softened strength'). However, the process described is only an interesting hypothesis, which has never been confirmed by well-documented data.

Typically, shear zones involve only thin parts of the subsoil, thus they cannot be easily found and represent a hidden threat for slopes, as they can trigger progressive slope failure. This matches the experience of researchers engaged in sampling or *in situ* tests, who sometimes recover, from apparently non-deteriorated deposits, soil samples that display highly fissured zones or an abnormally high water content (Milligan 1981), or find, during penetration tests, zones characterized by a much lower strength.

A good example of apparently non-persistent shear zones found in a stable slope has been described by Demers *et al.* (1993), who carried out piezocone tests (CPTU) in the bank of a river in the slightly OC Champlain clay, beside a failed slope. The profiles of the tip resistance showed the presence of weak zones, which have been interpreted as shear zones generated by stress changes induced by river erosion (Fig. 6). It is noteworthy that such zones were located at the same depth as a slip surface that had been found in the failed slope near the investigated site.

Unfortunately, data on the properties of soil taken from shear zones are rather scarce. Cruden *et al.* (1989) reported some tests performed on samples taken from a shear zone found at the base of an ice-thrust deposit in the Wabanum Lake area, Alberta. Triaxial tests revealed a ductile behaviour compared with the brittle behaviour of samples taken in fresh zones in the same deposit. Similar data were provided by Bertini *et al.* (1986) on the shear zone found at the base of a slide in Italy.

Well-documented data for shear zones found at the base of mudslides in highly tectonized clay shales in Southern Italy have been reported by Guerriero (1995) and Comegna (2005). In general, these show a water content that is much higher than that of the parent formation and of the uppermost deteriorated mudslide body (Picarelli 1993). This is due to the dilation that accompanies the formation of the shear zone, but could also be an effect of the change (decrease) in the mean effective stress occurring during the phase of increase in the deviatoric stress (Urciuoli 2002) and/or of physical–chemical processes discussed below.

Figure 7 compares the limit state surface (LSS) of samples taken from a mudslide body and from the shear zone found at its base (Comegna & Picarelli 2008). All specimens, in both the landslide body and the shear zone, were taken at practically the same depth by pushing samplers into the wall of a pit dug into the landslide, just above and within the shear zone. The LSS was obtained through isotropic and consolidated drained (CID) triaxial tests. Figure 7 also shows the range of suction measured on good quality saturated specimens taken from the shear zone. The effects of deterioration are clearly shown by comparing the size of the two LSSs. Assuming that suction is comparable with the mean *in situ* effective stress, the material of the shear zone appears to be normally consolidated or slightly overconsolidated (OCR = 1–2). Naturally, much higher differences might be observed by comparing the two LSSs with that of the parent formation, which is not available. In fact, the mudslide body above the shear zone also experienced strong deterioration, although it was less pronounced.

These considerations are confirmed by the results of triaxial tests showing a ductile behaviour of both the mudslide

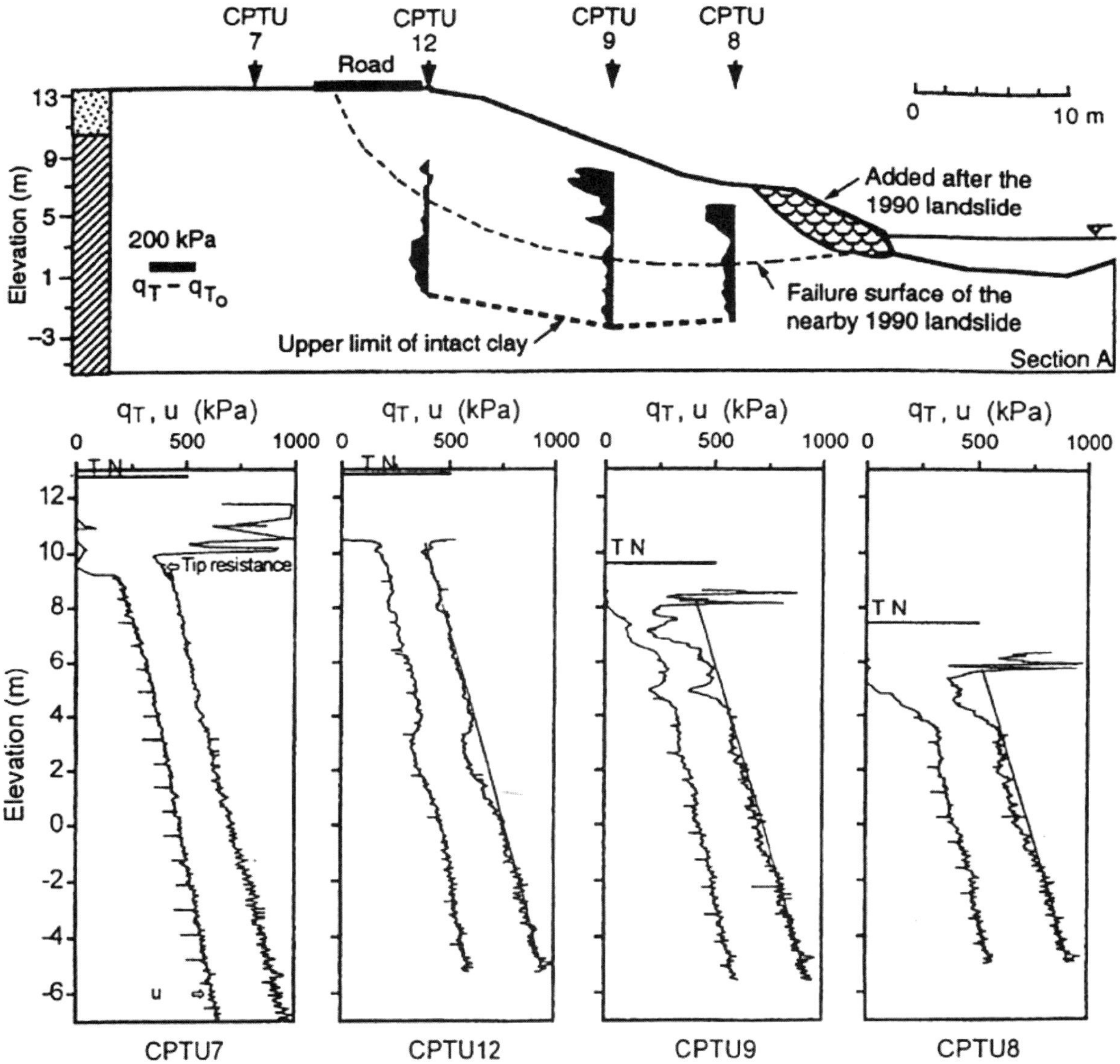

Fig. 6. The Maskinongé site and profiles of tip resistance and excess pore pressures obtained in CPTU tests (Demers *et al.* 1993).

body and the shear zone, even though their critical state line (CSL) has a slope higher than that of the reconstituted specimens (Guerriero 1995). Similar results are provided by direct shear tests, which show a residual strength higher than that of the reconstituted specimens (Fig. 8); however, Figure 8 also suggests a very slow progressive decrease in the residual strength of undisturbed specimens. These systematic differences could be due to the progressive breaking of small lithorelicts having the same role as the gravelly component in sand, or to physico-chemical effects, as discussed below.

Deterioration associated with prevailing volumetric strains
Deterioration caused by volumetric strains and consequent increase in water content is termed softening. According to Terzaghi (1936), the fundamental factors that govern softening are: (1) stress decrease; (2) availability of free water; (3) presence (or formation) of fissures that open as a

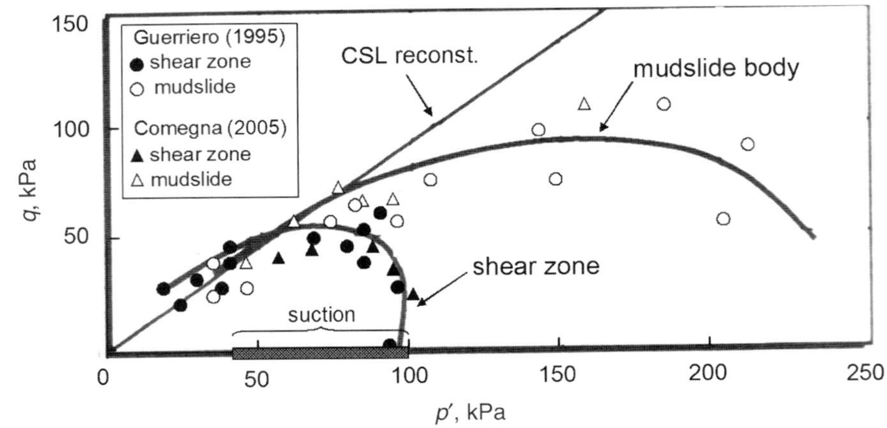

Fig. 7. Limit state surfaces of samples taken from a mudslide body and the shear zone (Comegna & Picarelli 2008).

consequence of unloading. Softening typically involves cuttings and slopes in highly fissured clays subjected to erosion. Fissures play a significant role, as they make water inflow on unloading easier and faster. As a consequence, softening can be relevant at the engineering time scale.

The concept of softening has dominated part of the geotechnical literature until now. Even recently, Terzaghi's arguments have been more or less reiterated in the same terms as they were formulated. Bearing on investigations on landslides in the Pierre shale, South Dakota, Botts (1986) described a process that practically reproduces the one suggested by Terzaghi (Fig. 9): opening of fissures as a result of stress decrease causes infiltration of free water, swelling of the clay adjacent to fissures under very low confining stress and reconsolidation of clay under its own weight. It is noteworthy that the soil structure shown in Figure 9b and c is typical of many softened and weathered formations (Chandler 1969), but it is also the same as in Figure 3, although the same structure is not necessarily due to the same mechanical process. In the intermediate stage of softening (Fig. 9b, c) the strength of the soil mass as a whole has an intermediate value between the strength of each component (Botts 1986), but in the long term (Fig. 9d), it drops to a value that is very close to the critical strength.

To explain the delayed failure of cuttings in the London Clay, Skempton (1948) adopted the same term, softening, proposed by Terzaghi, to indicate the decrease in the undrained strength caused by unloading as a result of excavation. However, such an idea has now been abandoned. For example, in one of his latest papers on the subject, Skempton (1977) suggested that long-term failures of

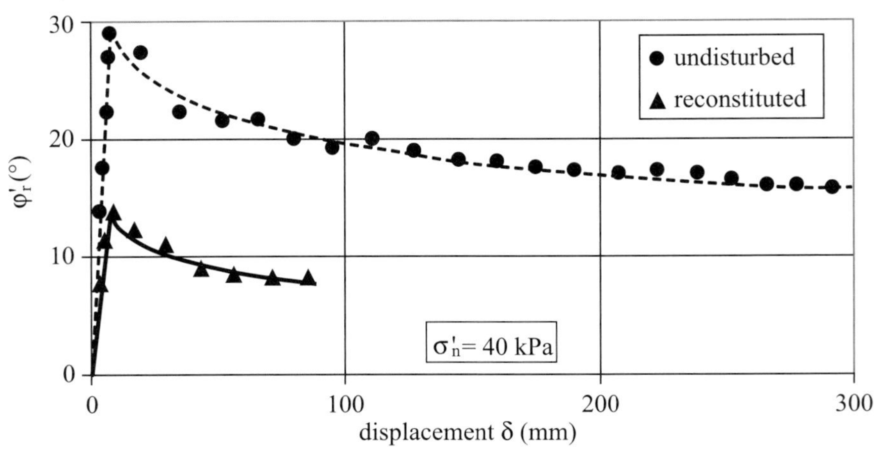

Fig. 8. Results of direct shear tests on a undisturbed and a reconstituted specimen taken from the same mudslide body (Picarelli 1993).

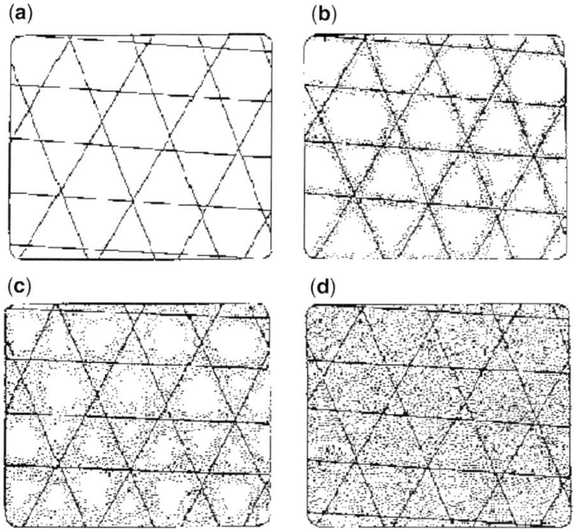

Fig. 9. Successive stages of softening according to Botts (1986).

cuttings in the London Clay are caused by the dissipation of negative excess pore pressures triggered by the excavation and the consequent decrease in the drained strength, a process that should not affect the strength parameters, c' and φ', but only the effective state of stress. However, he did not exclude that during the long-lasting phase of pore pressure equalization, some decay in the parameters of shear strength can take place.

The ideas reported above are essentially based on conceptual considerations suggested by the results of back-analyses of failed slopes (as in Fig. 2) whose rupture may have been triggered by other factors, such as geological factors or unpredictable pore pressures acting at failure, that are not related to deterioration.

Unfortunately, well-documented tests aimed at reproducing in the laboratory the mechanical process imagined by Terzaghi are scarce and often do not help in the clarification of doubts. However, in recent decades some studies have been focused on this topic (Leroueil & Hight 2002).

Based on the results of oedometer tests on undisturbed highly plastic OC clays and clay shales, some workers have noticed that unloading can cause very high volumetric strains leading to a final value for the void ratio that can be higher than the initial one. Such behaviour is shown by slightly fissured high-plasticity clay shales that crop out in the USA, Central America (Fig. 10) and other parts of the world. Similar data have been reported by Hight *et al.* (2002) and Takahashi *et al.* (2005) for the highly plastic London Clay, which displays a pronounced expandibility under effective stresses below 50–100 kPa. Yudhbir (1969) suggested that high swelling may be the consequence of passive failure induced by unloading; others have assumed that it is a consequence of destructuring occurring in the stage of compression or even in that of swelling, giving rise to a contraction of the LSS (Bjerrum 1967; Leroueil & Vaughan 1990).

Based on such remarks, Calabresi & Scarpelli (1985) investigated the mechanical effects of unloading through consolidated undrained (CIU) triaxial tests carried out on the lacustrine highly fissured medium-plastic OC Tody clay ($I_P = 28\%$) and the marine moderately fissured highly plastic OC Ancona clay ($I_P = 60\%$). They noted that pre-swelling under zero confining stress can bring about some decrease in the cohesion as a result of the increase in the water content. However, the soil behaviour remains dilative and the cohesion does not completely vanish (Fig. 11). The magnitude of the drop of cohesion depends on the water content increase and can be interpreted in the light

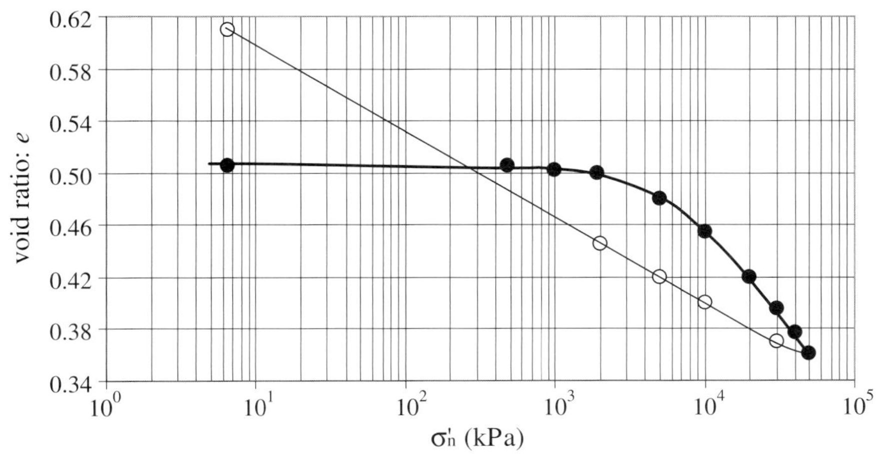

Fig. 10. Results of a 1D compression test on the Culebra shale (Banks *et al.* 1975).

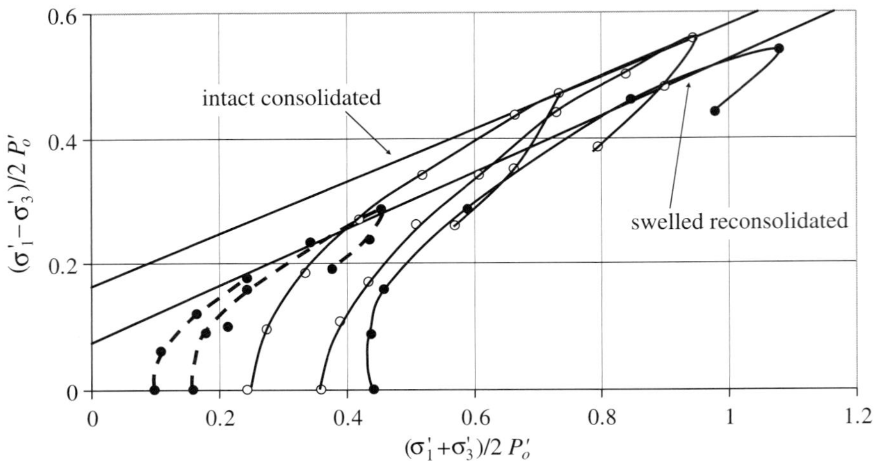

Fig. 11. Effects of pre-swelling on the OC Todi clay (Calabresi & Scarpelli 1985).

of the critical state theory. Therefore, swelling does not completely cancel the effects of overconsolidation and the soil continues to behave as a non-bonded OC clay whose strength depends on its void ratio at the end of the phase of swelling–reconsolidation.

Similar experiments were carried out by Graham & Au (1985), who performed a series of drained and undrained triaxial tests on the lacustrine moderately overconsolidated plastic Winnipeg clay ($I_P = 51\%$). Prior to the stage of rupture, some specimens were subjected to swelling under a very low confining stress (2–4 kPa) and others were subjected to freeze–thaw cycles, then they were reconsolidated to the established normal stress. All tests showed a decrease in the shear strength, but not a complete vanishing of the cohesion, the greater effects being induced by freeze–thaw cycles. Once again, swelling causes a decrease in strength, but not a full softening. It is worth mentioning that cycles of wetting–drying, which are responsible for the so called slaking process (Botts 1986), are notoriously responsible for very similar effects as freeze–thaw.

Tests carried out by Takahashi *et al.* (2005) on pre-swelled specimens of the marine OC London Clay ($I_P = 40\%$) revealed a similar soil behaviour.

The effects of swelling seem even more pronounced in marine highly tectonized and fissured OC clay shales cropping out along the Apennines chain, Italy, whose common feature is generally a very high plasticity. Bilotta *et al.* (1985) noted that the swelling index, C_s, of these formations may be much higher than that of expansive clays (Fig. 12). Also, Cicolella & Picarelli (1990) showed that the yield

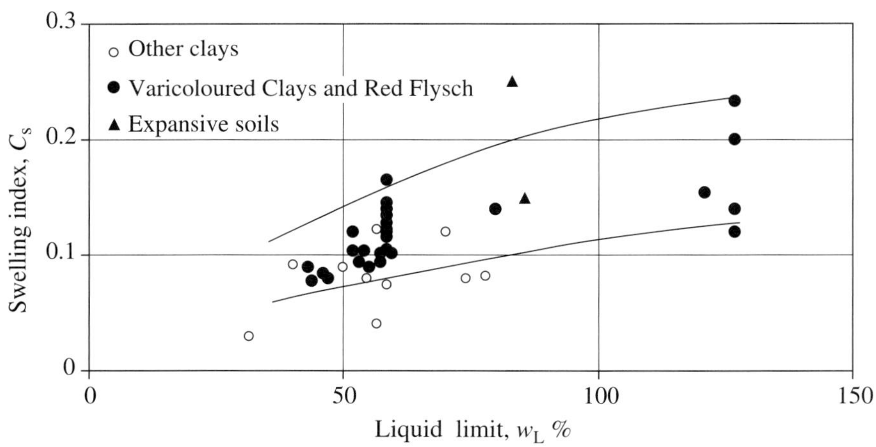

Fig. 12. Swelling index of highly plastic tectonized clay shales (Bilotta *et al.* 1985).

stress and the swelling index of fresh undisturbed specimens are higher than those of softened specimens taken from the same deposit at very similar depths. In addition, the swelling index of both is higher than that of reconstituted specimens. Finally, repeated cycles of compression–swelling with maximum stress increasing at each cycle can induce a continuous increase in the value of C_s from cycle to cycle. All these data seem to indicate destructuring as a fundamental mechanism of deterioration.

Cicolella & Picarelli (1990) investigated the effects of swelling on the shear strength of the marine highly fissured highly plastic Bisaccia clay shale ($I_P = 50–100\%$) through special triaxial and direct shear tests on specimens taken from a cutting.

Figure 13 reports the results of conventional direct shear tests performed on 2 cm thick specimens imposing normal stresses less than the swelling pressure, which was around 0.6 MPa. The soil was allowed to swell for a period of 10–100 days, during which distilled water was continuously renewed, then it was sheared. The role of the time of swelling, and thus of the increase in water content, is clear.

Further data for the same material are reported in Figure 14. Figure 14a shows the results of tests in which swelling under the established normal stress was allowed for 48 h only before shearing; Figure 14b shows the results of tests performed after a stage of swelling in the same box for 40–90 days under a very low normal stress of 10 kPa (pre-swelling), and a successive stage of consolidation under a normal stress less than the swelling pressure. A comparison between the data reveals the strong effect of pre-swelling, which alters the soil behaviour from dilative and brittle to contractive and perfectly ductile.

All the test results have been plotted in the Mohr plane (Fig. 15). The differences in the shear strength are very clear. The lower strength is attained after pre-swelling (Fig. 14b), which reduces the effects of overconsolidation. However, the shear strength is significantly affected by the magnitude of the volumetric strain induced by swelling even when the soil remains brittle, as shown by data for conventional tests carried out after different swelling times (Figs 13 & 14a). Therefore, although the single points cannot be clearly fitted by a straight line through the origin of the axes because of some inhomogeneity between specimens and because of the influence of fissures, which complicates the interpretation of the results, the role of swelling is clear in all cases.

Therefore, in this case more than in the previous ones, deterioration by swelling is a very clear process. As discussed above, destructuring seems to be a likely mechanism, but there are some clues suggesting that, at least in this last case which concerns highly plastic clay shales, physico-mechanical processes might play a major role. This point is discussed in the next section.

Physico-chemical processes of deterioration

Any change in pore fluid composition can greatly influence the mechanical behaviour of clay soils. In particular, a reduction in pore water salinity can produce a dramatic decrease in shear strength. Remarkable effects of such a decrease are exhibited by landslides in quick clays. These are marine clays deposited at the end of, or after, the last glaciation. As a result of the uplift caused by the retreat of glaciers, some deposits were subjected to freshwater leaching and, consequently, to a decrease in ion concentration of pore solution. In some cases, the decrease occurred at almost constant porosity and water content. As a result of the ion concentration decrease, the liquid limit dropped dramatically to values much lower than the *in situ* water content. This implies that these clays, which form firm solid deposits while undisturbed, can change into a liquid mass when remoulded.

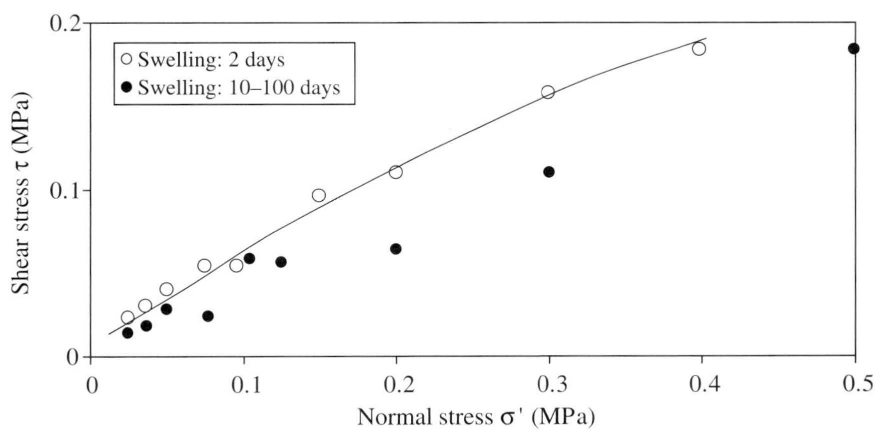

Fig. 13. Results of direct shear tests on the highly plastic Bisaccia clay shale, allowed to swell for different time spans under a normal stress lower than the swelling pressure (Picarelli 1991).

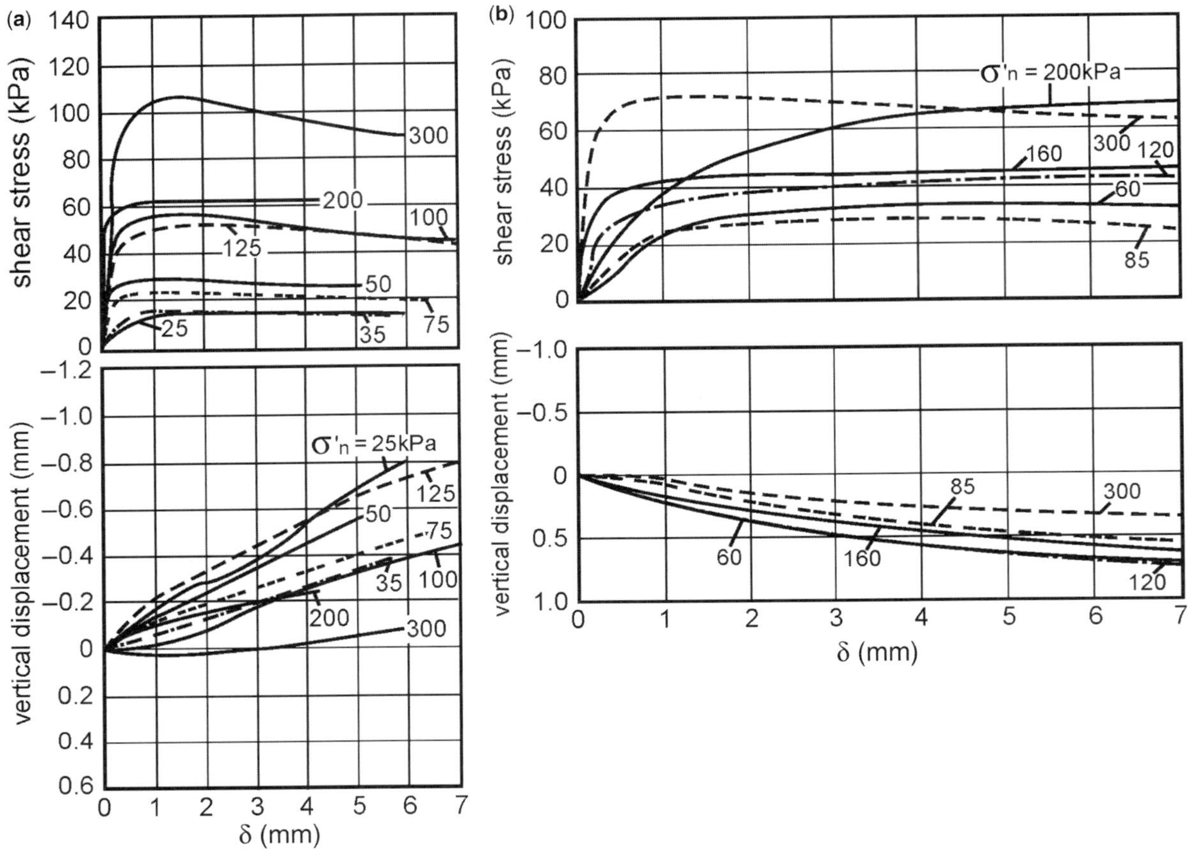

Fig. 14. Results of (**a**) conventional direct shear tests on the marine highly fissured highly plastic Bisaccia clay shale, and (**b**) tests carried out after a cycle of swelling in distilled water under a normal stress of 10 kPa (Cicolella & Picarelli 1990).

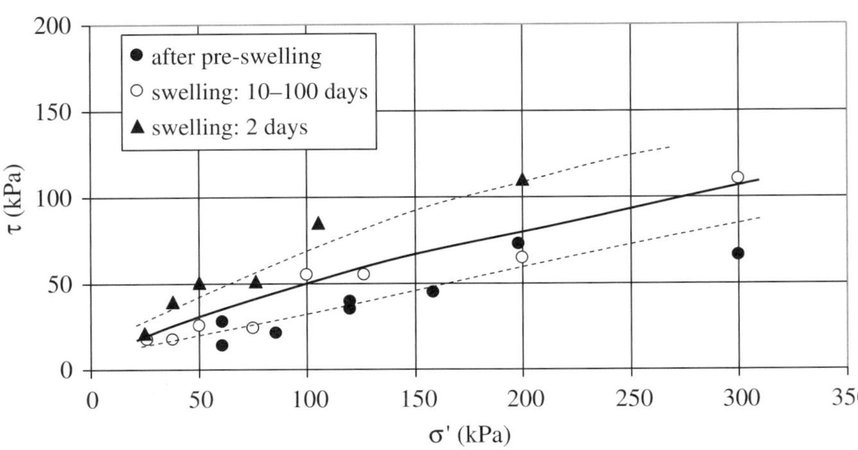

Fig. 15. Results of direct shear tests carried out in a stress range less than the swelling pressure imposing different stress histories and swelling times prior to shearing (modified after Cicolella & Picarelli 1990).

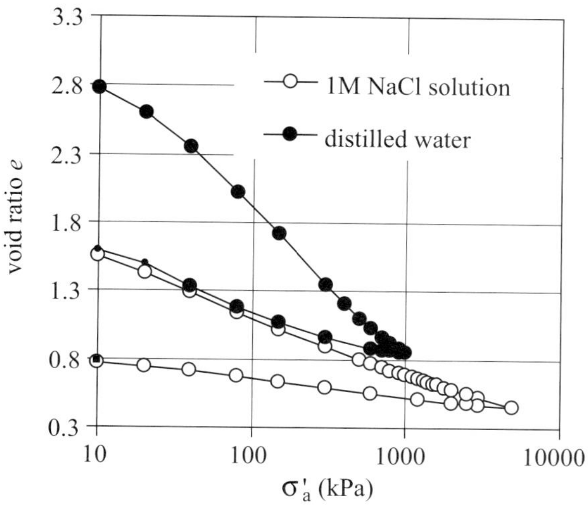

Fig. 16. One-dimensional compression curves of two reconstituted specimens of Bisaccia clay (Di Maio & Onorati 2000*a*).

A number of landslides in quick clays have been studied, described and interpreted on the basis of physico-chemical deterioration (Bjerrum 1954, 1955; Bjerrum & Rosenqvist 1956). Quick clays contain non-expanding phyllosilicate minerals (e.g. illite), whose behaviour is controlled by shear resistance at the particle contacts. Their fabric after sedimentation in a marine environment is of a 'house of cards' type, with edge-to-face contacts. As a result of the leaching of ions in the pore water, the shearing resistance at the particle contacts decreases and the fabric becomes metastable. Any disturbance, even small, may produce a fabric collapse. If the water content is higher than the liquid limit, the collapse can turn the solid into a liquid (Rosenqvist 1966). It is noteworthy that leaching does not always occur at almost constant volume in non-expanding clays such as illite or kaolinite. In fact, depending on the boundary conditions, the decrease in pore solution concentration may cause consolidation (Sridharan & Ventakappa Rao 1973).

The behaviour of soils containing expanding clay minerals (e.g. smectite) is very different. Experimental results on montmorillonitic soils show that, because of the high specific surface, their behaviour is mostly controlled by double-layer effects (Bolt 1956; Olson & Mesri 1970; Sridharan & Ventakappa Rao 1973). As ion concentration decreases, the thickness of the double layer increases, causing volume increase and shear strength reduction.

The influence of pore liquid composition on volume change behaviour can be clearly seen in Figure 16, which shows oedometric compression and swelling curves of two reconstituted specimens of the active Bisaccia clay shale, whose composition is reported in Table 1 together with other characteristics. One of the two specimens was obtained by mixing the powdered clay with distilled water and the other by mixing it with a 1M NaCl solution at about their respective liquid limits (110% and 50%). The specimens were submerged in the same liquid as their pore liquid, thus avoiding chemically induced water or ion flows. It can be seen that, at any value of axial stress, the void ratio and the slopes of the compression and swelling curve of the water-saturated material are higher than those of the solution-saturated material.

Pore fluid composition greatly influences the shear strength too, as can be seen in Figure 17, which shows the results of conventional CIU triaxial tests on specimens reconstituted with and exposed to distilled water and on specimens reconstituted with and exposed to 1M NaCl solution. All the specimens were normally consolidated. The tests were carried out with constant cell pressure during the strain-controlled deviatoric stress increase, and at a rate of axial displacement of 0.03 mm min^{-1}. The results show that the material reconstituted with the salt solution has a much higher shear strength than the material reconstituted with distilled water. In terms of strength parameters, $c' = 15$ kPa and $\varphi' = 18^\circ$ for the clay prepared with the solution, and $c' = 0$ and $\varphi' = 10^\circ$ for the material prepared with distilled water.

In nature, the clay can be in contact with a fluid different from the pore fluid. The difference can induce diffusion, advection, osmotic water flow and so on. As the hydraulic conductivity decreases, ion diffusion and osmotic water flow become the most probable conduction phenomena, which, in turn, cause changes in pore fluid composition. If such changes produce a decrease in the pore fluid ionic force (or an increase in its dielectric constant), then the clay undergoes deterioration. The occurrence of such

Table 1. *Characteristics of Bisaccia clay shale*

G_s	w_L (%)	*Kaolinite* (%)	*Illite* (%)	*Smectite* (%)	*Other minerals*	*A*	A_s (m^2 g^{-1})
2.78	110	10	20	(Ca-smectite) 30	Chlorite 10% Quartz 15% Calcite 10% Feldspars 5%	1.25	190

G_s, specific weight; w_L, liquid limit measured with distilled water; A, activity; A_s, specific surface determined by an approximate method based on methylene blue adsorption (Di Maio *et al.* 2004).

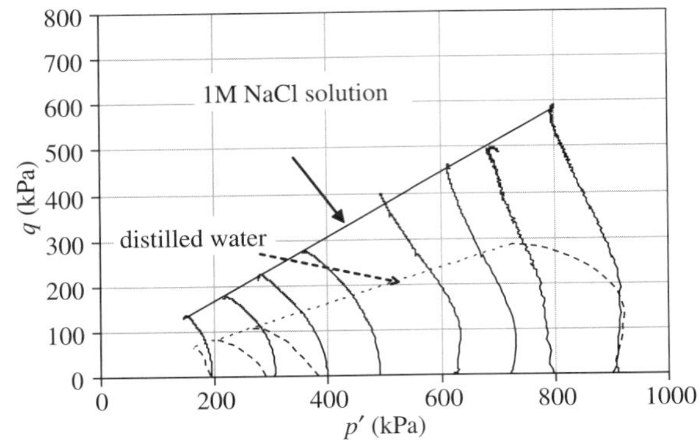

Fig. 17. Results of CIU triaxial tests on normally consolidated specimens of Bisaccia clay (Di Maio & Onorati 2000*b*; Di Maio 2004*a*).

phenomena can be simply checked in the laboratory by exposing specimens reconstituted with a given solution to a less concentrated solution.

Figure 18 shows the oedometer curves of two pairs of specimens of the Bisaccia clay shale carved from two undisturbed samples. Sample C1 was taken at a depth of about 2.5 m from ground level, and C5bis was taken at about 21 m depth. Sample C1 has an initial void ratio much higher than that of sample C5bis. As it will be shown below, this difference can be interpreted as the result of chemical processes of water adsorption. During the tests, one specimen of each sample was immersed in a bath of distilled water and another in a bath of 1M NaCl solution. The figure shows that during unloading the specimens exposed to water swelled much more than those exposed to the NaCl solution. Once swelling at 10 kPa had ended, the specimens that had been immersed in the salt solution were exposed to distilled water (Fig. 19). This exposure caused further high swelling, under constant external load, represented by the segments normal to the abscissa. A noticeable extrusion required the test on specimen C5bis to be interrupted.

Swelling induced by exposure to distilled water strongly depends on the stress level, as shown by Figure 20a. This figure shows the oedometer curves obtained for a series of specimens of the Bisaccia clay reconstituted with 1M NaCl solution. The specimens were initially compressed to 5000 kPa while submerged in the same solution, and subsequently unloaded. During unloading, they were exposed to distilled water at different values of axial stress; exposure to distilled water caused swelling under constant Terzaghi's effective stresses. Figure 20b shows the increase in void ratio Δe produced by exposure to distilled water v. axial stress. The figure shows that the void ratio gradients increase dramatically with decrease in axial stress.

Similar gradients of void ratio with depth are found *in situ*, as can be seen in Figure 21, which shows data for samples from four boreholes in the Bisaccia subsoil. The uppermost 2.5 m cover, where $e > 1$, is a mudslide body. From 2.5 to 5 m there is a soft soil. At greater depths the material is hard. Assuming that the highest void ratio of the uppermost materials depends on the exposure to freshwater at low vertical stresses, a simplified analysis has been carried out using previous data. Based of the results of oedometer tests carried out on the reconstituted material, two theoretical rebound curves have been obtained, based on a simplified hypothesis

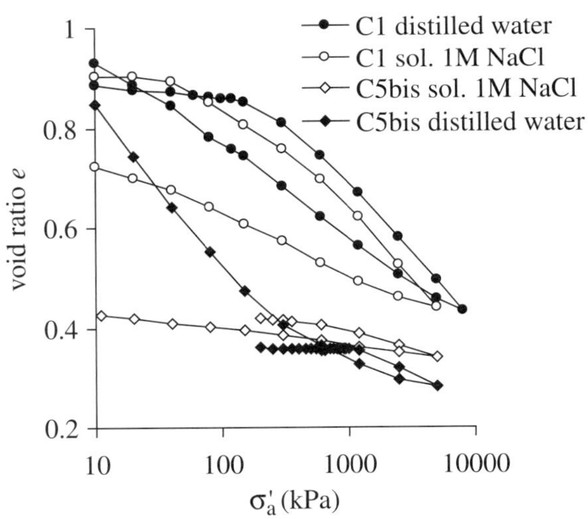

Fig. 18. Oedometer curves of specimens of Bisaccia clay shale exposed to distilled water and to a salt solution (Di Maio & Onorati 2000*a*).

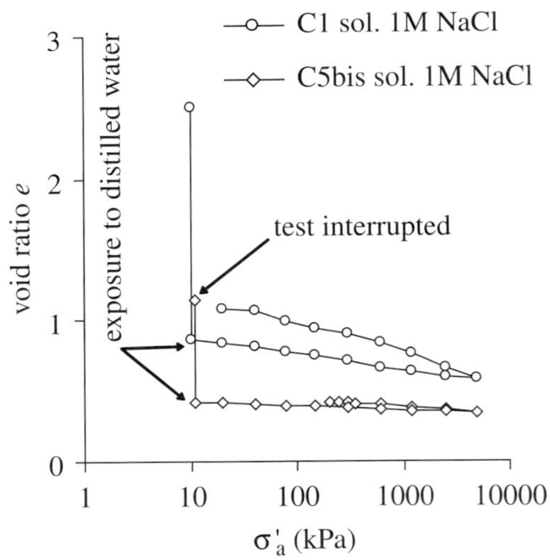

Fig. 19. Omotic swelling of Bisaccia clay shale caused by exposure to distilled water (Di Maio & Onorati 2000*a*).

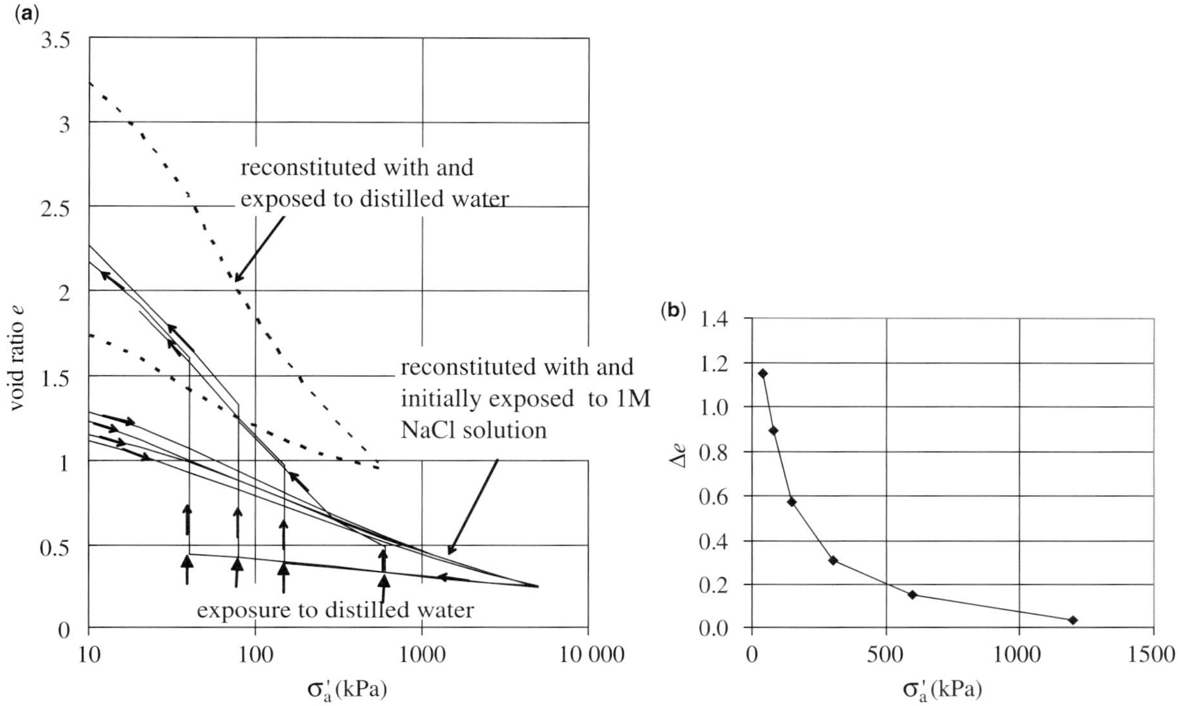

Fig. 20. (**a**) Swelling caused by exposure of the Bisaccia clay reconstituted with 1M NaCl solution to distilled water at several different axial stress values on the unloading curve (Di Maio 2004*b*); (**b**) chemically-induced void ratio increase Δe vs. effective axial stress.

of 1D virgin compression and parallelism of the swelling curves. The fine line represents the theoretical rebound curve of the Bisaccia clay sedimented in a 1M NaCl solution and eroded while in contact with the same solution. The bold curve is the theoretical curve for the material sedimented and eroded being always in contact with distilled water. Both curves have been obtained based on the reasonable hypothesis that erosion corresponds to the removal of an overburden of 1000 kPa, as indicated by geological data (Picarelli & Urciuoli 1993*b*).

The field data virtually overlap the first theoretical curve for depths greater than about 6 m, whereas for smaller depths, they gradually approach the second theoretical curve. A reasonable explanation of this trend is that exposure to rainwater caused an osmotic swelling in the uppermost soil of the type shown in Figure 20 for specimens exposed to distilled water. The long time required for osmotic swelling (Di Maio 1996; Di Maio & Onorati 2000*c*), as well as its dependence on the stress level, probably explains why it could have occurred only in the uppermost clay.

Such data suggest that physico-chemical phenomena may couple with the mechanical processes discussed above. In particular, swelling caused by unloading, as it causes the inflow of water from nearby zones, can favour leaching or freshwater adsorption, especially at small depths below the ground surface. This can explain the decrease in strength of tectonized plastic marine clay shales, such as the Bisaccia clay shale, discussed above. Based on data reported in this section, the dramatic change in void ratio caused by swelling seems to be a consequence of exposure to, and adsorption of, freshwater. Because of the consequent change in physico-chemical forces at the interparticle contacts, such a phenomenon could produce the observed vanishing of the dilation angle (Fig. 14b), and possibly, a decrease in the critical friction angle as in Figure 17. Similar processes might affect other marine plastic clays described in the literature, at least at low confining stresses, explaining their high expansibility. In those clays that are characterized by minor increases in the void ratio, the role of physico-chemical processes might be less significant than that of mechanical processes.

Finally, it cannot be excluded that physico-chemical phenomena interact also with shear as a consequence of water inflow associated with dilation. They could play some role in the progressive deterioration of mudslide bodies, which generally are relatively thin and undergo continuing opening of vertical cracks, which favours infiltration of freshwater from the ground surface. The same slow decrease in shear strength displayed by natural specimens

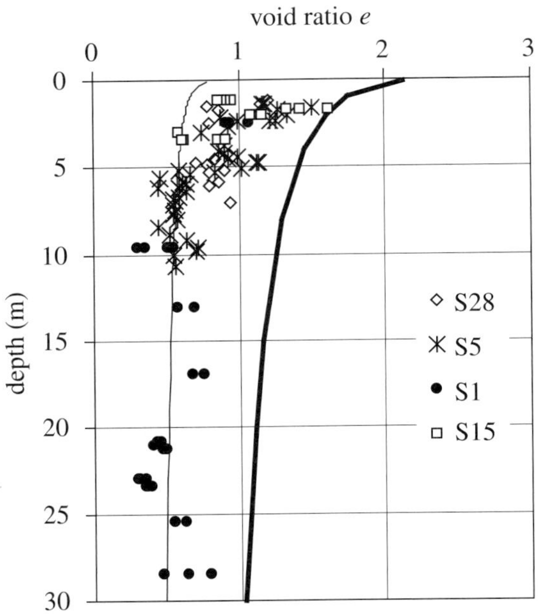

Fig. 21. Theoretical and experimental void ratio vs. depth from the ground level for the Bisaccia clay shale (Di Maio & Onorati 2000*a*).

taken from mudslides (Fig. 8) might be explained not only by the progressive alignment of clay mineral aggregates and the breaking of lithorelicts but also by slow adsorption of distilled water as a consequence of osmotic water flow or of pore ion outward diffusion.

Summary and conclusions

Deterioration is one of the major factors governing the stability of slopes in stiff overconsolidated clays and clay shales. It is always a predisposing factor that is responsible for a time-dependent weakening of the soil, which can combine with other phenomena to cause slope collapse, but in some cases it is the only factor that can explain sudden failures that are otherwise unexplainable.

The role of soil deterioration in slope failure has been well known for a long time, but some uncertainties often still cover its real causes. Often deterioration is associated with softening or strain-softening processes, but the real phenomena that lead to soil failure are not clearly identified. This favours the use of terms whose meaning is not clear and not representative of what really occurs.

In principle, mechanical deterioration can reflect: (1) localized deterioration involving only relatively thin soil bands at whatever depth, which spread along well-defined directions; (2) general deterioration involving large soil volumes starting from the ground surface down to a given depth. However, physico-chemical processes as a result of changes in pore fluid composition can also develop and interact with the mechanical processes, with magnitude and form varying for the different types of clays.

The first type of mechanical process of deterioration is generally associated with strain-softening. It is caused by plastic shear strains inducing macroscopic damage to soil through sets of non-persistent (minor shears) and persistent shear discontinuities (principal shears). It causes a dramatic change in soil structure and fabric, which includes breaking of bonds (destructuring), increase in the water content and change in the arrangement of clay particles in very thin zones around the shear discontinuities. As a consequence of strain-softening, first true cohesion is lost, then the dilation angle decreases to zero, producing a loss of the apparent cohesion (or a decrease in the peak friction angle), and finally, at high deformation, the friction angle can locally decrease to its residual value. In principle, in marine smectitic (expanding) soils, the inward flow of freshwater as a result of the opening of fissures induced by shear or simply as a result of dilation, could add physico-chemical effects to mechanical effects. However, this can occur only in the vicinity of sources of freshwater, and thus only at small depths below the ground surface or at the contact with more permeable formations fed with freshwater.

The second type of mechanical deterioration, which is generally called softening, is essentially caused by a generalized increase in the void ratio. It may be induced by either monotonic or cyclic stress changes, causing breaking of bonds and increase in the water content. If deterioration is due to cyclic changes in the state of stress, its effects can be described by the term fatigue. Breaking of bonds causes a decrease in the true cohesion, whereas increase in the water content can produce a decrease in the dilation angle. If swelling is associated with adsorption of freshwater or ion diffusion outward from the pore fluid of marine smectitic soils, even the critical friction angle can decrease.

Finally, in non-expanding clays, the second type of deterioration can also occur at constant volume. For instance, in the case of quick clays, leaching can produce physico-chemical effects through a decrease of shear resistance at the particle contacts, affecting both the dilation angle and the critical friction angle.

Although deterioration is one of the main causes of landslides in fine-grained soils, systematic studies have not as yet been carried out, and only the results of partial investigations can be found in the technical literature. In particular, a general framework of the processes that bring about deterioration of fine-grained materials is missing and the terminology itself is still ambiguous. An improvement of the experimental procedures to predict at least the effects of those phenomena that are important at an engineering scale is necessary. On this last point, the advice of Burland (1990) to perform laboratory tests using liquids similar to the natural soil liquid appears a wise one, at least in those cases in which major and rapid changes in soil properties are likely. This is not easy because of the difficulty in the investigation of the composition of natural pore water.

References

BANKS, D. C., STROHM, W. E., DE ANGULO, M. & LUTTON, R. J. 1975. *Study of clay shale slopes along the Panama Canal. Report 5. Engineering Analyses of Slides and Strength Properties of Clay Shales along the Gaillard Cut.* US Waterways Experiment Station, Technical Report **S-70-9**.

BERTINI, T., CUGUSI, F., D'ELIA, B. & ROSSI-DORIA, M. 1986. Lenti movimenti di versante nell'Abruzzo Adriatico: caratteri e criteri di stabilizzazione. *In*: *Proceedings of 16th Convegno Nazionale di Geotecnica, Bologna*, Vol. 1, 91–100.

BILOTTA, E., PELLEGRINO, A. & PICARELLI, L. 1985. Geotechnical properties and slope stability in structurally complex clay soils. *In*: *Geotechnical Engineering in Italy. An Overview, Cap. 3: Physical and Mechanical Properties*. AGI, Rome, 195–214.

BJERRUM, L. 1954. Geotechnical properties of Norwegian marine clays. *Géotechnique*, **4**, 49–69.

BJERRUM, L. 1955. Stability of natural slopes in quick clay. *Géotechnique*, **5**, 1–101.

BJERRUM, L. 1967. Progressive failure in slopes of overconsolidated plastic clays. *International Journal of the Soil Mechanics Engineering Division, ASCE*, **SM5**, 1–49.

BJERRUM, L. & ROSENQVIST, I. TH. 1956. Some experiments with artificially sedimented clays. *Géotechnique*, **6**, 124–136.

BOLT, G. H. 1956. Physico-chemical analysis of the compressibility of pure clays. *Géotechnique*, **6**, 86–93.

BOTTS, M. E. 1986. *The effects of slaking on the engineering behavior of clay shales*. PhD thesis, University of Colorado, Boulder.

BURLAND, J. B. 1990. On the compressibility and shear strength of natural clays. *Géotechnique*, **40**, 329–378.

CALABRESI, G. & SCARPELLI, G. 1985. Effects of swelling caused by unloading in overconsolidated clays. *In*: *Proceedings of International Conference on Soil Mechanics and Foundation Engineering, San Francisco*, Vol. 1, 411–414.

CHANDLER, R. 1969. The effect of weathering on the shear strength properties of Keuper Marl. *Géotechnique*, **19**, 321–334.

CHANDLER, R. J. 1984. Recent European experience of landslides in over-consolidated clays and soft rocks. *In*: *Proceedings of IV International Symposium on Landslides, Toronto*, Vol. 1, 61–81.

CICOLELLA, A. & PICARELLI, L. 1990. Decadimento meccanico di una tipica argilla a scaglie di elevata plasticità. *Rivista Italiana di Geotecnica*, **XXIV**, 5–23.

COMEGNA, L. 2005. *Proprietà e comportamento delle colate in argilla*. PhD thesis, Seconda Università degli Studi di Napoli.

COMEGNA, L. & PICARELLI, L. 2008. Anisotropy of a shear zone. *Géotechnique*, **58**, 737–742.

CRUDEN, D. M., THOMSON, S. & TSUI, P. C. 1989. The geotechnical characteristics of an ice-thrust mudstone, Wabamun Lake area, Alberta. *Canadian Geotechnical Journal*, **26**, 227–234.

DEMERS, D., LEROUEIL, S. & D'ASTOUS, J. 1993. *In situ* testing in a landslide area at Maskinongé, Québec. *In*: *Proceedings of 46th Canadian Geotechnical Conference, Saskatoon*, 465–474.

DI MAIO, C. 1996. Exposure of bentonite to salt solution: osmotic and mechanical effects. *Géotechnique*, **46**, 695–707.

DI MAIO, C. 2004*a*. Shear strength of clays and clayey soils: the influence of pore fluid composition. *In*: LORET, B. & HUYG-HE, J. M. (eds) *Chemo-mechanical Couplings in Porous Media-Geomechanics and Biomechanics*, Springer, Berlin, 45–55.

DI MAIO, C. 2004*b*. Consolidation, swelling and swelling pressure induced by exposure of clay soils to fluids different from the pore fluid. *In*: LORET, B. & HUYG-HE, J. M. (eds) *Chemo-mechanical Couplings in Porous Media Geomechanics and Biomechanics*, Springer, Berlin, 19–43.

DI MAIO, C. & ONORATI, R. 2000*a*. Swelling behaviour of active clays: the case of an overconsolidated, marine origin clay. *In*: *Proceedings of 8th International Symposium on Landslides, Cardiff*, Vol. 1, 469–474.

DI MAIO, C. & ONORATI, R. 2000*b*. Influence of pore liquid composition on the shear strength of an active clay. *In*: *Proceedings of 8th International Symposium on Landslides, Cardiff*, Vol. 1, 463–468.

DI MAIO, C. & ONORATI, R. 2000*c*. Osmotic softening of overconsolidated clays. *In*: *Proceedings of International Conference on Geotechnical and Geological Engineering, Melbourne* [CD-ROM publication].

DI MAIO, C., SANTOLI, L. & SCHIAVONE, R. 2004. Volume change behaviour of clays: the influence of mineral composition, pore fluid composition and stress state. *Mechanics of Materials*, **36**, 435–451.

EIGENBROD, K. D., GRAHAM, J. & BURAK, J. P. 1992. Influence of cycling pore-water pressures and principal stress ratios on drained deformations in clay. *Canadian Geotechnical Journal*, **29**, 326–333.

GRAHAM, J. & AU, V. C. S. 1985. Effects of freeze–thaw and softening of a natural clay at low stresses. *Canadian Geotechnical Journal*, **22**, 69–78.

GUERRIERO, G. 1995. *Modellazione sperimentale del comportamento meccanico dei terreni in colata*. PhD thesis, Università di Napoli Federico II.

HIGHT, D. W., MCMILLAN, F., POWELL, J. J. M., JARDINE, R. J. & ALLENOU, C. P. 2002. Some characteristics of London Clay. *In*: TAN, T. S., PHOON, K. K., HIGHT, D. W. & LEROUEIL, S. (eds) *Proceedings of International Workshop on Characterisation and Engineering Properties of Natural Soils, Singapore*, Vol. 2. Balkema, Rotterdam, 851–907.

HUTCHINSON, J. N. 1988. Morphological and geotechnical parameters of landslides in relation to geology and hydrogeology. *In*: *Proceedings of 5th International Symposium on Landslides, Lausanne*, Vol. 1. Balkema, Rotterdam, 3–35.

LEROUEIL, S. & HIGHT, D. W. 2002. Behaviour and properties of natural soils and soft rocks. *In*: TAN, T. S., PHOON, K. K., HIGHT, D. W. & LEROUEIL, S. (eds) *Proceedings of International Workshop on Characterisation and Engineering Properties of Natural Soils, Singapore*, Vol. 1. Balkema, Rotterdam, 255–360.

LEROUEIL, S. & VAUGHAN, P. R. 1990. The general and congruent effects of structure in natural soils and weak rocks. *Géotechnique*, **40**, 467–488.

LEROUEIL, S., TAVENAS, F., BRUCY, F., LAROCHELLE, P. & ROY, M. 1979. Behaviour of destructured natural clays. *International Journal of Soil Mechanics Engineering Division, ASCE*, **105**, 759–778.

MILLIGAN, V. 1981. Panel contribution 'Prediction and Performance'. *In*: *Proceedings of 10th International Conference on Soil Mechanics and Foundation Engineering*, Vol. 4, 581–585.

MORGENSTERN, N. R. 1977. Slopes and excavations in heavily overconsolidated clays. *In*: *Proceedings of 9th International Conference on Soil Mechanics and Foundation Engineering, Tokyo*, Vol. II, 567–581.

OLSON, R. E. & MESRI, G. 1970. Mechanisms controlling the compressibility of clays. *Journal of the American Society of Civil Engineers*, **96**, 1853–1878.

PICARELLI, L. 1991. Discussion of paper 'The general and congruent effects of structure in natural soils and weak rocks' by S. Leroueil & P. R. Vaughan. *Géotechnique*, **41**, 281–284.

PICARELLI, L. 1993. Structure and properties of clay shales involved in earthflows. *In*: ANAGNASTOPOULOS, A., SCHLOSSER, F., KALTEZIOTIS, N. & FRANK, R. (eds) *Proceedings of International Symposium on the Geotechnical Engineering of Hard Soils–Soft Rocks, Athens*, Vol. 3. Balkema, Rotterdam, 2009–2019.

PICARELLI, L. & URCIUOLI, G. 1993*a*. Parametri geotecnici rilevanti per l'analisi e la progettazione di gallerie in argilla. *In*: *Proceedings, XVIII Convegno Nazionale di Geotecnica, Rimini*, Vol. II, 95–109.

PICARELLI, L. & URCIUOLI, G. 1993*b*. Effetti dell'erosione in argilliti di alta plasticità. *Rivista Italiana di Geotecnica*, **27**, 29–47.

PICARELLI, L., LEROUEIL, S., URCIUOLI, G., GUERRIERO, G. & DELISLE, M. C. 1997. Occurrence and features of shear zones in clay. *In*: *Proceedings of International Symposium on Localisation and Bifurcation Theory for Soils and Rocks, Gifu*, 259–270.

PICARELLI, L., URCIUOLI, G. & RUSSO, C. 2004. Effect of groundwater regime on the behaviour of clayey slopes. *Canadian Geotechnical Journal*, **41**, 467–484.

PICARELLI, L., COMEGNA, L. & TOMMASI, P. 2006. Discussion of the paper 'Evolution of shear-zone structure in undrained ring-shear tests' by M. W. Agung, K. Sassa, H. Fukuoka and G. Wang. *Landslides*, **3**, 265–268.

RIEDEL, W. 1929. Zur mechanik geologischer Brucherscheinungen. *Zentralbl att für Mineralogie, Geologie und Palaeontologie, B*, 354–368.

ROSENQVIST, I. TH. 1966. The Norwegian research into the properties of quick clay – a review. *Engineering Geology*, **1**, 445–450.

SKEMPTON, A. W. 1948. The rate of softening of stiff fissured clays with special reference to London Clay. *In*: *Proceedings of II International Conference on Soil Mechanics and Foundation Engineering, Rotterdam*, Vol. 2, 50–53.

SKEMPTON, A. W. 1964. Long term stability of clay slopes. *Géotechnique*, **14**, 77–102.

SKEMPTON, A. W. 1967. Some observations on tectonic shear zones. *In*: *Proceedings of 2nd International Conference on Rock Mechanics, Belgrade*, Vol. 1, 329–335.

SKEMPTON, A. W. 1970. First time slides in overconsolidated clays. *Géotechnique*, **20**, 320–324.

SKEMPTON, A. W. 1977. Slope stability of cuttings in brown London Clays. *In*: *Proceedings of IX International Conference on Soil Mechanics and Foundation Engineering, Tokyo*, Vol. 3, 261–270.

SKEMPTON, A. W. & PETLEY, D. J. 1967. The strength along structural discontinuities in stiff clays. *In*: *Proceedings of Geotechnical Conference, Oslo*, Vol. 2, 55–69.

SRIDHARAN, A. & VENTAKAPPA RAO, G. 1973. Mechanisms controlling volume change of saturated clays and the role of the effective stress concept. *Géotechnique*, **23**, 359–382.

TAKAHASHI, A., FUNG, D. W. K. & JARDINE, R. J. 2005. Swelling effects on mechanical behaviour of natural London clay. *In*: *Proceedings of 16th International Conference on Soil Mechanics and Geotechnical Engineering, Osaka*, Vol. 2, 443–446.

TCHALENKO, J. S. 1970. Similarities between shear zones of different magnitudes. *Geological Society of America Bulletin*, **81**, 1625–1640.

TERZAGHI, K. 1936. Stability of slopes in natural clay. *In*: *Proceedings of 1st International Conference on Soil Mechanics, Harvard*, Vol. I, 161–165.

TSUI, P. C., CRUDEN, D. M. & THOMSON, S. 1988. Mesofabric, microfabric and submicrofabric of ice-thrust bedrock, Highvale mine, Wabamun Lake area, Alberta. *Canadian Journal of Earth Sciences*, **25**, 1420–1431.

URCIUOLI, G. 2002. Strains preceding failure in infinite slopes. *International Journal of Geomechanics*, **2**, 93–112.

VALLEJO, L. 1989. An extension of the particulate model of stability analysis for mudflows. *Soils and Foundations*, **29**, 1–13.

YUDHBIR, Y. 1969. *Engineering behaviour of heavily overconsolidated clays and clay shales with special reference to long-term stability*. PhD thesis, Cornell University, Ithaca, NY.

Mass movement and landscape evolution in weathered granite and gneiss terrains

Piotr Migoń

Department of Geography and Regional Development, University of Wrocław, pl. Uniwersytecki 1, 50–137 Wrocław, Poland (e-mail: migon@geogr.uni.wroc.pl)

Abstract: The paper assesses the contribution of mass movement to the development of landforms in weathered granite and gneiss terrain, using examples from SE Brazil, southern Mexico, SE China, and other locations. The style of slope failures varies depending on local relief and the grade of weathering, which mutually influence each other. Shallow debris slides and boulder falls are common in less weathered rock and steep terrain, whereas rotational slides appear to dominate in highly weathered rock and in more subdued topography. In certain topographic circumstances, slides may turn into debris flows or earthflows. Extreme rainfall is a typical trigger of mass movement in weathered terrain. Mapping of past landslides may considerably assist in hazard assessment, as they usually leave a clear geomorphological record.

Geomorphological evolution of a terrain may proceed through a variety and combination of surface processes, whose relative roles vary spatially depending on local relief, bedrock characteristics and environmental conditions, mainly temperature and rainfall patterns. One important control on the patterns of hillslope evolution is the presence or absence of a thick mantle of weathered material (saprolite) near the topographic surface.

Global distribution of thick saprolites reflects a multitude of factors. On the one hand, it is climate dependent and positively correlated with high temperature and high humidity (Pedro 1968; Thomas 1994; Tardy & Roquin 1998). Hence, the tropical zone is characterized by the extensive occurrence of deeply weathered rock, in subdued and hilly to moderately mountainous settings (Thomas 1994). However, the distribution of deeply weathered mantles extends well beyond the tropics, reaching in Europe a latitude of 50–60°N (Migoń & Lidmar-Bergström 2001). Much of this weathered mantle has been inherited from the early Cenozoic and Mesozoic, and has survived as a result of low long-term denudation rates in shield settings or under a protective sedimentary cover maintained over very protracted periods of geological history. In contrast, some types of weathered mantles, such as the thick mantles of grus (i.e. disaggregated rock, with little geochemical change), which are widely present in mountains and uplands of Central Europe and North America, are not necessarily very ancient materials and form in present-day temperate environments (Migoń & Thomas 2002). Consequently, deep weathering is not restricted to tropical regimes, although the depth of weathering and degree of mineralogical and textural change are indeed at their maximum in humid low latitudes. Inevitably, therefore, the significance of weathering lies at the very core of tropical geomorphology (Thomas 1994, 1997).

This paper reviews the evidence for the role of mass movement in landform development in weathered granite and gneiss terrains, using examples from different geographical settings. The review emphasizes geomorphological aspects of landsliding *sensu lato* in saprolitic materials, and the importance of zoning in the weathered mantle (see Ruxton & Berry 1957; Geological Society Engineering Group Working Party 1990) in understanding the variability of gravity-driven phenomena, and for demonstrating the relationships between weathering grade and types of mass movement. Finally, a range of landforms and their implications will be presented.

Deep weathering profiles and patterns in basement rocks

Deep weathering, which is both chemical and physical, can affect all types of bedrock, and depending on the environmental setting produces saprolitic materials that vary in thickness, texture, strength, and geochemistry (Irfan 1996, 1999; Taylor & Eggleton 2001). Although the majority of saprolites show significant mineralogical and fabric change in relation to the parent rock, some may be predominantly disaggregated rock masses, with minor effects of mineral alteration. However, thick weathered mantles derived from granite and gneiss are less dependent on climate or environment and are distributed widely from equatorial terrains to high-latitude plateaux and hillslopes (Migoń 2006). The thickness of saprolitic cover on granite may be substantial. Grus, which is a poorly decomposed but thoroughly disintegrated material, is typically 5–20 m thick, whereas more decomposed clay-rich saprolites may be up to 30–40 m

From: CALCATERRA, D. & PARISE, M. (eds) *Weathering as a Predisposing Factor to Slope Movements*. Geological Society, London, Engineering Geology Special Publications, **23**, 33–45.
DOI: 10.1144/EGSP23.4 0267-9914/10/$15.00

thick. Thicknesses as great as 70–90 m have been reported from several low-latitude settings such as Hong Kong (Shaw 1997) and SE Brazil (Thomas 1995). Despite this considerable variability in thickness it has long been observed that weathering mantles derived from granitoid rocks show clear vertical differentiation. These observations led Ruxton & Berry (1957) to propose a formal subdivision of a granite weathering profile, which included five principal classes (Table 1; Fig. 1). Later work refined the concept and a six-fold division was proposed (Dearman *et al.* 1978), which has been widely adopted (Geological Society Engineering Group Working Party 1990; Table 1).

A notable feature of many deeply weathered granites is the presence of corestones. These are unweathered, or significantly less weathered compartments of parent rock set amidst an otherwise more extensively disintegrated and decomposed rock mass (Fig. 2). Corestones may be typically from 20–30 cm to more than 5 m across, depending primarily on joint spacing. They are typically rounded, although near the base of a profile cubic shapes may prevail. The sharp boundary between a corestone and the surrounding weathered granite is the so-called 'weathering front'. A similar structural feature may separate the saprolite from the underlying rock mass. Finally, although original fabric is normally much less distinct in the saprolite, original discontinuities such as joint planes may be traced even in the completely weathered rock (Au 1996; Ehlen 1999; Aydin 2006). Weakening of fabric bears on the significant reduction of shear strength of the saprolite, which may be 10–15 times less than in the solid rock (Irfan 1996).

Weathering profiles need not be complete, in the sense that all six weathering grades are present. 'Incomplete' profiles are fairly common, particularly in hilly terrains, and these may include only divisions I–III (i.e. from solid rock to clayey grus). The interpretation of such saprolites is not straightforward. For example, weathering profiles may be truncated by erosion, typically associated with recent uplift or changing environmental conditions, both leading to acceleration of surface processes, including landslides. Alternatively, weathering may not have reached a more advanced grade because of the local hillslope setting. The rate of slope erosional processes, compared with the rate of weathering, may be high enough to prevent the saprolite from reaching mineralogical maturity.

The progress of weathering vertically is controlled by the pre-existing structure of the rock, its mineralogy and the presence of discontinuities. Granites in particular are far from uniform even at the outcrop scale (hence the differences in weathering extent between heavily jointed and massive granites), and so the thickness of the saprolite may be highly variable. In one of the early studies Thomas (1966) showed that saprolites derived from fine-grained granites in the Jos Plateau, Nigeria, vary in depth from a few metres to more than 50 m over distances of hundreds or tens of metres. Many subsequent studies, drawing for example on detailed side investigations and large-scale

Table 1. *Weathering grades for weathered basement rocks*

Ruxton & Berry (1957)			*Geological Society Engineering Group Working Party (1990)*		
Description	*Formal name*	*Weathering zone code*	*Weathering zone code*	*Formal name*	*Description*
Structureless sandy clay or clayey sand, up to 30% of clay, dominance of quartz and kaolin	Residual debris	I	VI	Residual soil	All rock material converted to soil; original structure and fabric destroyed; large change in volume
Few rounded corestones, <10% of the section area, <5% of clay		IIa	V	Completely weathered	All rock decomposed and disintegrated but original structure and fabric still largely intact
Much of the original rock structure preserved, corestones occupy 10–50% of the section area	Residual debris with corestones	IIb	IV	Highly weathered	>50% of rock decomposed and disintegrated; fresh or weakened rock locally present, including as corestones
Dominance of rectangular corestones (50–90%), surrounded by residual matrix	Corestones with residual debris	III	III	Moderately weathered	<50% of rock decomposed and disintegrated; rock discoloured and partly weakened; corestones may be common
>90% is solid rock, minor amounts of debris along fractures, iron staining may be present	Partially weathered rock	IV	II	Slightly weathered	Discoloration as the main sign of change, mainly along discontinuities; some weakening of fabric
No visible signs of rock weathering	Bedrock		IB	Faintly weathered	Slight discoloration along major discontinuities
			IA	Fresh	No visible signs of weathering

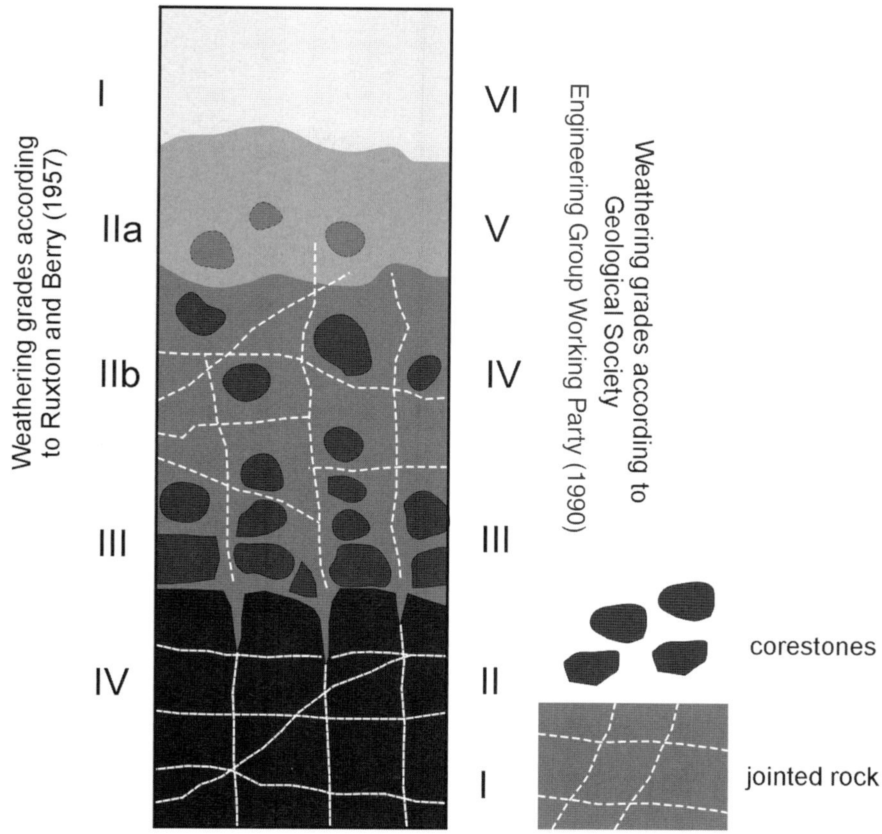

Fig. 1. An ideal weathering profile developed on basement rock, particularly granite.

excavations, have confirmed the great variability in saprolite thickness. Shaw (1997) in Hong Kong has documented the range of depth to the solid rock from surface to 90 m. In high-relief areas, the thickness of the weathered mantle typically increases towards the footslope, but in some hilly terrains the relationships may be reversed. A multi-convex topography, identified by Thomas (1974) as a distinctive category of basement terrains, is often associated with great depth of weathering under hilltops. Similar patterns have been observed in the humid tropical mountains of Papua New Guinea (Pain & Ollier 1981).

Mass movement in weathered granite and gneiss terrains

SE Brazil case study

The coastal ranges of SE Brazil, and the section of the Serra do Mar between São Paulo and Rio de Janeiro in particular, offer a multitude of examples of relationships between deep weathering and mass movement. The ranges represent a section of the great escarpment running along most of Brazil's Atlantic coast and are formed of Precambrian basement rock, comprising mainly granite and gneiss (da Silva & de Andrade Ramos 2002). They act as a prominent orographic barrier to moisture-laden air masses advancing from the Atlantic and receive considerable rainfall throughout the year. In summer, daily totals are not uncommonly around 300 mm, with intensities in excess of 50 mm h^{-1}. Abundant rainfall, coupled with high biological productivity and tectonically induced structural weakening of the basement, creates conditions for efficient weathering. Granite and gneiss are weathered to great depth, locally as much as 80 m (Thomas 1994). In steep terrain, the bulk of saprolite is composed predominantly of poorly to moderately decomposed rock, with original structure largely preserved, but in more subdued topographic settings highly decomposed materials are widespread (Fig. 3).

Within the Serra do Mar, several major types of geomorphological terrains may be distinguished. These include the following.

Fig. 2. Corestones in deeply weathered granite on the Meghalaya plateau, NE India. Two people in the upper right provide the scale (photograph courtesy of A. Latocha).

Fig. 3. A thick zone of weathering over Precambrian gneiss in Serra do Mar, SE Brazil, exposed in a road cut through a cupola-shaped hill. Rills and gullies are associated with collapsed pipes and efficient surface runoff.

(1) Massive steep-sided domes of solid rock, rising as much as a few hundred metres above their surroundings. They may occur in isolation or in clusters, or project from mountain crests. Their upper slopes are usually bare, with occasional patches of regolith cover. Rio de Janeiro itself is a classic location for huge granite and gneiss domes (da Silva & de Andrade Ramos 2002).

(2) Highly dissected mountain massifs, with steep forested slopes underlain by a discontinuous mantle of weathered rock of variable thickness. However, the weathering grade is not high and decomposition is limited. The Tijuca massif [1012 m above sea level (a.s.l.)] towering above Rio de Janeiro serves as an example.

(3) Hilly, multi-convex relief typified by associations of cupola-like hills 50–100 m high separated by flat-floored valleys. In contrast to the previous type, weathering is far more advanced and many single hills are weathered throughout, with others having a rock core mantled by a thick saprolitic cover. Arcuate slope hollows, incised into many hills, are a notable feature here. Low-angle planar surfaces occur in the front of the hollows, accounting for the general slope concavity (de Meis & Monteiro 1979; Coelho Netto 1999) This type of morphology exists, for example, in the intermontane depression of the Paraiba do Sul.

The types of relief distinguished above are associated with different types of mass movement phenomena. Massive domes are subject to long-term unloading, resulting in the origin of surface-parallel partings (sheeting joints). Hence, they become affected by slab slides and rock falls, particularly near overhangs produced by selective weathering. Scars within the rock slopes, and extensive talus around the perimeter, usually hidden in dense forest, constitute geomorphological evidence for slope failures. In highly dissected terrains, mass movement is triggered by heavy rainfall and particularly affects slope hollows and first-order basins, which act as runoff collectors (Fernandes *et al.* 2004; Vieira & Fernandes 2004). The majority of failures are shallow translational slides and debris avalanches involving the entire weathering mantle, so that fresh bedrock surfaces become exposed in many places. After reaching valley floors, the sliding masses may transform into debris flows and travel far beyond the steep catchments. The reasons for this transformation are complex and include breakdown of the internal structure of the sliding mass, the surplus of water, available from streams occupying the valleys, and the confining effect of relief. Subsequent slides may affect colluvium derived from previous landslides (De Ploey & Cruz 1979).

In the multi-convex hilly relief the patterns of slope instability are different. Deep-seated rotational slides are most common and these, if not confined by topography, transform into earthflows spreading across the footslopes and building up ramps. The location of the landslides is not uncommonly controlled by relict joint planes, which focus subsurface runoff and become subject to piping and tunnel erosion (Fig. 3), often antecedent to more widespread slope failures (Coelho Netto 1999). Regionally repetitive associations of slope hollows and ramps are interpreted as the geomorphological record of past landslides, covering the period of the late Pleistocene and the Holocene (de Meis & Monteiro 1979; Coelho Netto 1999).

South Mexico case study

The Veladero massif in southern Mexico is a granite stock of Palaeogene age, an outlier of the Sierra Madre del Sur mountain range (Ducea *et al.* 2004). Topographically, it rises steeply above the Bay of Acapulco, with relative relief of >900 m and a slope gradient in excess of 25° (Migoń & Alcántara-Ayala 2008). Despite its location in the humid subtropical belt both the depth and grade of weathering are moderate. Disintegrated to moderately decomposed granite ubiquitously underlies the hillslopes, but the thickness of the saprolite is normally only a few metres. Densely packed corestone horizons and weakly disintegrated rocks occur near the base of many exposures. These visual observations are consistent with the mineralogy of the weathered mantle and the colluvium. A mixture of smectite, illite and kaolinite has been identified in the saprolite (Migoń & Alcántara-Ayala 2008).

Evidence for the role of mass movement is threefold. First, the slope morphology is corrugated, with an abundance of hollows of various shapes (Fig. 4). These are scars left by past landslides, which often affected the entire weathered mantle, exposing jointed bedrock. Bedrock exposures in turn shed boulders, which accumulate in the newly created hollows. Likewise, exposed corestone horizons are susceptible to further instability and fail as a result of selective water erosion of the fine saprolitic material. Second, stacked colluvial packages occur along the footslopes, comprising large boulders, as much as 2 m long, chaotically set within a sandy–silty matrix. The observed thickness of the colluvial packages is up to 4 m (Migoń & Alcántara-Ayala 2008). The combined geomorphological and sedimentological evidence points to shallow debris slides, and occasional wedge failures, as the prevalent mass movement mechanisms shaping the lower slopes of the Veladero massif. Third, valleys dissecting the massif are affected by debris flows. The main sources of debris are the primary weathered mantle and colluvium produced by past landslides (Lugo *et al.* 2002). In addition, lag deposits from ancient flows may be remobilized. Major debris flows are triggered by rainfall associated with hurricanes, which may bring as much as 300–400 mm of rain in 24 h, as was the case in October 1997. The transport capacity of flows is enormous. During hurricane Pauline in 1997, boulders as much as 9 m long were transported, completely choking the streets of Acapulco and causing much damage to infrastructure.

SE China case study

SE China, including Hong Kong, is a classic area for studying deep weathering patterns in granite rocks (Ruxton & Berry 1957; Berry & Ruxton 1960; Irfan 1996) and mass

Fig. 4. Topographic hollows (remnants of past debris slides) incised into a weathered granite slope, Veladero massif, Mexico. Boulders seen within the central hollow are products of rock fall from exposed faces of jointed granite. Colluvium packages are hidden from the view by residential buildings.

movements in weathered terrain (e.g. So 1971; Au 1998; Chen & Lee 2005). Much of the latter work is written from the viewpoint of engineering geology, but nevertheless these papers contain valuable observations about geomorphology of weathered slopes affected by landslides and debris flows.

The territory of Hong Kong is a mountainous area, with a highest elevation of 957 m a.s.l. and ubiquitous regolith-covered slopes that are locally very steep (>45°), in places subvertical. The hills comprise two principal rock types: granite of Jurassic–Cretaceous age and volcanic materials of similar age. Deep weathering is pervasive and the depth of granite-derived weathered mantle is typically 30–60 m, although highly variable spatially (Shaw 1997). All possible weathering grades are recorded, from disintegrated granite with corestones at depth through decomposed material to clay-rich 'residual soil' in the topmost part (Ruxton & Berry 1957; Au 1998). However, the last grade is relatively uncommon, contrary to what might be expected in the humid subtropical zone, and the majority of granitic saprolites contain mainly sand and silt fraction, with only subordinate clay. Relict joint planes are frequent in the moderately weathered granite and volcanic rocks (Au 1996) and these, especially if significant amount of clay is present along them, typically act as sliding planes because of much reduced strength (Chen & Lee 2005; Aydin 2006).

The relationships between weathering and the patterns of mass movement have been comprehensively analysed by Au (1998), and later by Ng *et al.* (2002). The following picture emerges from these studies.

(1) Shallow debris avalanches and debris flows are most common, whereas debris slides are less frequent and deep-seated slumps are very rare. Many debris avalanches are small and occur within artificial slope undercuts, affecting colluvium rather than weathered rock *in situ*, but they are frequent on natural slopes as well. Debris slides are the dominant type of mass movement affecting weathered bedrock, although these may transform into debris avalanches.

(2) Significant variations in physical properties of granite saprolites over short distances cause differential saturation and pore-water build-up, therefore most failures are local.

(3) The presence of solid rock compartments within saprolites, or less weathered veins, reduces hydraulic conductivity and leads to the rise of pore-water pressure and, eventually, failure above the solid rock.

(4) Piping is widespread within colluvium and upper saprolites, particularly where corestones are present, and is an important factor contributing to slope instability.

(5) Shallow landslides, especially if located on valley sides, turn into long run-out channelized debris flows after reaching drainage lines.

The situation in Hong Kong can be contrasted with that observed further inland in the Guangdong province (Xu 1996). The topography is more subdued and consists of a multitude of hills that are weathered throughout. Some resemble fairly symmetrical cupolas, but there are also elevations with distinct hemispherical hollows cut into their slopes. These amphitheatre-like features narrow down into funnel-like outlets and widen again into broad footslope fans. Xu (1996) introduced the Chinese word 'benggang', meaning 'collapsed hill', to describe this landform association. This term also provides the clue to the origin.

Hollows are produced by rotational slides within deeply weathered granite, which liquefy and flow towards lower ground, occasionally leaving a transitional furrow behind. Single benggang may reach 300 m in length, but they typically occur in clusters, reaching a density of a few tens per 1 km^2. Apparently, Chinese benggang are similar in appearance to hollows and ramps described from SE Brazil.

Complementary observations

In addition to the examples discussed above, slope instability phenomena in weathered basement rocks have been recorded in a multitude of other locations. Shallow debris slides in moderately weathered granite (grus or Japanese 'masa') are frequent in the mountains of Japan and the Korean Peninsula. They are triggered by heavy rainfall of daily intensities of 100–300 mm, related to passing typhoons (Chigira 2001; Wakatsuki *et al.* 2002; Onda *et al.* 2004). It has been observed that the density of landslides is often greater in granite areas in comparison with gneiss or shale terrain, but the average thickness of the failed mass is smaller. In the adjacent shale, deep-seated landslides are much more common. Onda *et al.* (2004) pointed to the scalloped topography of granite hillsides and perched groundwater tables in thin grus as factors conducive to widespread slope instability. Their findings are thus parallel to the conclusions reached by Fernandes *et al.* (2004) in SE Brazil.

Considerable extent of slope instability has also been demonstrated in the deeply weathered Serre granites in Calabria, southern Italy (Calcaterra *et al.* 1996; Calcaterra & Parise 2005). Again, shallow slides within the near-surface horizons of weathered mantle occurred first, but the steep terrain and sparse vegetation favoured transformation of these slides into debris flows. Altogether, hundreds of landslides of various types and different ages have been recorded in two small drainage basins, each a few tens of square kilometres in area, emphasizing the great role of mass movement in shaping weathered terrains.

Mass movements in various weathered basement rocks are common in the African highlands, for example in Malawi (Shroder 1976) or Zambia (Thomas 1999). Upland morphology with relatively little steep terrain bears on the type of mass movement. Rotational slides and short run-out earthflows are dominant, leaving amphitheatre-shaped scars in the upper slope and colluvial fans at the hillside base. In contrast, steep mountain slopes of subtropical Puerto Rico (De Graff *et al.* 1989) and of humid tropical coastal ranges of Venezuela (Larsen & Wieczorek 2006), where the weathered mantle is relatively thin, are mainly affected by shallow earth and debris slides, which may incorporate corestones. In the latter area bedrock is gneiss and schist and weathered schistosity planes provide excellent sliding surfaces. During the flash flood disaster in northern Venezuela in 1999 colluvial packages were eroded and ancient debris-flow deposits undermined, supplying more boulders. Gneiss blocks as large as 11 m long were recorded to have moved (Larsen & Wieczorek 2006). On reaching valley floors, slides turned into devastating debris flows and flash floods, carrying large boulders further away, and eventually depositing them as boulder fans in low-gradient, densely built-up areas.

Weathering–mass movement relationships

It becomes evident that weathered basement terrains, and granites in particular, host frequent slope failures. These vary in style, depending on the geology, relief and environmental factors. However, it appears that specific weathering grades tend to be associated with specific types of mass movement. Older miscellaneous observations were assembled by Durgin (1977), who suggested the relationships listed in Table 2. More recent work, including that reviewed above, is generally consistent with his proposal, although another summary relationship is offered here, related to the classification scheme by Ruxton & Berry (1957) and the Geological Society Engineering Group Working Party (1990) (Fig. 5).

Early stages of weathering, which involve preferential alteration along discontinuity planes and opening of major fractures as a result of tensile stress (stages I and II in the Geological Society Engineering Group Working Party (1990) scheme), are associated with rock slides and rock falls, depending on local slope gradient. In either case, shear strength of the rock along a parting is significantly reduced (Au 1996). Secondary clay accumulations within dilation joints, which may be very thin (Ehlen 1999) but laterally very extensive (Campbell & Parry 2002), also contribute to slope failures in an otherwise massive rock. Slope collapses on the hillsides of granite or gneissic domes are typical gravitational phenomena causally related to incipient weathering.

At more advanced stages of deep weathering (stage III), with a mantle of grus (arène) present over a hilly topography, shallow debris slides are preferentially generated. Often, indurated joint planes (Ehlen 1999), or slickensides in

Table 2. *Weathering – mass movement relationships according to Durgin (1977)*

Stage of weathering	*Mass movements*
Saprolite (thoroughly decomposed rock)	Rotational slides, slumps
Decomposed granite (<15% of fresh rock left)	Debris flows, debris avalanches, debris slides
Core stone stage (15–85% of fresh rock left)	Rock fall avalanches, rolling rocks
Fresh rock (<15% of weathered material along joints)	Rock falls, rock slides, block glides; debris avalanches and slides over sheeting surfaces

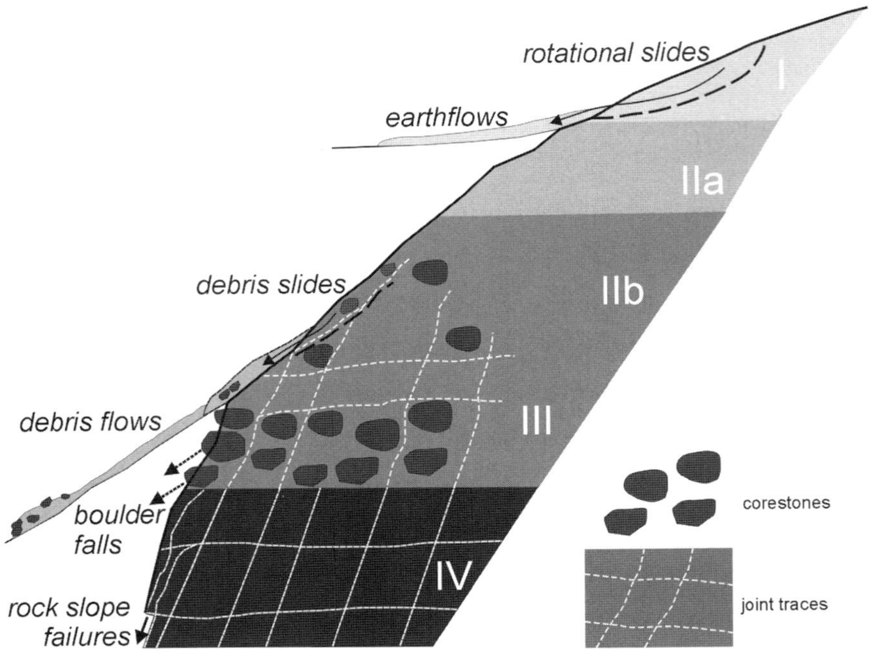

Fig. 5. Weathering grade–mass movement relationships in basement rocks.

tectonized basement rocks turned to grus, act as sliding planes. In terrains where the mantle of grus is thin (a few metres), and the weathering front separating grus from solid rock is sharp, slides may occur at this interface (Chigira 2001; Onda 2004). This is because the weathering front provides an efficient barrier to groundwater movement, resulting in the development of ephemeral perched water tables during rainstorms. Hence, the thin grus above the interface becomes saturated and the pore-water pressure rises above the limit of material stability.

Another type of mass movement associated with disintegrated rock, and with weathered granites in particular, is corestone fall. If a grus section with frequent unweathered boulders is exposed, selective washing of the encapsulating finer material by rainwater may excavate and undermine corestones, so that they project out of the outcrop face. Further erosion, or an external trigger such as a seismic event, may lead to their fall and movement downslope by rolling or bouncing. This type of movement is probably comparatively rare in natural landscapes, but is likely to be common in man-made slopes, especially where they are poorly designed and maintained (Durgin 1977; Grigg & Wong 1987; De Graff *et al.* 1989).

Highly and completely weathered rock (stages IV and V), with few corestones left and significant percentages of secondary clays, is typically affected by rotational slides. The development of curved failure planes is facilitated by the homogenization of the saprolite and disappearance of discontinuity planes. Au (1996) demonstrated, using the example of saprolites from Hong Kong, that even if joints persist, the ratio of joint strength to intact strength diminishes toward unity.

The relationships of mass movements with weathering grades III to V are further influenced by local relief. If a hillside affected by mass movement is long and steep enough, and topographic and hydrological conditions are conducive, then an initial slide may transform into a flow-type phenomenon. Geomorphological settings favouring the transformation are hillslope hollows and valley floors occupied by streams present at the time of failure. After a debris slide reaches a local watercourse, the content of water in the sediment increases, the pore-water pressure builds up and the sliding mass liquefies (Iverson *et al.* 1997). The events described from the Tijuca Massif in SE Brazil (Coelho Netto 1999; Fernandes *et al.* 2004), when local debris slides transformed into debris flows travelling along the valleys for a few kilometres, provide a very good example. Numerous instances of similar transformation have been recorded in the steep mountainous terrain of Hong Kong (Ng *et al.* 2002). In more weathered terrains, colluvium from rotational slides may liquefy and continue moving as earthflows, spreading wide over the available footslope space. The wide depositional surfaces below the landslide scars in multi-convex terrain are the product of this type of transformation (Xu 1996; Coelho Netto 1999).

Durgin (1977), in his review of landsliding in weathered granite, asserted that slope instability phenomena are most common in 'decomposed' rock rather than in 'saprolite', and hence at stage III rather than IV or V. Regional analysis of mass movement patterns in the Hiroshima area, Japan, resulting from extremely high precipitation of 250 mm on 29 June 1999 (Chigira 2001) confirms this view. The majority of landslides affected those granite slopes where surface-parallel micro-sheeting was present and the characteristics of weathered granite conformed with sandy grus. The likely reasons suggested by Durgin (1977) included a significant decrease in bulk density, practically no cohesion of either saturated or dry material, and efficient seepage. In addition, sliding may exploit indurated joint planes. However, relief remains a key control of failure mechanism. Slightly and moderately weathered rock, by virtue of its 2–3 times higher strength in comparison with highly weathered rock (Irfan 1996), can support steeper slopes than those developed in fine-grained, clayey saprolites. These steep, metastable slopes may be more prone to abrupt yielding after rapid disturbance, whether by brief but intense rainfall (Chigira 2001) or an earthquake, than gentler slopes, which release stress continuously through creep and shallow soil flow. In hilly to mountainous terrains steep slopes underlain by thick clay-rich saprolites may simply not exist.

Geomorphological consequences of landslides in weathered terrains

The relationships between deep weathering patterns, mass movement and relief are both complex and mutually interdependent, involving a range of feedbacks. The pre-existing topography dictates the pattern of mass movements, including the location of slope failures, the travel path, and the run-out distance. However, landslides and other types of mass movement play a major role themselves in shaping the topography of weathered terrain, creating new landforms and providing new avenues for sediment transfer. In turn, some of these landforms have considerable lifetimes, constituting thereby a readily available geomorphological record of mass movement in the past (e.g. Thomas 1999; Migoń & Alcántara-Ayala 2008).

Geomorphological consequences of slope instability events in the weathered terrain may be considered in terms of both form and process. Mass movement effects are essentially erosional in the upper slope and depositional in the lower slope and/or in the valleys below. In highly weathered rocks, hemispherical slope hollows may become the dominant landform and terrain unit, changing a convex relief compartment into a scalloped hill. The density of the hollows is locally considerable. Shroder (1976) mapped landslides in the deeply weathered and moderately dissected Nyika Plateau in Malawi, Africa, and recorded 235 landslides over an area of 277 km^2. In the Bananal area of SE Brazil slope hollows cover up to 20% of the surface of headwater drainage basins (Coelho Netto 1999; Fig. 6). In Wuhua County in SE China more than 10 000 rotational slides of various size were triggered by heavy rainfall associated with a typhoon in 1986, in an area of *c.* 2800 ha (Xu 1996). The importance of slope hollows has also been noted in deeply weathered granitoids in Calabria (Calcaterra & Parise 2005). In less weathered terrains, where the thickness of the saprolite is usually lesser, complete stripping down to solid rock may take place (De Ploey & Cruz 1979; Coelho Netto 1999; Fernandes *et al.* 2004). As a result, a rock slope facet becomes exposed and this, if sufficiently jointed, may shed blocks and boulders to the lower slope, where they come to rest on colluvial fans

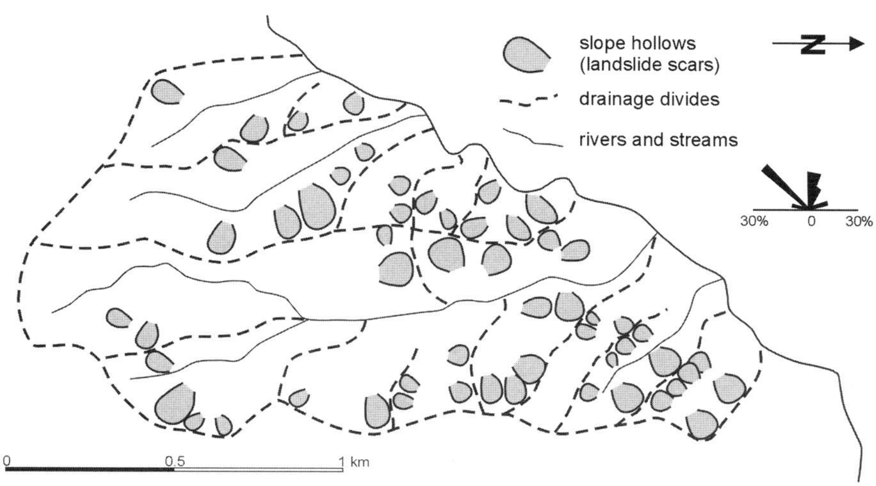

Fig. 6. Distribution of slope hollows in the deeply weathered gneissic terrain near Bananal, SE Brazil. It should be noted that elongation of many of the hollows coincides with major joint directions (shown at centre right) (after Coelho Netto 1999, simplified by the author).

from preceding debris slides (Migoń & Alcántara-Ayala 2008). Another class of landforms associated with mass movement is funnel-shaped depressions connecting slope hollows to footslopes below. These are mainly erosional facets moulded by earthflows or debris flows, subsequent to slides. Landforms produced by debris flows do not seem to be particularly distinctive compared with those related to debris flows in other landscape settings, for instance in high-mountain terrain. However, the structure of associated levees reflects the nature of the saprolite. If corestones were abundant, then boulder-rich levees would easily form. In the Veladero granite massif, Mexico, levees are densely packed boulder ridges, with single blocks up to 4–5 m long (Lugo *et al.* 2002; Migoń & Alcántara-Ayala 2008).

Depositional landforms are less obvious than the erosional ones, particularly after some time elapses and revegetation occurs. Low-angle (<5°) footslope fans and aprons built of landslide material are common and these may grade down to alluvium-covered valley floors, local relief permitting. The size of these terrain units and their relative height are highly variable, depending on the magnitude and frequency of landslides in the source area. Around the Jos Plateau, Nigeria, the thickness of colluvial units is locally as much as 20 m (Hill & Rackham 1978), whereas depositional units in hilly lands of SE Brazil may extend for a few hundred metres, probably indicating a long and complex erosional and depositional history (de Meis & Monteiro 1979).

The occurrence of landslides paves the way for further geomorphological processes. Among these, gully erosion is worth particular attention, as it works side-by-side with landsliding. Usually, funnel-shaped transitional zones of landslides and footslope fans built of poorly consolidated material favour the subsequent development of gully networks (Coelho Netto 1999). Gullied slope hollows and valley floors are a common sight in SE Brazil (Fig. 7), SE China, and elsewhere. However, extending gullies promote seepage and tunnel erosion, and undermine the slope, triggering new landslides, albeit on a rather local scale.

The geomorphological evolution of landslide-affected weathered terrains is best described as episodic, with long intervals of apparent stability punctuated by brief periods of considerable geomorphological change. The latter are typically associated with heavy rainfall (e.g. So 1971; Xu 1996; Coelho Netto 1999; Chigira 2001; Lugo *et al.* 2002). Daily totals well in excess of 100 mm appear to be required to initiate mass movement at a large scale, whereas major catastrophes, such as those affecting Hong Kong in 1966 (So 1971) and Serra das Araras near Rio de Janeiro in 1967 (Jones 1973), were associated with at least 100 mm of rain in 1 h. Critical rainfall thresholds are as important as intrinsic thresholds of saprolite stability beneath a slope of given gradient. For example, Wakatsuki *et al.* (2002) noted that gradients between 27 and 33° are sufficient for shallow landsliding on gneiss, whereas 35–40° is required to trigger slips in weathered granite.

Stripping of a saprolite by landslide may be complete or incomplete. This appears to depend on both local relief and weathering grade. Complete stripping, typical of steep terrains underlain by less weathered mantles, means complete exhaustion of the reservoir for any subsequent

Fig. 7. Heavy gullying in the alluviated first-order valley floor, near Bananal, SE Brazil. Scalloped outlines of gully walls indicate the importance of piping in the enlargement of this feature. Exits of pipes are visible on the exposed gully wall.

movements. Exposed rock slopes may be very long-lived features; their steep gradients hamper the encroachment of vegetation and renewed development of the weathering mantle. Episodic landsliding has long been considered as one of the processes leading to the excavation of massive domed inselbergs, which typically rise from weathered footslopes in many low-latitude terrains (Hurault 1963; Thomas 1978). However, even after some saprolite is re-established, high gradients will favour mass movement and constant removal of loose material, retarding the formation of thicker clay-rich saprolites. The Veladero massif in Mexico is one example of a terrain where this pathway of long-term slope development has been hypothesized (Migoń & Alcántara-Ayala 2008). In more subdued, multi-convex topography, rotational slides (slumps) are common and their failure plane is seldom anywhere near the weathering front. Referring to the general concept of extrinsic and intrinsic limits to slope evolution, in moderately weathered steeplands the slopes appear to be weathering limited, whereas in the deeply weathered, multi-convex terrains, it is the efficacy of hillslope transport that sets the limits to how the slopes evolve.

Initial slope failures usually trigger a cascade of processes that extend down the drainage basin and may propagate well beyond the hilly, landslide-prone terrain. Extensive alluviation of valleys recorded in SE Brazil (Coelho Netto 1999), Hong Kong (Berry & Ruxton 1960) and the Guangdong province in SE China (Xu 1996) has its roots in equally extensive mass movement phenomena in upper catchment area.

Finally, a number of feedback relationships are evident from the study of landslides in weathered terrains, emphasizing the major role of feedbacks in the geomorphological evolution in general. For example, there are spatial and temporal linkages between landsliding and gullying. Analysis of slope hollow hydrology suggests a similar feedback (Fernandes *et al.* 2004; Vieira & Fernandes 2004). Slides leave concavities of variable depth in the upper slope zone of depletion. These topographic hollows are invaded by tropical vegetation relatively quickly, suggesting a re-establishment of stability, but their existence actually sets the stage for future movements. This is because the presence of hollows significantly affects slope hydrology, with most runoff being directed to the hollows. Elevated pore-water pressure typifies these landscape scars, which become prone to renewed landsliding. It seems that unless a very large event takes place and the regolith is completely stripped, hillside hollows and landslides will develop concurrently.

Conclusions

Weathered terrains, especially in low latitudes, are characterized by frequent landslides, debris flows and earthflows, and cannot be adequately understood without considering the role of episodic gravity-driven processes. In basement terrains, where structural differentiation of weathering profiles is often fairly clear, there are clear relationships between weathering grade, relief and mass movement. Less weathered rock is able to support steeper terrain and hence debris slides turning not uncommonly into flows, sometimes followed by boulder movement, are frequent. In highly weathered terrain, dominated by clayey saprolites, the relief is more subdued and the dominant type of slope failure is rotational sliding, which may turn into earthflows.

With respect to the etching–stripping model of landscape evolution, which is particularly useful to explain the long-term geomorphological evolution of the tropics (Thomas 1994), landslides appear to be the most potent agents of weathering mantle removal and provide valuable insights into the nature of the stripping component and its temporal dimension. Their high magnitude, wide extent, and episodic occurrence indicate that much of long-term geomorphological change is accomplished during ‘events’, capable of substantially affecting a large tract of terrain in a brief episode, rather than over the protracted time intervals between these events. Landslides and debris flows in weathered basement bedrock are clearly the ‘formative events’ *sensu* Brunsden (1993), which dictate the pattern of landform evolution for many subsequent years.

Geomorphology must play a major role in the landslide hazard assessment in weathered terrains. First, pre-existing relief is a key control of the location of failure zones and pathways taken by moving saprolitic masses. Geomorphological mapping is therefore indispensable in predicting where subsequent events may take place and how far their effects might extend down the valley. For example, Ng *et al.* (2002) concluded that many landslide sources are located near the pronounced breaks of slope that are common on valley sides or at the head of hollows. Thus, mapping the extent of slope breaks will help to locate potential places of future landslides, whereas an analysis of valley morphology will help to determine possible run-out distances. Second, landslides and subsequent debris flows or earthflows usually leave clear geomorphological and sedimentological evidence, in both the zones of depletion and deposition, although ground check may be necessary to confirm landslide origin. Their identification may considerably enlarge our database beyond the historical record, and assist in a more complete understanding of mass movement patterns in a given area. The relationships between terrain, weathering grade and mass movement explored in this paper may guide the search for traces of past slope instability at different scales.

Acknowledgements. This paper is part of the long-term research project focused on geomorphological evolution of basement terrains, funded by the Ministry of Science and Higher Education in Poland via the University of Wrocław. The research project in Mexico, referred to in this paper, was supported by the Instituto de Geografía and Programa de Intercambio Académico, CTIC, UNAM. My special thanks go to the researchers who in various ways contributed to the development of ideas presented in this paper. These are, in alphabetical order, I. Alcántara-Ayala, N. Fernandes, C. Limeira Mello, M. Naise de Oliveira Peixoto and M. Thomas. Finally, my sincere thanks go to C. Ollier and S. Campbell, whose comments and suggestions helped to improve the original manuscript.

References

AU, S. W. C. 1996. The influence of joint-planes on the mass strength of Hong Kong saprolitic soils. *Quarterly Journal of Engineering Geology*, **29**, 199–204.

AU, S. W. C. 1998. Rain-induced slope instability in Hong Kong. *Engineering Geology*, **51**, 1–36.

AYDIN, A. 2006. Stability of saprolitic slopes: nature and role of field scale heterogeneities. *Natural Hazards and Earth Science Systems*, **6**, 89–96.

BERRY, L. & RUXTON, B. P. 1960. The evolution of Hong Kong Harbour. *Zeitschrift für Geomorphologie, Neue Folge*, **4**, 97–115.

BRUNSDEN, D. 1993. Tablets of stone: toward the ten commandments of geomorphology. *Zeitschrift für Geomorphologie, Neue Folge, Supplementband*, **79**, 1–37.

CALCATERRA, D. & PARISE, M. 2005. Landslide types and their relationships with weathering in a Calabrian basin, southern Italy. *Bulletin of Engineering Geology and the Environment*, **64**, 193–207.

CALCATERRA, D., PARISE, M. & DATTOLA, L. 1996. Debris flows in deeply weathered granitoids (Serre Massif – Calabria, Southern Italy). *In*: SENNESET, K. (ed.) *Proceedings of the 7th International Symposium on Landslides, Trondheim*. Balkema, Rotterdam, 171–176.

CAMPBELL, S. D. G. & PARRY, S. 2002. *Report on the investigation of kaolin-rich zones in weathered rocks in Hong Kong*. Geotechnical Engineering Office, Report **132**.

CHEN, H. & LEE, C. F. 2005. Geohazards of slope mass movement and its prevention in Hong Kong. *Engineering Geology*, **76**, 3–25.

CHIGIRA, M. 2001. Micro-sheeting of granite and its relationship with landsliding specifically after the heavy rainstorm in June 1999, Hiroshima Prefecture, Japan. *Engineering Geology*, **59**, 219–231.

COELHO NETTO, A. L. 1999. Catastrophic landscape evolution in a humid region (SE Brazil), Inheritances from tectonic, climatic and land use induced changes. *Supplementi di Geografia Fisica e Dinamica Quaternaria*, **3**, 21–48.

DA SILVA, L. C. & DE ANDRADE RAMOS, A. J. L. 2002. Pão de Açúcar, RJ. Cartão postal geológico do Brasil. *In*: SCHOBBENHAUS, C., CAMPOS, D. A., QUEIROZ, E. T., WINGE, M. & BERBERT-BORN, M. L. C. (eds) *Sítios Geológicos e Paleontológicos do Brasil*. DNPM/CPRM–Comissão Brasileira de Sítios Geológicos e Paleobiológicos (SIGEP), Brasília, 263–268.

DE MEIS, M. R. M. & MONTEIRO, A. M. F. 1979. Upper Quaternary 'rampas', Doce river valley, Southeastern Brazilian plateau. *Zeitschrift für Geomorphologie Neue Folge*, **23**, 132–151.

DEARMAN, W. R., BAYNES, F. J. & IRFAN, T. Y. 1978. Engineering grading of weathered granite. *Engineering Geology*, **12**, 345–374.

DE GRAFF, J. V., BRYCE, R., JIBSON, R. W., MORA, S. & ROGERS, C. T. 1989. *Landslides, their extent and significance in the Caribbean*. *In*: BRABB, E. E. & HARROD, B. L. (eds) *Landslides. Extent and Economic Significance*. Balkema, Rotterdam, 51–80.

DE PLOEY, J. & CRUZ, O. 1979. Landslides in the Serra do Mar, Brazil. *Catena*, **6**, 111–122.

DUCEA, M. N., GEHRELS, G. E., SHOEMAKER, S., RUIZ, J. & VALENCIA, V. A. 2004. Geologic evolution of the Xolapa Complex, southern Mexico: evidence from U–Pb zircon geochronology. *Geological Society of America Bulletin*, **116**, 1016–1025.

DURGIN, P. B. 1977. Landslides and the weathering of granitic rocks. *Reviews in Engineering Geology*, **3**, 127–131.

EHLEN, J. 1999. Fracture characteristics in weathered granites. *Geomorphology*, **31**, 29–45.

FERNANDES, N. F., GUIMARÃES, R. F., GOMEZ, R. A. T., VIEIRA, B. C., MONTGOMERY, D. R. & GREENBERG, H. 2004. Topographic controls of landslides in Rio de Janeiro, field evidence and modeling. *Catena*, **55**, 163–181.

GEOLOGICAL SOCIETY ENGINEERING GROUP WORKING PARTY 1990. Tropical residual soils. Engineering Group Working Party Report. *Quarterly Journal of Engineering Geology*, **23**, 1–101.

GRIGG, P. V. & WONG, K. M. 1987. Stabilization of boulders at a hillslope site in Hong Kong. *Quarterly Journal of Engineering Geology*, **20**, 5–14.

HILL, I. D. & RACKHAM, I. J. 1978. Indications of mass movement on the Jos Plateau, Nigeria. *Zeitschrift für Geomorphologie Neue Folge*, **22**, 258–274.

HURAULT, J. 1963. Recherches sur les inselbergs granitiques nus en Guyane Française. *Revue de Géomorphologie Dynamique*, **14**, 49–61.

IRFAN, T. Y. 1996. Mineralogy, fabric properties and classification of weathered granites in Hong Kong. *Quarterly Journal of Engineering Geology and Hydrogeology*, **29**, 5–35.

IRFAN, T. Y. 1999. Characterization of weathered volcanic rocks in Hong Kong. *Quarterly Journal of Engineering Geology and Hydrogeology*, **32**, 317–348.

IVERSON, R. M., REID, M. E. & LAHUSEN, R. G. 1997. Debris-flow mobilization from landslides. *Annual Review of Earth and Planetary Sciences*, **25**, 85–138.

JONES, F. O. 1973. *Landslides of Rio de Janeiro and the Serra das Araras escarpment, Brazil*. US Geological Survey, Professional Papers, **697**.

LARSEN, M. C. & WIECZOREK, G. F. 2006. Geomorphic effects of large debris flows and flash floods, northern Venezuela, 1999. *Zeitschrift für Geomorphologie Neue Folge, Supplementband*, **145**, 147–175.

LUGO, J., VÁZQUEZ CONDE, M. T., ZAMORANO OROZCO, J. J., MATÍAS RAMÍREZ, G. & GÓMEZ ARIZMENDI, A. 2002. El huracán Pauline en Acapulco, octubre de 1997. *In*: LUGO, J. & INBAR, M. (eds) *Desastres naturales en América Latina*. Fondo de Cultura Económica, México City, 267–288.

MIGOŃ, P. 2006. *Granite Landscapes of the World*. Oxford University Press, Oxford.

MIGOŃ, P. & ALCÁNTARA-AYALA, I. 2008. Weathering and landform development in a subtropical mountainous terrain, Veladero massif, Mexico. *Zeitschrift für Geomorphologie Neue Folge*, **52**, 1–16.

MIGOŃ, P. & LIDMAR-BERGSTRÖM, K. 2001. Weathering mantles and their significance for geomorphological evolution of central and northern Europe since the Mesozoic. *Earth Science Reviews*, **56**, 285–324.

MIGOŃ, P. & THOMAS, M. F. 2002. Grus weathering mantles – problems of interpretation. *Catena*, **49**, 5–24.

NG, K. C., PARRY, S., KING, J. P., FRANKS, C. A. M. & SHAW, R. 2002. *Guidelines for natural terrain hazard studies*. Geotechnical Engineering Office, Report **138**.

ONDA, Y. 2004. Hillslope hydrology and mass movement in the Japanese Alps. *In*: OWENS, P. N. & SLAYMAKER, O. (eds) *Mountain Geomorphology*. Arnold, London, 153–164.

ONDA, Y., TSUJIMURA, M. & TABUCHI, H. 2004. The role of subsurface water flow paths on hillslope hydrological processes, landslides and landform development in steep mountains of Japan. *Hydrological Processes*, **18**, 637–650.

PAIN, C. F. & OLLIER, C. D. 1981. Geomorphology of a Pliocene granite in Papua New Guinea. *Zeitschrift für Geomorphologie Neue Folge*, **25**, 249–258.

PEDRO, G. 1968. Distribution des principaux types d'alteration chimique à la surface du globe, présentation d'une esquisse géographique. *Revue de Géographie Physique et Géologie Dynamique*, **10**, 457–470.

RUXTON, B. P. & BERRY, L. 1957. Weathering of granite and associated erosional features in Hong Kong. *Geological Society of America Bulletin*, **68**, 1263–1282.

SHAW, R. 1997. Variations in sub-tropical deep weathering profiles over the Kowloon Granite, Hong Kong. *Journal of the Geological Society, London*, **154**, 1077–1085.

SHRODER, J. F. JR. 1976. Mass movements on the Nyika Plateau, Malawi. *Zeitschrift für Geomorphologie Neue Folge*, **20**, 56–77.

SO, C. L. 1971. Mass movements associated with the rainstorm of June 1966 in Hong Kong. *Institute of British Geographers, Transactions*, **53**, 55–66.

TARDY, Y. & ROQUIN, C. 1998. *Derive dês continents. Paléoclimats et alterations tropicales*. BRGM, Orleans.

TAYLOR, G. & EGGLETON, R. A. 2001. *Regolith Geology and Geomorphology*. Wiley, Chichester.

THOMAS, M. F. 1966. Some geomorphological implications of deep weathering patterns in crystalline rocks in Nigeria. *Institute of British Geographers, Transactions*, **40**, 173–193.

THOMAS, M. F. 1974. Granite landforms, a review of some recurrent problems of interpretation. *Institute of British Geographers, Special Publication*, **7**, 13–37.

THOMAS, M. F. 1978. The study of inselbergs. *Zeitschrift für Geomorphologie Neue Folge, Supplementband*, **31**, 1–41.

THOMAS, M. F. 1994. *Geomorphology in the Tropics. A Study of Weathering and Denudation in Low Latitudes*. Wiley, Chichester.

THOMAS, M. F. 1995. Models for landform development on passive margins. Some implications for relief development in glaciated areas. *Geomorphology*, **12**, 3–15.

THOMAS, M. F. 1997. Weathering and landslides in the humid tropics, A geomorphological perspective. *Journal of the Geological Society of China*, **40**, 1–16.

THOMAS, M. F. 1999. Evidence for high energy landforming events on the central African plateau: Eastern Province, Zambia. *Zeitschrift für Geomorphologie Neue Folge*, **43**, 273–297.

VIEIRA, B. C. & FERNANDES, N. F. 2004. Landslides in Rio de Janeiro, the role played by variations in soil hydraulic conductivity. *Hydrological Processes*, **18**, 791–805.

WAKATSUKI, T., TANAKA, Y. & MATSUKURA, Y. 2002. The effects of bedrock lithology on soil layer structure, slip depth and slope angle of shallow soil slips on granite and gneiss slopes in Korea. *Transactions, Japanese Geomorphological Union*, **23**, 223–236.

XU, J. 1996. Benggang erosion, the influencing factors. *Catena*, **27**, 249–263.

Degradation of marls; two case studies from the Iberian Peninsula

E. E. Alonso*, J. A. Pineda & R. Cardoso

Department of Geotechnical Engineering and Geosciences, Universitat Politècnica de Catalunya, Jordi Girona 1–3, Building D-2, 08034 Barcelona, Spain

*Corresponding author (e-mail: eduardo.alonso@upc.edu)

Abstract: The strength degradation of two marl formations (Pamplona marls in Spain and Abadia marls in Portugal) is analysed in the paper. In the case of Pamplona marls, a number of large-scale direct shear tests on natural and saturated specimens of the undisturbed material were performed. Tests on marl–concrete interfaces were performed and are also presented. Both cases analysed show that the saturation of the marl leads to a strong reduction of friction and effective cohesion. Negative effects of suction changes on the mechanical behaviour of the marls were investigated in more detail in suction controlled tests on Abadia marls, a more plastic material. A numerical simulation of wetting effects on a marl fragment of finite size helps to identify the nature of degradation mechanisms, which are associated with the development of tensile surfaces inside the marl matrix. They, in turn, are due to the expansive behaviour of the clay fraction of the marl relative to suction changes.

A frequent observation in most civil engineering projects where clayey rocks are present is their high degradation potential as a result of weathering processes, which takes place, for instance, when they are excavated. 'Degradation' is a general term used to describe a series of phenomena, such as strength reduction, volume changes, stiffness loss, and, quite often, loss of mass continuity as a result of the development of fissures that can cause instability phenomena of finite volumes of rock (Alonso & Alcoverro 2002).

Experience indicates that degradation is frequently associated with material unloading and exposure to atmospheric conditions, which is typically the case in underground and surface excavations. Unloading changes effective stresses and leads to swelling of rocks if active minerals are present. In addition, wetting and drying cycles induce plastic deformations producing fissuring in the rock. The induced cracks are an easy way for the water to enter the rock.

Field profiles of water content of exposed clayey rock formations often show the presence of a well-defined 'weathering front'. Below the front the water content remains at its undisturbed low value. Above the front, the material evolves rapidly into a clay formation with a substantial increase in water content. Increasing water content (which is also an indication of increased porosity) leads to a fast reduction in mechanical properties (Pinyol *et al.* 2007).

Materials that exhibit these characteristics are very common in the Iberian Peninsula. Design provisions for engineering works (tunnelling, foundation of structures and slope stability, e.g.) in these materials must consider their evolving behaviour. El Bierzo clay, Lilla claystone, Pamplona marls, all from Spain, and Abadia marls, from Portugal, are examples of such materials. Figure 1 show the loss of strength and brittleness of red continental low-plasticity Tertiary clay from El Bierzo (Spain). When unweathered (i.e. in a massive state not affected by atmospheric action) this Tertiary clay may be described as a brittle material of medium strength (unconfined compression strength, $q_u = 1.0–3.0$ MPa). Two slow direct shear tests were performed on both unweathered and weathered specimens under identical confining stresses (100 kPa). The intact core, having a water content of 17%, was taken at a depth of 12 m and exhibited a marked brittle behaviour. A shallower specimen, having a slightly higher water content (19.2%), reacted as a ductile material. Its shear strength was close to the 'residual' strength determined in the shear box at saturated conditions, after a full-shearing cycle. The point to note in this case is that there are two alternative circumstances leading to the low strength: either shear straining of the unweathered material or natural weathering, which resulted in a (relatively small) increase in water content. However, the identification of these separate mechanisms is not simple in this case because the cores were recovered from a natural slope with symptoms of shallow instability. Therefore some previous natural strain in the upper core specimen cannot be disregarded.

Experience in tunnelling through these materials indicates that, in general, clayey rocks maintain relatively good supporting properties when mechanically excavated. However, when exposed to environmental conditions in the tunnel (relative humidity changes) the rock degrades in a relatively short time. This is the case for the Lilla claystone, an argillaceous rock containing anhydrite and gypsum veins that occurs in the Tertiary Lower Ebro Basin in NE Spain. This material was excavated by the Lilla tunnel of the

From: Calcaterra, D. & Parise, M. (eds) *Weathering as a Predisposing Factor to Slope Movements*. Geological Society, London, Engineering Geology Special Publications, **23**, 47–75.
DOI: 10.1144/EGSP23.5 0267-9914/10/$15.00

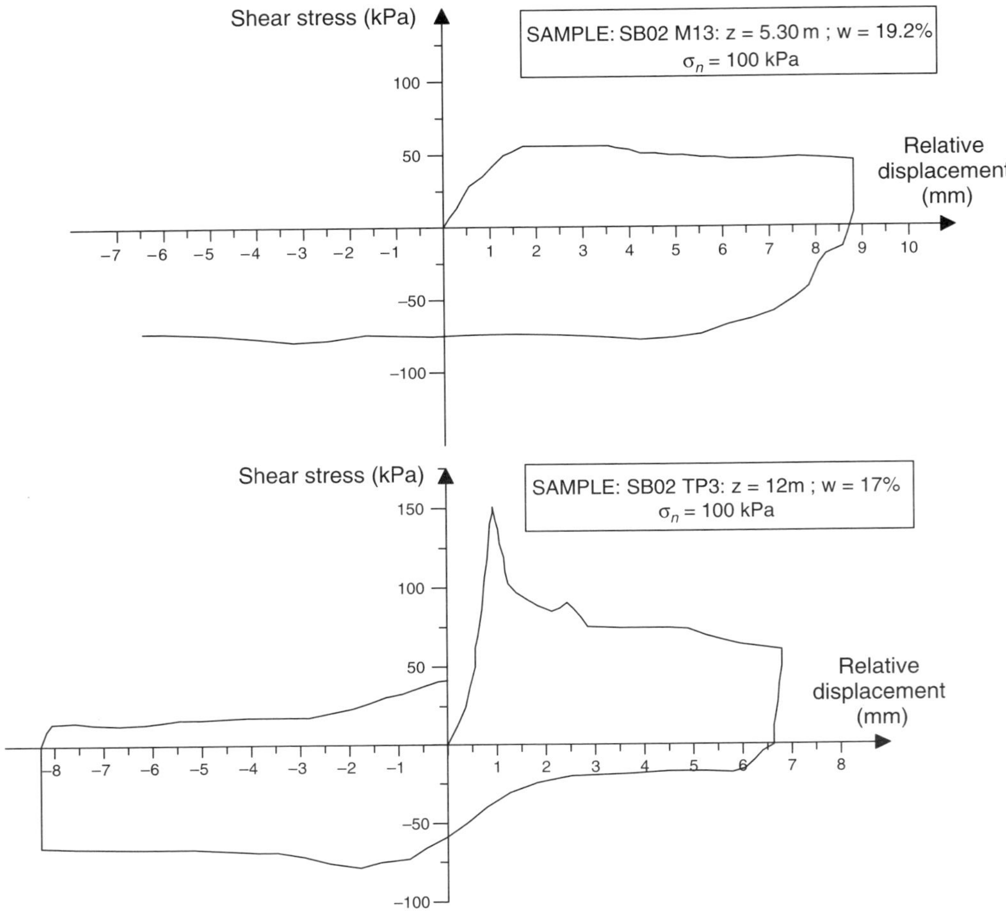

Fig. 1. Shear stress–relative displacement curves measured in two cores from El Bierzo Tertiary clay formations (modified from Pinyol *et al.* 2007).

Madrid–Zaragoza–Barcelona high-speed railway. Figure 2 shows the unconfined compression strength measured in rock cores taken in boreholes located in the floor of the tunnel (Berdugo 2007). The plot shows the fast reduction in strength as water content increases. Visual identification of cores as 'fresh' or weathered is also shown in the figure. Strong differences in unconfined compression strength were observed for similar water content values. For instance, the maximum strength for unweathered samples at 4% water content was about 50 MPa whereas the maximum value for weathered material was about 6 MPa. For samples tested, the degree of saturation was in the range of 50–75% and 50–100% for unweathered ('fresh') and weathered samples, respectively. A more detailed description of this case has been given by Alonso & Berdugo (2005).

These two examples serve to introduce the geotechnical and degradation features of two natural formations: Pamplona marls (Spain) and Abadia marls (Portugal). Geotechnical properties and weathering processes affecting the mechanical behaviour of these two marls are discussed below in more detail.

In the next section degradation phenomena of Pamplona marls are presented. The main geological and geotechnical characteristics are described. In addition, the degradation of the material is discussed in connection with the stability conditions of a concrete dam (Monreal dam). It will be shown that the selection of strength parameters needed in stability analysis is a complicated task.

The influence of water content changes (suction changes) on the hydro-mechanical properties of Abadia Marls will be investigated. A numerical analysis carried out to simulate hydraulic effects on degradation phenomena of this material will be presented. The authors have no knowledge of any previous studies on Pamplona and Abadia marls other than those mentioned in the text.

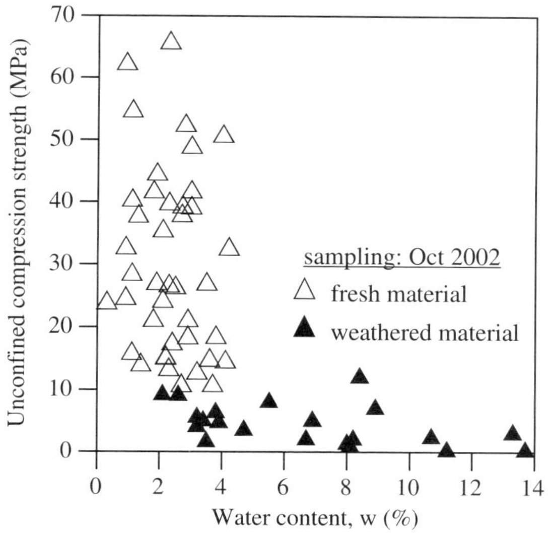

Fig. 2. Variation of unconfined compression strength with water content of Lilla claystone (Berdugo 2007).

The final section summarizes the most important topics discussed in this paper.

Pamplona marls

Geological characterization

Pamplona marl is an argillaceous rock of Northern Spain. This material can be defined as 'problematic' rock because of the extreme changes in behaviour when it is subjected to seasonal variations. In nature, a well-defined weathering front is usually observed. Within the Pamplona marl formation three main strata may be distinguished: a lower marly unit, an intermediate stratum, interbedded with sandstone layers, and a marly upper stratum. The lower stratum occupies most of the formation and it is composed of bluish grey massive marls that are prone to weathering. They decompose easily into centimetre sized fragments, which soon become a brown clayey material. Weathered superficial marls produce a brown clayey deposit, mixed with colluvial–alluvial Quaternary materials and terrace deposits located above. At deeper levels these brown clay levels are more compacted and their colour changes to grey, until they reach the fractured marls, and, at depth, the unweathered marls. This level contains frequent interlayers of sandstones with calcareous cement. The structure of these deposits shows N120–125°E orientation with a dip of 40–50° NE. In general, the depth of the weathering front ranges from 3 to 5 m (Anonymous 2002). Correspondence between different colours and weathering degrees is common in marls.

Weathered Pamplona marls show an average fines content of 90%. They are low-plasticity clays (CL) with liquid limit values ranging between 27 and 50% and plastic limits between 11 and 28%. The density of solids shows a variationbetween 2.60 and 2.76 Mg m^{-3} within the altered area. Within the weathered upper stratum the natural water content ranges from 2.6 to 24.0%, with a mean value of 16.5%. Figure 3 shows the variation of water content and Atterberg limits with depth. The unconfined compressive strength of the weathered layer ranges from 80 to 800 kPa. In terms of soil mechanics concepts, the soil consistency may be defined as intermediate–hard. It may be defined also as stable to rigid soil (Hoek & Bray 1981).

The mineral composition of unweathered marl has a significant proportion of calcite mineral aggregates (*c.* 50%), quartz (*c.* 20%), phyllosilicates (clays, mainly illite and chlorite) and mica. The dry density of unweathered marls ranges between 1.78 and 2.59 Mg m^{-3}. Natural water content below the weathering front shows very uniform values with a mean of 3% (Fig. 3). Unconfined compressive strength varies from 0.44 to 73.1 MPa, which classifies the rock from highly weathered to moderately hard material following Hoek & Bray (1981) (Fig. 4). Standard penetration test (SPT) may be performed in the upper weathered material and a significant scatter in values is found (21 to >50). SPT cannot be performed in the unweathered material.

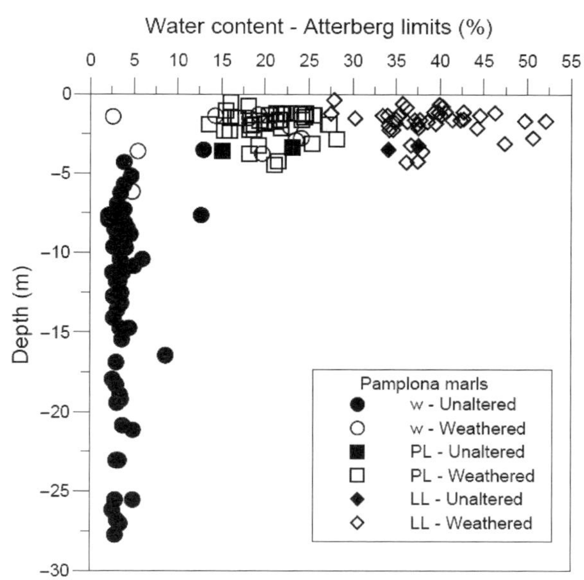

Fig. 3. Variation in water content and Atterberg limits with depth (modified from Mapa Geotécnico del Área de Pamplona, Anonymous 2002).

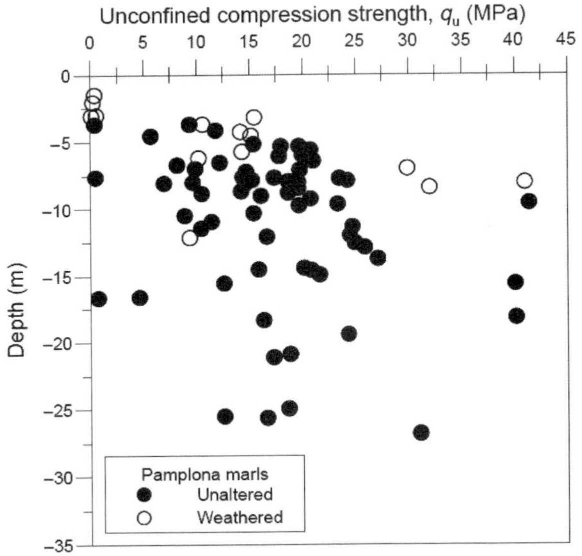

Fig. 4. Variation of unconfirmed compression strength of Pamplona marls with depth (modified from Mapa Geotécnico del Área de Pamplona, Anonymous 2002).

Stability problems on Pamplona marls

The general structure of Pamplona marls is determined by folds oriented NW–SE. These have a subhorizontal dip in extensive areas. However, this is not the case for outcrops located NE of Pamplona, where layers of marls tend to be vertical: dip values change between 50° and 70°, as shown in Figures 5 and 6. These figures show cuts in Pamplona marls on the northern Navarra highway at Burria. The layers show a dip similar to the angle of the excavated slope. Surface weathering may lead, however, to shallow slides and debris flows, as shown in Figure 5. These instability phenomena are more marked when the dip of the slopes is similar to the dip of the marl layers (Fig. 6). The sedimentation planes are preferential sliding surfaces as a result of the reduced shear strength that can occur in the low-roughness contacts.

Debris flows are frequently observed on natural slopes with heights in excess of 30 m and slopes in the range 1H:1V to 1H:3V. Weathering and erosion are commonly observed in these slopes. In artificial cuts the short-term stability is directly linked to the inner structure of the rock mass; that is, the relationship between the angle of the slope and the layer's dip (see Fig. 5). In general, cuts higher than 15–20 m and slopes varying between 1H:1V and 1H:2V could show debris flows, creep-slides, and even circular sliding (Anonymous 2002).

Strength parameters reported in the literature, obtained from drained and undrained triaxial tests, show some scatter probably related to the degree of weathering of the samples tested. Figure 7 shows the relationship between internal friction angle and cohesion for weathered samples of Pamplona marls. Also included are the strength parameters of Llundain marls, which are present in the same area. They show a similar composition and mechanical response.

Fig. 5. Steep natural slopes at Burria site (Pamplona). Debris flows on weathered materials.

Fig. 6. Pamplona marl; road cut on northern Navarra highway at Burria. The cut is stable because the slope angle corresponds to the dip of marl layers.

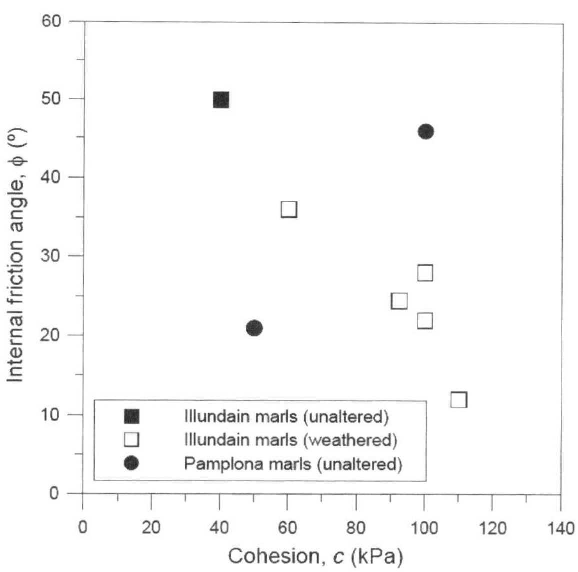

Fig. 7. Variation of internal friction angle with apparent cohesion for weathered Pamplona marls (modified from Mapa Geotécnico del Área de Pamplona, Anonymous 2002).

Drained triaxial tests indicate that the internal friction angle ranges between 22° and 36°, and cohesion ranges between 60 and 100 kPa. However, lower (12°) and higher friction values have been reported (Fig. 7).

Sliding stability on weathered surfaces: the case-study of Monreal Dam

General overview

A typical weathering profile of Pamplona marls is presented in Figure 8. This profile was observed at Monreal Dam site, which will be described below. The figure shows a stratigraphic section of the area, in which an upper clay layer containing gravels (average depth 3.90 m) may be observed. This stratum is followed by a 0.80 m gravel layer, and deeper marls (weathering intensity II–III and average thickness of 1.50 m). Below these materials, a marl layer (weathering intensity IB and average thickness of 1.50 m) reaches the foundation level. It was at this level that unweathered marl samples were collected to investigate the foundation properties of Monreal dam. Weathering intensity was evaluated using the criteria proposed by Bieniawski (1989). Further information about the sampling procedure has been given by Alonso *et al.* (2005).

Monreal dam is a gravity dam (massive concrete) with an upstream vertical face, a downstream slope of 0.8H:1V, and a maximum height from foundation level of 21.90 m. The dam is 153 m long and the concrete blocks of the dam, separated by joints, are founded at different levels. The concrete blocks were designed to be founded on unweathered marls (Fig. 9).

Mechanical borings and geophysical tests were carried out *in situ* to classify available materials. The geotechnical sequence was defined as follows: topsoil (thickness 0.20–0.80 m); alluvium and/or colluvium (thickness 2–5 m); weathered marls (thickness 2–3 m); cracked

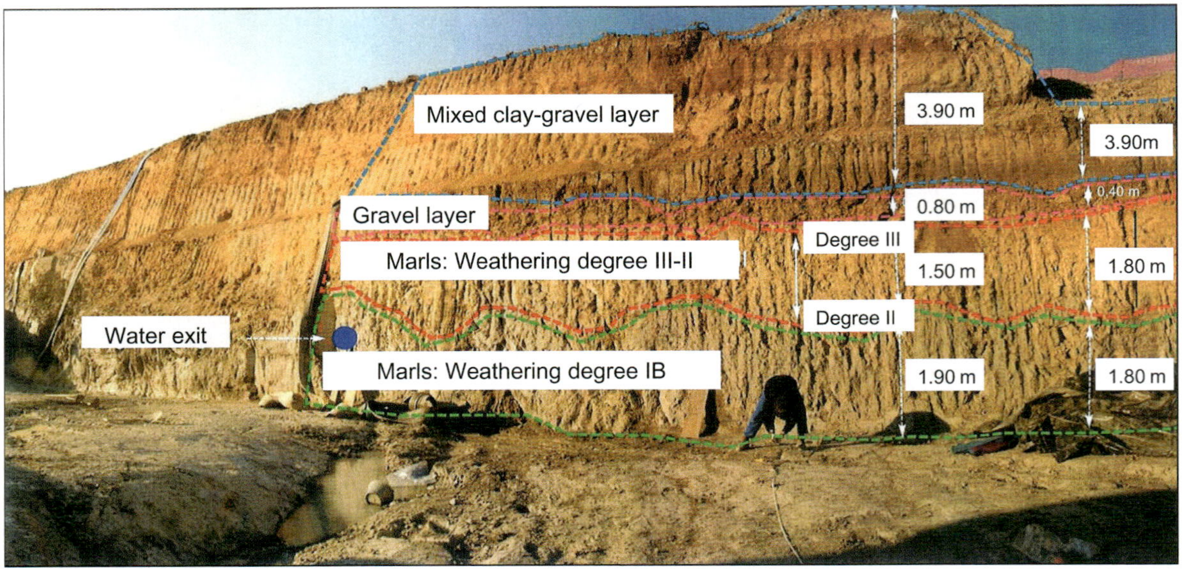

Fig. 8. Stratigraphic profile of Pamplona marls at Monreal Dam site.

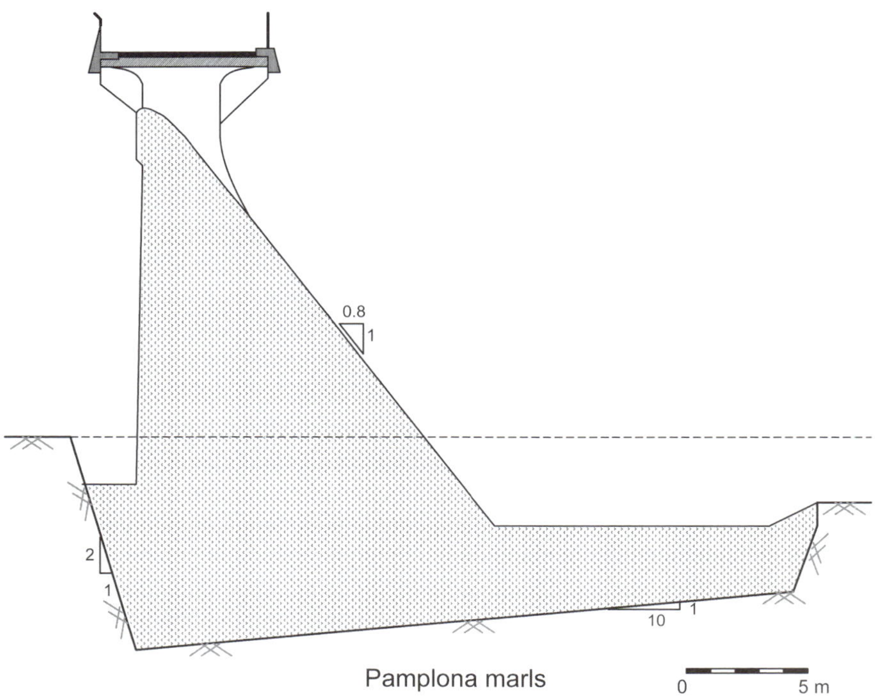

Fig. 9. Central cross-section of Monreal dam.

marls and sound marls at greater depths. The unconfined compressive strength for unweathered marls at this location varies from 15 MPa to more than 20 MPa.

Small increments in water content result in large drops in strength. Thus, marls appear either as unweathered materials ($q_u \approx 10$ MPa) or as weathered materials ($q_u \approx 0.4$ MPa). The dam designer adopted the following strength parameters to perform stability calculations: effective cohesion $c' = 290$ kPa and internal friction angle $\phi' = 30°$.

During excavation of the dam's foundation the natural weathering profile (Fig. 9) could be observed. Only three layers were identified: an upper mixed gravel–clay layer, followed by a weathered marl layer in which weathering processes have transformed the stiff and strong rock into stiff weathered clay, and finally the unweathered and unfissured marls. However, a few hours after excavation, the unweathered layer developed cracks as a result of stress relief and exposure to the local environment. This situation is illustrated in Figure 10, where four layers can be easily observed, including the layer of fissured marls at the foundation plane. Crack formation immediately after excavation is shown in Figure 11. Strong suction changes caused by drying as a result of exposure to the environment were the main factor responsible for the physical degradation observed. Once excavated for the first time the material evolves rapidly from a rock-type material towards a soil-type material. It is a common characteristic of these 'evolutive' materials to be very sensitive to exposure to environmental conditions. Small changes in water content at low stress levels lead to dramatic changes in mechanical properties such as loss of stiffness, swelling (if active minerals are present) and loss of material continuity because of cracking. This behaviour has been previously reported by various researches on several argillaceous soft rocks (e.g. Olivier 1987; Botts 1998; Bossart *et al.* 2002, among others).

The fast degradation of Pamplona marls raised some doubts about the strength parameters to be used in the analysis of Monreal dam stability against foundation sliding. The main issue is how to select the strength parameters for stability analysis. A test programme was carried out to check the validity of the values adopted in the design. Some of the tests performed were aimed at determining the strength of concrete–marl contact interfaces.

Laboratory tests performed on blocks of marl from Monreal site

Block samples were taken at five points of the foundation area (samples M1, M2, M3, M4 and M5) for laboratory tests performed at the Universitat Politècnica de Catalunya (Barcelona, Spain). Heavy-duty saws were used to cut the block samples. Two joint families were observed *in situ* (N120–160°, dip 40–50° NE; N35–80°, dip 40–50° NE). They complicated the extraction of block samples because the samples tended to fissure following the existing joints. Cracks developed soon once natural stresses were released

Fig. 10. Weathering profile and foundation plane.

Fig. 11. Cracking caused by unloading and exposure to atmospheric conditions.

and the samples were exposed to the atmosphere. The samples were covered with paraffin wax to limit cracking and to reduce water losses.

Table 1 summarizes the index properties measured on the fine fraction (<425 μm, No. 40 ASTM), where w_L is the liquid limit, w_P the plastic limit, PI the plasticity index, and ρ_s the density of solids. Table 2 summarizes the values corresponding to the natural density (ρ_n), dry density (ρ_d), void ratio (e) and degree of saturation (S_r), determined in seven 'unweathered' samples from blocks M3-E2 and M1-A3. To determine the natural density, the paraffin wax method was employed. it should be noted that the marl *in situ* appears unsaturated (S_r ranges from 0.73 to 0.95). Given the low porosity of the marl a significant field suction (a few megapascals) is expected.

Table 1. *Index properties for unweathered marl*

Sample	*Classification*	w_L (%)	w_P (%)	PI (%)	ρ_s (Mg m^{-3})
M3-E2	CL	37.0	18.7	18.3	2.60

Mineralogical analysis was performed to determine the presence of expansive minerals. The results showed that the predominant mineral phases are calcite and quartz.

Table 2. *Initial conditions for unweathered samples*

Block	*Sample*	w (%)	ρ_n (Mg m^{-3})	ρ_d (Mg m^{-3})	e	S_r (%)
M3-E2	1	5.14	2.31	2.20	0.18	72.9
M3-E2	2	6.51	2.32	2.18	0.19	87.4
M3-E2	3	8.13	2.29*	2.12*	0.23*	92.8
M3-E2	4	8.41	2.25*	2.08*	0.25*	86.5
M1-A3	5	6.94	2.31	2.16	0.20	88.6
M1-A3	6	7.24	2.29	2.14	0.22	86.5
M1-A3	7	6.87	2.34	2.19	0.19	95.3

*Values calculated for nearly saturated samples, after shear testing.

Significant proportions of mica and chlorite were also detected. The presence of mica can explain the drop in strength once the marl cementation (which seems to be weak) disappears.

Direct shear (DS) tests were carried out at two shearing conditions: (1) at natural water content (called here 'dry' conditions); (2) at 'saturated' conditions (i.e. achieved by adding water to the shear plane, immersed in a bath of water). To preserve as much as possible the water content and density conditions, shearing was carried out directly on the block sample. Each block (sealed and paraffined) was placed in a Perspex mould, into which mortar concrete was poured at the bottom to form the base plate that contains the sample.

A Wykeham Farrance WF 25505 shearing device with a 30 cm × 30 cm box was used. Shearing tests were carried out at three vertical stress levels, ranging from 32 to 163 kPa. The material, either 'saturated' or 'dry', was subjected to a progressive increase in tangential stress at a 5 μm min^{-1} constant horizontal displacement rate, until it reached a maximum displacement of 50 mm. The shearing time for each test was of 1 day. Figure 12a shows the set-up for direct shear tests.

The shear strength properties of the concrete–unweathered marl contact was also investigated in direct shear tests performed at three vertical stress levels (50, 100 and 183 kPa). The direct shear test was carried out under 'dry' conditions (i.e. at natural water content) using the same 30 cm × 30 cm shear box. The material was subjected to the same test procedure as described above.

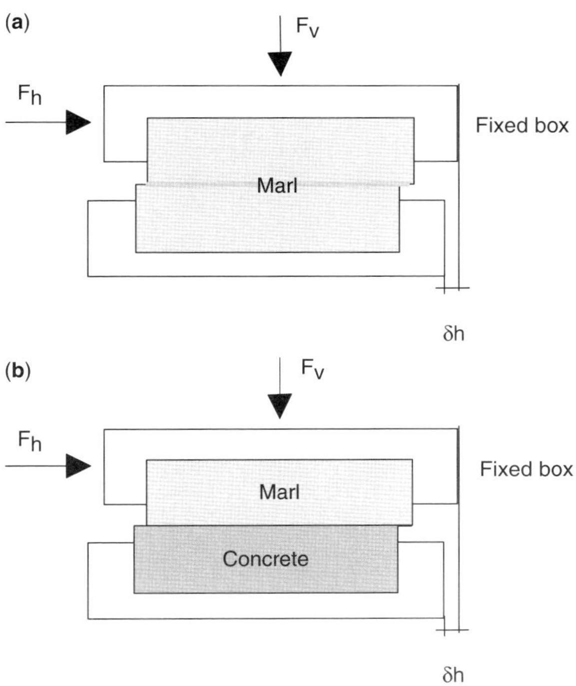

Fig. 12. Set-up of direct shear tests (**a**) on marls (**b**) on concrete–marl contact.

The set-up used for the marl–concrete contact was as follows. Surfaces of the future shear plane were adjusted horizontally within the shear box, keeping the concrete at the bottom. The lower part of the box was filled with cement mortar to fix the rock sample. The contact area was surrounded by sand (minimum thickness 15 mm), and the upper part of the box was filled with more mortar. After a day of curing, the two boxes were installed in the direct shear apparatus. Figure 12b shows the test set-up.

Before preparing the marl–concrete contact, roughness profiles of marl samples were also measured using a profile measuring device. Vertical and horizontal displacements were measured using a linear variable differential transformer (LVDT). Readings were stored in a personal computer by means of a digital data acquisition system. The direction of the roughness profiles in each sample was the same as the shear direction of the direct shear tests. Roughness profiles were used to determine the joint roughness coefficient (JRC) following Barton & Choubey (1977). By comparing each roughness profile with standard profiles a numerical value of JRC is determined, which can be related to some mechanical and hydraulic properties of joints. This parameter was used to determine the empirical strength envelopes using the criteria proposed by Barton *et al.* (1985). JRC values obtained from the roughness profile are given in Figure 13 for samples M1, M2 and M4. Two profiles were measured for each sample and correspond to specimens M1, M1-A, M2, M2-A, M4 and M4-A. The estimated closer reference profiles are also shown. Two widely different values of JRC (11 and unity), were approximated by this method.

Finally, residual shear strength conditions were evaluated by means of a ring shear test performed on remoulded material having a gravimetric water content close to the liquid limit ($w_L = 36\%$). Shearing was applied under three applied normal stress levels ($\sigma'_v = 40$, 80 and 150 kPa).

Strength

Figure 14 shows the shear stress–horizontal displacement relationship obtained from two samples subjected to similar vertical stresses (32 and 42 kPa). The only difference between them was the condition of shearing. Sample A ($\sigma_v = 32$ kPa) was sheared at natural water content whereas sample B ($\sigma_v = 42$ kPa) was sheared at saturated conditions. The results indicate that the saturation process leads to a significant decrease of the peak strength and to brittleness of the marl. At large displacements the strengths obtained in both cases are very similar and tend to the residual strength value. Special care was taken in all tests to maintain the same orientation of the strata in relation to the direction of shearing so that the natural anisotropy of the material would not affect the results.

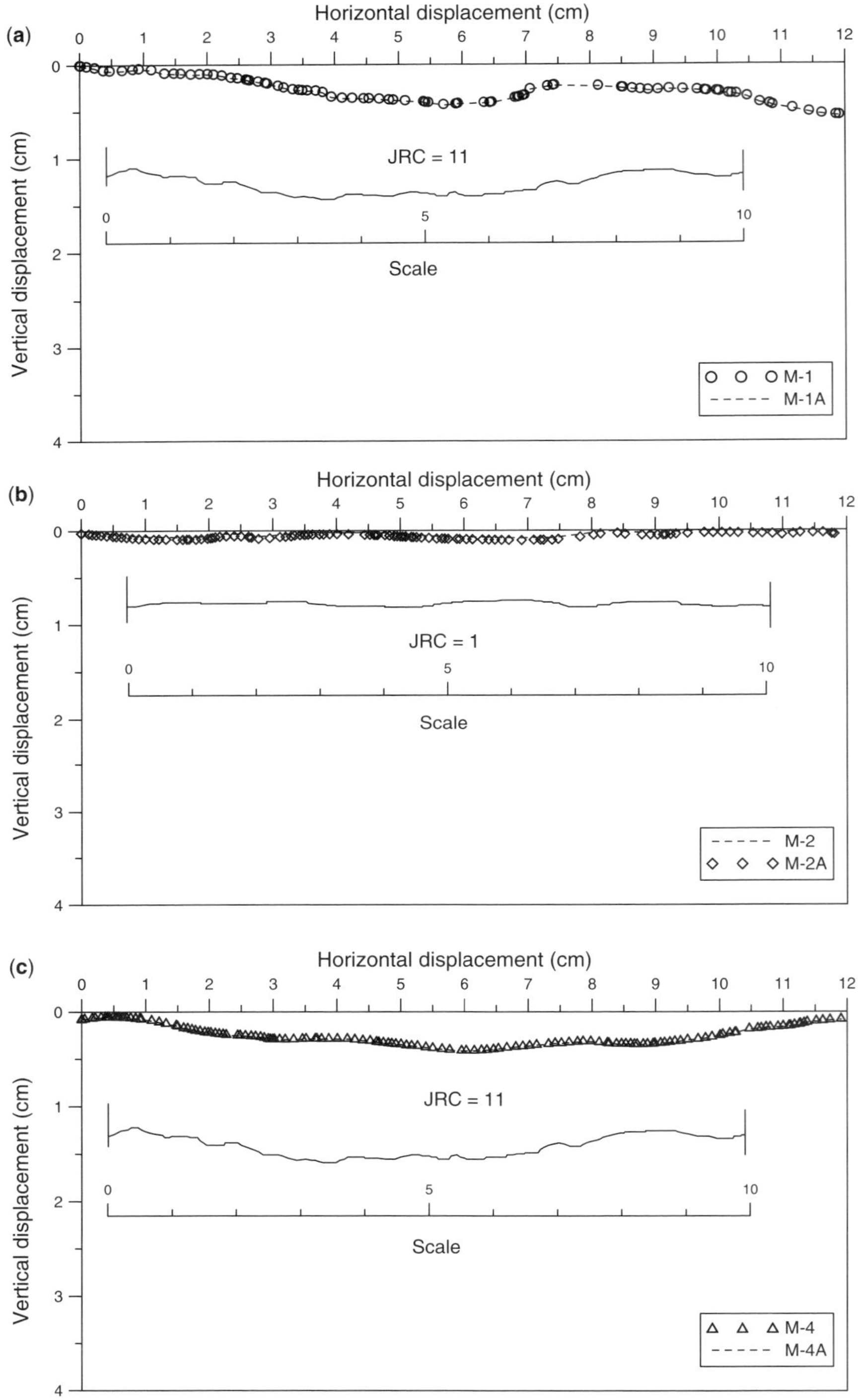

Fig. 13. Roughness profiles measured on three samples and the corresponding JRC value: (**a**) sample M1 (JRC = 11); (**b**) sample M2 (JRC = 1); (**c**) sample M4 (JRC = 11).

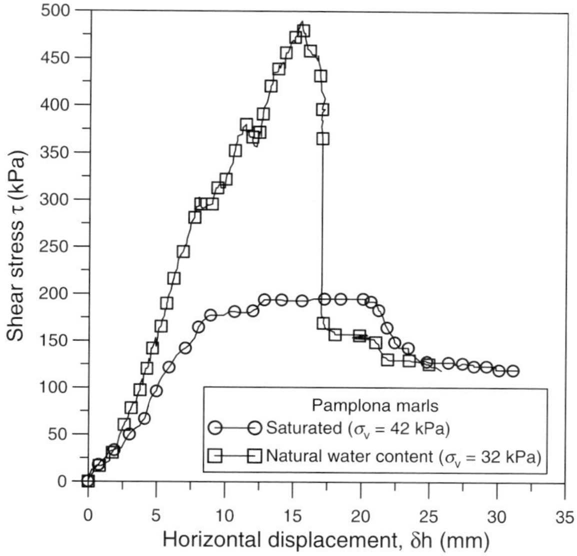

Fig. 14. Stress–displacement relationship for two unaltered samples sheared at different states (natural water content and saturated conditions).

A summary of the strength envelopes obtained from the shear tests is presented in Figure 15. The figure shows extremely large variations in strength envelopes, which are attributed to shearing conditions (natural and saturated state) and to the type of test. For unweathered samples in both natural and saturated conditions, a linear relationship between shear stress and vertical stress was obtained. The strong drop in strength associated with wetting indicates the high weathering susceptibility of this material. The friction angle decreased from $\phi' = 41^\circ$ (for 'natural' conditions) to $\phi' = 31^\circ$ (for saturated conditions). The drop in apparent drained cohesion was large: $c' = 452$ kPa to $c' = 168$ kPa for the dry and saturated shear tests respectively. The strong drop in cohesion of this material has a great influence on the stability of the dam, as discussed below.

A residual friction angle close to 16° was obtained in the ring shear apparatus (Fig. 15). This value is consistent with the absence of active clay minerals. However, it is lower than expected in view of the general mineralogical content and the low plasticity of the material. Most probably, the mica content explains this result.

Figure 15 also presents the peak conditions obtained from the large shear box on the concrete–marl contacts. The figure also includes data obtained by Romero (2005) on the concrete–marl contact under saturated conditions. The envelope obtained for the unweathered samples under natural water content conditions was bilinear with a friction angle (ϕ') of 38° and a cohesion intercept of 80 kPa for normal stresses below 100 kPa. For vertical stresses greater than 100 kPa the strength parameters can be approximated by a friction angle of 11° and $c' = 140$ kPa. This bilinear shape is an indication of the nonlinear character of the envelope. Under saturated conditions the concrete–marl contact showed a friction angle value of $\phi' = 20^\circ$ and a cohesion value of $c' = 36$ kPa. The low strength parameters obtained by Romero (2005) were explained by the higher degree of weathering of the samples, compared with the block samples used in the large shear test at natural conditions. Again, saturating the contact results in a significant drop of strength.

Based on JRC values reported earlier, the strength envelopes could be obtained as follows (Barton *et al.* 1985):

$$\tau = \sigma_n \tan\left[\mathrm{JRC}\log_{10}\left(\frac{\mathrm{JCS}}{\sigma}\right) + \phi_r\right] \quad (1)$$

where JRC is the roughness coefficient (11 or unity for this case), JCS is the unconfined compression strength of the marl ($q_u \approx 9$ MPa, average from three samples at similar water content and density), ϕ_r is the residual friction angle, which has been approximated by the value obtained in ring shear tests ($\phi_r = 16^\circ$) and σ is the normal stress. Figure 15 shows the strength envelopes obtained using equation (1) for the two extreme values of JRC determined in samples. Also included is the experimental strength envelope for concrete–marl (unweathered) contact. The strength parameters obtained from two JRC values were $\phi' = 33^\circ$, $c' = 14$ kPa and $\phi' = 18^\circ$, $c' = 12$ kPa, for JRC values equals to 11 and unity, respectively. The use of the Barton's formula for comparison with marl–concrete contact results is justified because, under the dam working

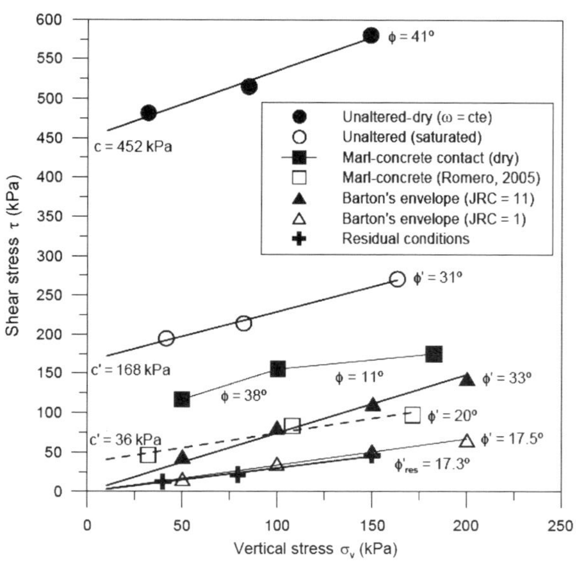

Fig. 15. Strength envelopes for unweathered and remoulded samples. Also shown is the residual strength envelope.

conditions, the marl–concrete contact will generate a joint playing a key role in the dam's stability. Strength developed along this joint is defined by the natural roughness of the marl caused by the excavation process. Concrete water and the pouring and curing process may reduce the available strength at the contacts. Laboratory tests on contacts have shown a reduction in strength compared with the intact material.

In summary, the internal friction angle for unweathered samples ranges from 31° to 40° whereas for the concrete–marl contact this parameter ranges from 20° to 38°. The internal friction angle based on the Barton *et al.* (1985) approximation varies from 18° to 33°.

Cohesion exhibited a strong dependence on saturation conditions and type of test. In general, cohesion values decreased markedly with the saturation process. Saturation of the samples reduced the apparent effective cohesion from 452 kPa to low values around 36 kPa for the concrete–marl contact.

Table 3 summarizes the strength parameters obtained from each of the tests performed on different samples of Pamplona marls. This information was used to evaluate the sliding stability of the Monreal dam.

Stability against foundation sliding of Monreal dam

The factor of safety (FS) is defined as the ratio between current strength and the sliding stresses in equilibrium with design conditions. Strength is calculated from the soil strength parameters (c' and ϕ'). Thus, the FS can be expressed as follows:

$$FS = \frac{\tau_{current}}{\tau_{sliding}} = \frac{[c' + \sigma' \tan \phi']}{\tau_{sliding}} \qquad (2)$$

where the sliding stress on the foundation plane is given by the tangential (T) and normal (N′) force components, which can be obtained from a conventional analysis of dam equilibrium [which will not be presented here; details have been given by Alonso *et al.* (2005)]. The tangential stress component was 1580 kN m^{-1} whereas the normal stress component was close to 1989 kN m^{-1}.

Figure 16 shows the changes in the FS, obtained using equation (2), as a function of the cohesion (c'), for friction angles (ϕ') ranging from 10° to 40°. For any cohesion value, the maximum drop in FS for a change in friction angle from 40° to 10° is around unity. The effect of cohesion is paramount. For instance, the FS obtained for the unweathered samples at natural water content was FS = 5, whereas the FS for the saturated unweathered sample dropped to a half: FS = 2.4. The FS obtained using the design parameters is FS = 3.5. The FS for strength parameters derived from tests performed on dry concrete–marl contacts ranged from FS = 1.6 to FS = 1.8 (as a result of the non-linearity of the strength parameters shown in Fig. 15). Under saturated conditions (using values obtained by Romero 2005) the FS becomes lower than unity: 0.87. FS values obtained from Barton *et al.* strength envelope were smaller than unity.

The variability of strength parameters measured in the various tests performed is clear from the values in Table 3. The degree of weathering has a major influence on the results, as it strongly affects the cohesion of the marls. The effective friction angle is also reduced when the specimen is saturated, but to a lesser degree. The factor of safety of the dam against sliding is critical for the most unfavourable strength parameters, which, in this case, were obtained for saturated marl–concrete contact and by Barton's criteria (Fig. 16). In these cases, consideration of the downstream

Table 3. *Summary of strength parameters for Pamplona marls*

Sample	*Shearing*	σ_v (kPa)	c' (kPa)	ϕ' (°)
Unweathered	Dry	33–149	452	41
Unweathered	Saturated	42–163	168	31
Contact	Dry	50–100	80	38
Contact	Dry	100–163	140	11
Contact	Saturated	30–170	36*	20*
Residual	Saturated	40–150	0	17
Barton	(JRC = 11)	50–200	14	33
Barton	(JRC = 1)	50–200	2	18

*Data obtained from Romero (2005).

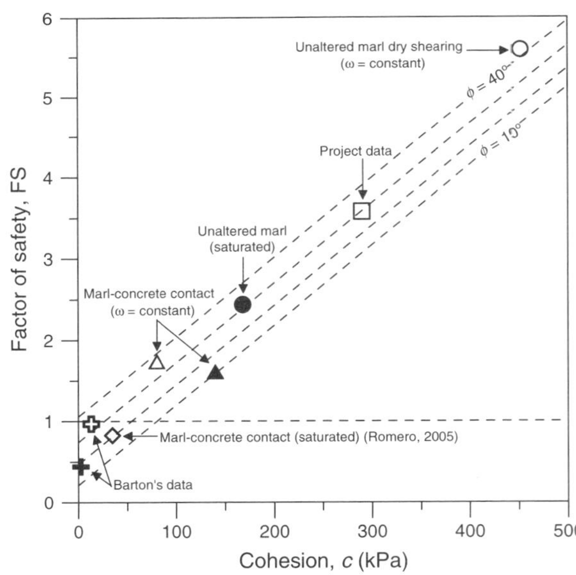

Fig. 16. Variation of factor of safety with cohesion and friction angle for different sets of strength parameters of Pamplona marls. w/c, water content.

passive resistance of the dam structure was sufficient to achieve an acceptable FS against forward sliding.

Concluding remarks, Pamplona marls

The influence of weathering processes such as stress relief and exposure to environmental conditions of Pamplona marls has been presented. This material shows a strong degradation in short time periods (a few hours), once it is unloaded, when water content increases. *In situ* Pamplona marls show a marked weathering profile (Figs 3, 8 & 10). In 3–4 m the marl evolves from unweathered (intact) states to weathered and fissured marls, with associated variations in water content (suction changes), Atterberg limits and density. These changes are reflected in changes of the mechanical properties (strength, stiffness and compressibility).

Strong degradation of Pamplona marls, in terms of shear strength loss, was observed in direct shear tests on unweathered samples under saturated and *in situ* water content conditions. Similarly, the state of saturation ('saturated' and 'dry' conditions) also had a major effect on the shear tests performed on concrete–marl interfaces (Table 3). A practical consequence is that the selection of the shear strength parameters for the evaluation of the stability of the dam is a complex matter. The factor of safety is strongly dependent on the cohesion of the material, which is drastically reduced by the saturation process. Effective friction decreases less markedly when specimens are saturated.

A way to reduce this degradation when the structure is being built is to minimize, after excavation, the time of exposure of the marl to the atmosphere (a few hours). Appropriate timing of construction work may help to achieve this goal.

Abadia marls

Characterization

Abadia marls from Arruda dos Vinhos (north of Lisbon, Portugal), are part of the Lusitanian Basin (Late Jurassic in age; Fig. 17). A detailed description of the geology has been given by Ferreira (1981), Leinfelder (1993) and Jeremias (2000). The complex reaches 800 m in thickness in some areas and the average slope gradient found in this formation is about 19° (Jeremias 2000; Zêzere 2002). Landslides are frequent and occur for slope gradients ranging between 10° and 21° (Zêzere *et al.* 1999). The material *in situ* is essentially saturated independently of depth and degree of weathering. This is due to high phreatic levels owing to the proximity to Tagus River and the low permeability of the material.

Figure 18 shows a core sample of Abadia marls and a detailed view of a core, in which the stratification can be identified. At the surface, the weathered material can be identified by its yellow colour. The bedrock (non-altered material) is grey to dark brown. The thickness of the weathered zone varies between 50 cm and 5.0 m. In recently cut slopes it is very common to see local landslides of disaggregated material (Fig. 19; photograph taken in the summer). These local instabilities are caused mainly by changes in the water content (or suction changes): drying leads to local disaggregation in steeper slopes during the summer; wetting after rain periods leads to mass movements (landslides) of some importance. Shallow translational slides sometimes with rotational geometry on steep slopes occur frequently and are characterized by large movements. Landslides occur especially after heavy rain periods, indicating that Abadia marls are very sensitive to wetting (Zêzere *et al.* 1999; Zêzere 2002).

Slaking durability tests (ISRM 1979; ASTM 1992) performed by Jeremias (2000) on several samples of Abadia marls at different depths show low durability for the weathered superficial layers and low to medium durability for the unweathered rock (Table 4). It is noteworthy that the anisotropy owing to stratification plays a very important role in the physical degradation observed for this material. Water adsorption tests performed by Jeremias at 95% relative humidity (RH) and 25 °C (96 h duration) showed average values of water content varying between 3.40 and 5.08%, indicating a good capacity for water absorption. This capacity increases with degree of weathering.

The evolution of the grain-size distribution measured in specimens dried after the test, of a sample of marl fragments having initially a range of sizes varying between 9 and 4.75 mm when subjected to eight wetting–drying cycles is presented in Figure 20. The Hardin index R_d (Hardin 1985) is the ratio between the difference of the areas above the grain-size distribution curves after and before wetting and the area above the original grading, before wetting, and can be used to classify the degradation observed. According to Figure 20, this index increased to 16% as a result of the first wetting. Almost half of the particles (49%) were reduced in size after the second cycle, an indication of a fast degradation process. The water content of the marl fragments before the first wetting was 11% ($s = 10$ MPa). Similar behaviour was observed for materials having initially different water contents. It was realized that the degradation process increases for drier fragments, and also for fragments with larger dimensions. This result agrees with those of the slaking tests results previously presented.

The mineralogical composition of the marls is presented in Table 5 (Jeremias 2000). Carbonates (calcite mainly) are present, and also clays and a very small percentage of organic matter. The relation between the percentages of clay and of carbonates justifies the designation of the material as marl. Among the clay minerals there is almost no smectite and the expansive minerals present are chlorite and gypsum (traces found in some samples). The presence of these minerals may explain the degradation (mainly

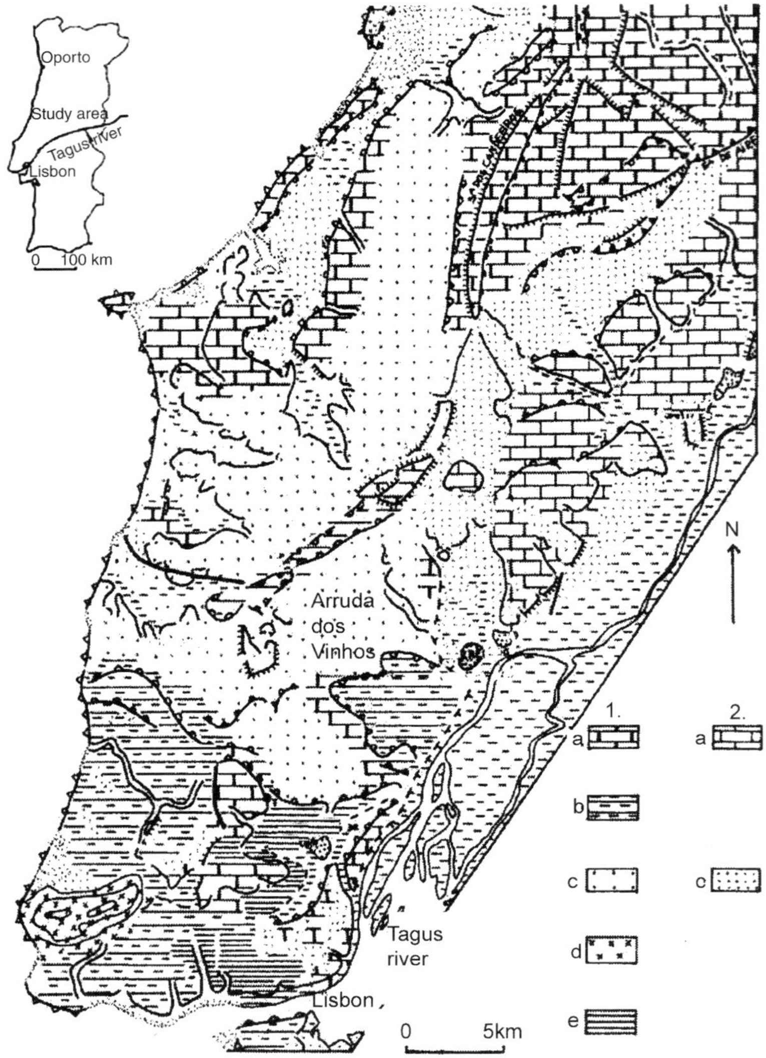

Fig. 17. Geological map of the region north of Lisbon (adapted from Ferreira 1981). 1, Secondary formations: a, marls and coralline limestones; b, marls from 'Abadia Beds'; c, disaggregated rocks, limestones and marls; d, volcanic rocks: granite, gabbro and syenite; e, volcanic complex of Lisbon: alternating basalts, ashes and cinders. 2, Tertiary and Quaternary formations: a, limestones; c, disaggregated rocks.

cracking and disaggregation) and the moderate expansive behaviour observed in this material when fully saturated, a feature discussed further in this paper.

The fine fraction of the marl (particle dimensions <0.20 mm) can be classified as low-plasticity clay (CL), which is in accordance with the nature of the clay minerals found. The Atterberg limits that allowed this classification and other relevant geotechnical properties of Abadia marls are presented in Table 6. The ranges of values presented in this table are for both weathered and non-weathered samples. The highest porosities and water contents were measured in samples from the most superficial layers (higher degree of weathering). By comparing the weathered superficial layers with the deepest layers (Jeremias 2000) it can be seen that there are no significant changes in the percentages of minerals, especially the clays. This indicates that the nature of the clay minerals is almost the same independent of the degree of weathering. Therefore

Fig. 18. Core sample of Abadia marls. (**a**) Core sample. (**b**) Core detail, showing stratification.

weathering phenomena are mainly physical (disaggregation), and chemical weathering is reflected mainly in the change of colour (mainly oxidation of the non-clayey minerals, less feldspar and the presence of calcite instead of dolomite).

No significant changes were found for the Atterberg limits of the material after eight wetting and drying cycles (wetting by full saturation at 20 °C followed by drying at 104 °C), as presented in Table 6. This indicates that weathering, artificially induced by wetting and drying cycles, did not lead

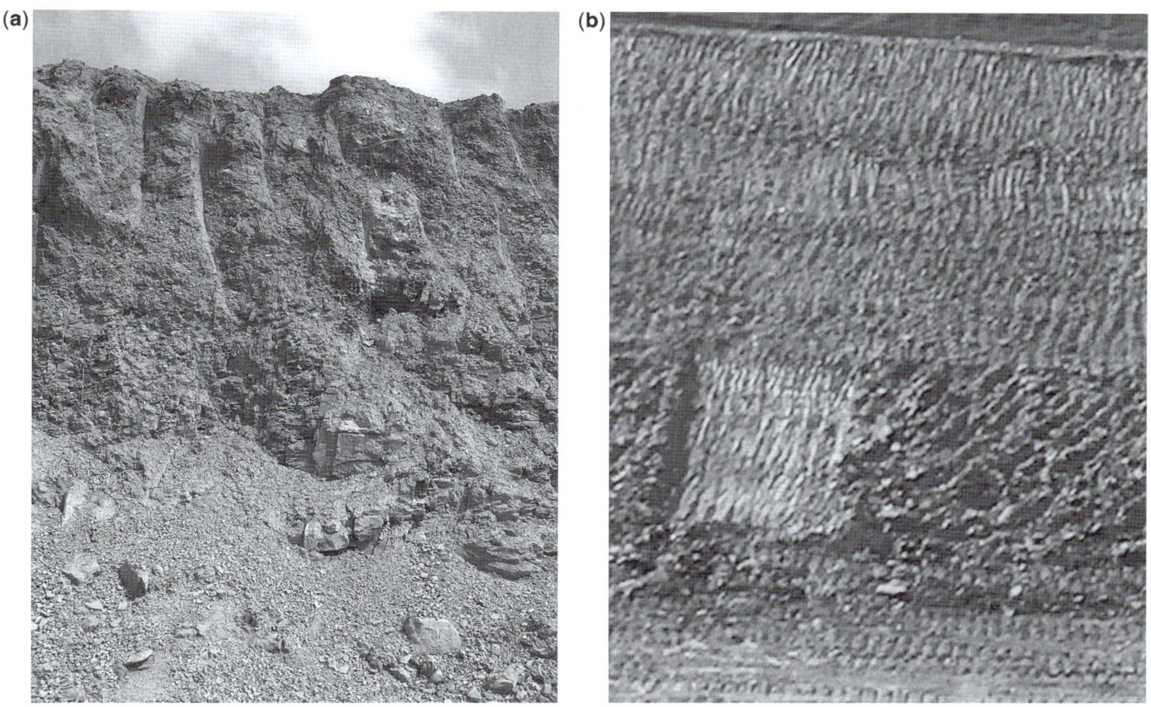

Fig. 19. General view of Abadia marls: (**a**) weathered steep slope; (**b**) non-weathered material (before being covered for protection).

Table 4. *Results of slake durability tests (Jeremias 2000)*

	I_{d1} (%)	I_{d2} (%)	I_{d5} (%)
Weathered samples	56–66	17–19	1–2
Unweathered samples	67–89	43–75	8–29

I_{dn}, percentage of material retained in the mesh after $n \times 1000$ rotations.

Table 5. *Mineralogical analysis of Abadia marls (Jeremias 2000), average values*

Mineral	(%)
Carbonates (calcite and dolomite)	16–23
Quartz	5–10
Other non-clay minerals (K-feldspar, mica, pyrite, apatite, rutile, etc.)	8–17
Organic matter	0–2
Clay constituents (no smectite detected)	54–64
Chlorite	1–5
Kaolinite	17–30
Illite	21–35
Smectite	0–1
Mixed layer clays	30–60

to changes in the mineralogical composition. Jeremias (2000) also confirmed that the degradation caused by wetting was mainly physical in artificial weathering tests carried out on several samples of this type of material, as there were no relevant changes in the mineralogical composition after the tests. The samples were subjected to four types of artificial weathering: (1) exposure to atmospheric conditions (300 days); (2) exposure to atmospheric conditions with a treatment with a 5% solution of sulphuric acid; (3) 300 wetting and drying cycles performed in an environmental chamber (wetting at 25 °C and RH = 95% and drying at 65 °C and RH = 0%); (4) tests similar to those in (3) but with spraying of the samples with a solution of NaCl (10%). Higher weathering degrees (mainly disaggregation) were achieved with method (2) because of the corrosive action of acid on carbonates and method (3) because of the physical degradation caused by wetting and drying cycles.

Results from mercury intrusion porosimetry (MIP) tests on marls samples presented in Figure 21a allowed identification of a fairly uniform pore size with a mean diameter of 148 nm. This uniformity can be seen also in the scanning electron microscopy (SEM) photograph presented below. The full disaggregation of this material [remoulded samples prepared with water content $w = 1.35w_L$, as suggested by Burland (1990)] shows similar

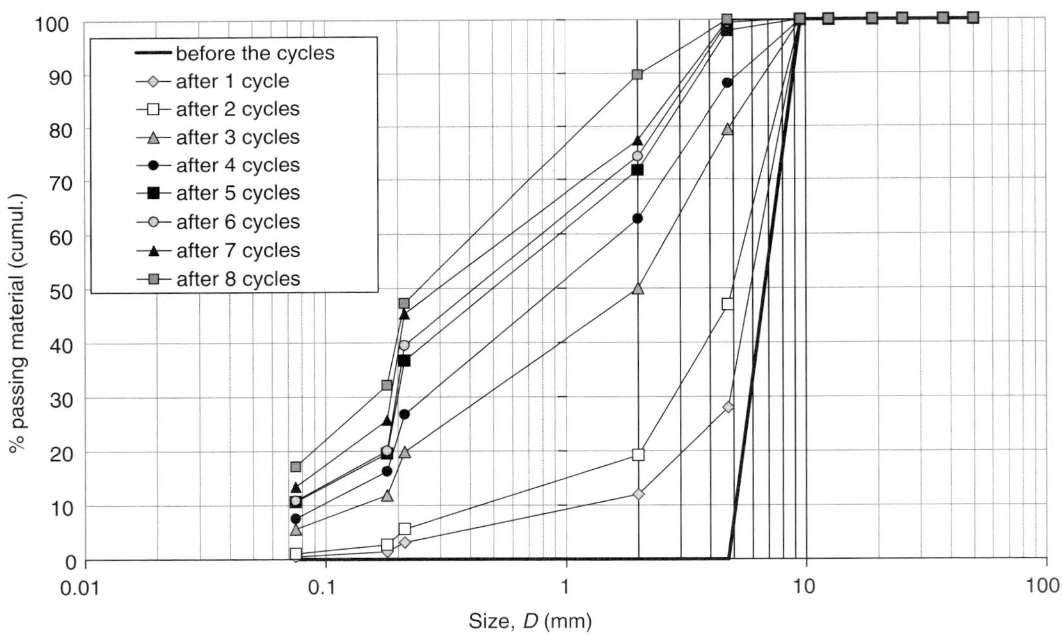

Fig. 20. Size evolution during eight wetting–drying cycles of a set of particles with uniform size (9 mm ≥ D > 4.75 mm).

Table 6. *Some geotechnical properties of Abadia marls*

Liquid limit, w_L	41–52%
Before the cycles	50%
After three wetting–drying cycles	51%
After eight wetting–drying cycles	49%
Plasticity index, PI	20–25%
Before the cycles	23%
After three wetting–drying cycles	26%
After eight wetting–drying cycles	24%
Solid weight density (average value)	27.5 kN m^{-3}
In situ void ratio/porosity (Jeremias 2000)	0.22–0.55/18–33%
In situ saturated water content, w_{sat} (Jeremias 2000)	8–22%

pore size distribution but now with a larger average diameter (305 nm), indicating that the disaggregation in the presence of water contributes to an increase in the porosity of the rock.

Figure 22 shows the variation with depth of dry unit weight measured in two core samples (CS1 and CS2) (Oliveira 2006) and the void ratio computed assuming the average saturated water content from Table 6 ($w_{sat} = 12\%$). This value for the water content was adopted because data available for the *in situ* values are not reliable (values corresponded to saturation degrees higher than 100%), as coring was performed with water. As expected, the superficial weathered layers showed lower dry density and higher void ratio.

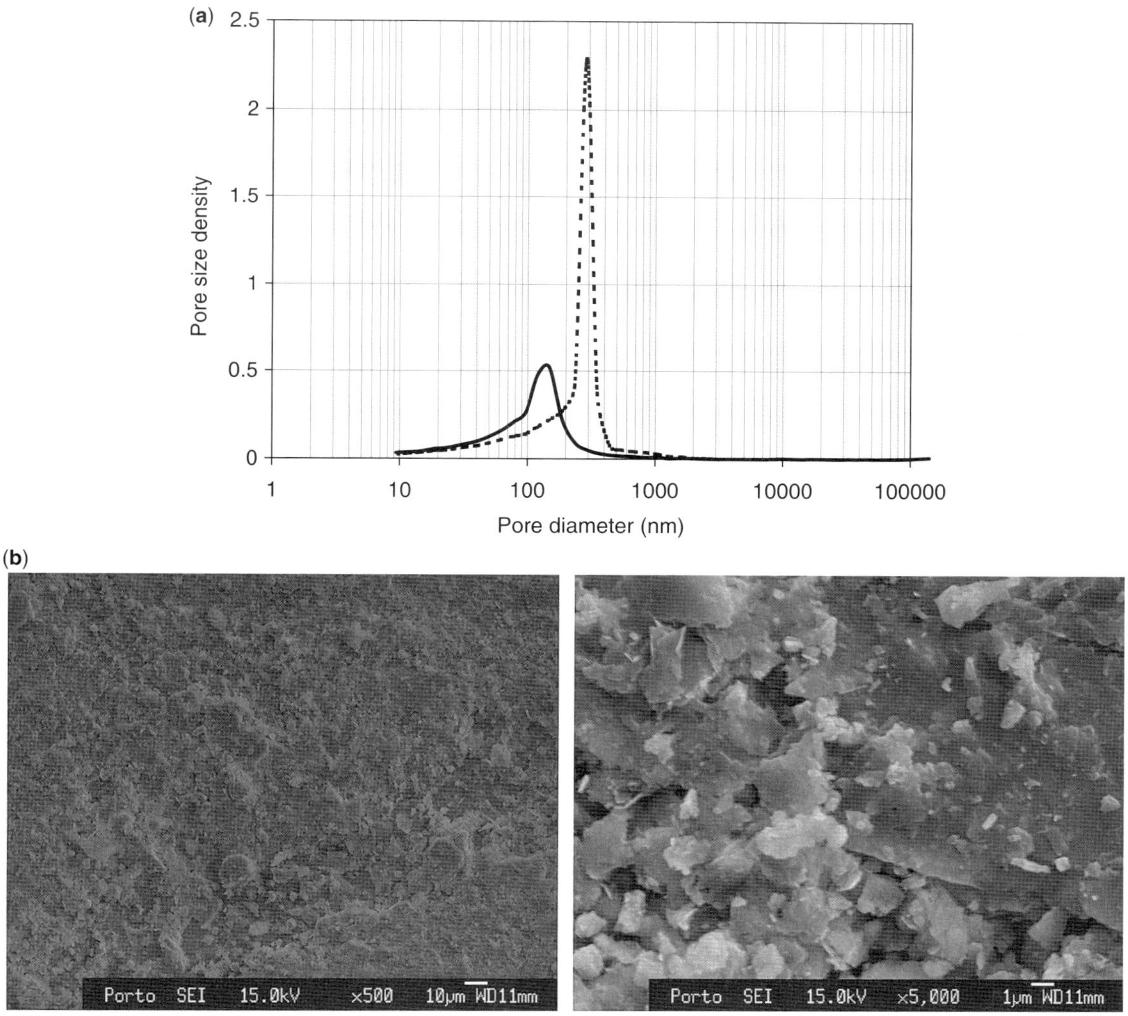

Fig. 21. Pore-size distribution of Abadia marl. (**a**) Results from mercury intrusion porosimetry. (**b**) SEM images (non-weathered rock).

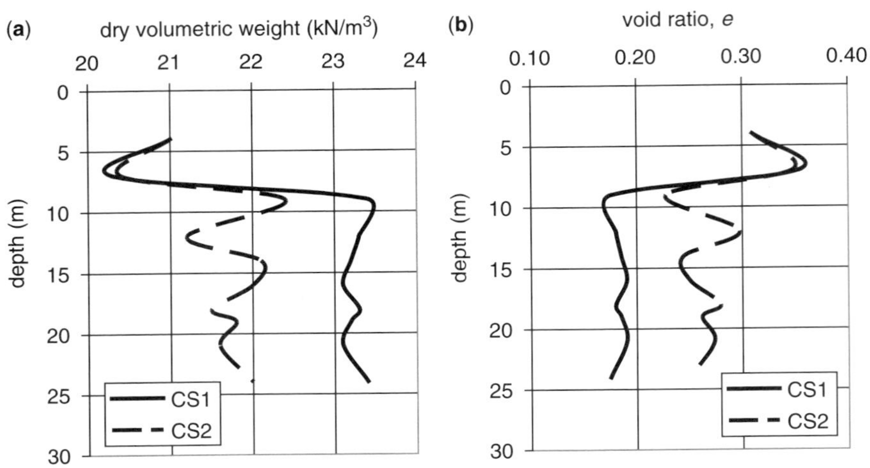

Fig. 22. Variation with depth of (**a**) the dry volumetric weight (adapted from Oliveira 2006) and (**b**) void ratio (assuming $w = 12\%$).

Weathering of Abadia marls is caused by wetting and drying cycles controlled by atmospheric changes. The associated suction changes are one of the main causes of rock physical degradation leading to the loss of mechanical properties (strength, cohesion–loss of cementation or bonding and stiffness). This degradation is one of the main factors responsible for landslides triggered by bank erosion and shallow translational slides on steeper slopes, all instability phenomena that occur frequently in Abadia formations.

Concepts of unsaturated soil mechanics were used to understand the degradation of Abadia marls and will be explained below. Laboratory tests were performed to characterize the hydromechanical behaviour of this material. Experimental data were used to calibrate a simple numerical model intended to clarify the degradation phenomena observed by simulating the wetting of a marl fragment.

The effect of suction

Abadia marls, when exposed to atmospheric actions (wetting and drying cycles), and undergo important physical disaggregation. The grain-size distribution of the weathered material includes a large percentage of material with dimensions less than 0.075 mm (passing ASTM sieve No. 200). Figure 23 shows a sequence of photographs of a fragment of marl (5 cm × 3 cm × 3 cm, $w = 4\%$, $s_{\text{initial}} = 100$ MPa) during wetting to full saturation in a laboratory environment ($T = 20\,°C \pm 5\,°C$, and RH $= 45\% \pm 10\%$). Significant cracking led to the complete breakage of the fragment into smaller fragments.

A possible explanation for the mechanism of degradation of marl fragments is schematically presented in Figure 24 assuming isotropic behaviour. When wetting occurs the particle boundary is first wetted and a suction gradient is created inside the rock fragment. This suction gradient induces water transfer and reduces in time until it reaches a zero value when full saturation is attained. As long as there are suction gradients, differential swelling deformations will be developed inside the rock fragment. The geometry, stratification and confinement of the fragment restrain swelling displacements and lead to tensile and shear stresses, which eventually result in cracking and loss of structure. A similar mechanism can be invoked to explain the degradation caused by drying (shrinkage instead of swelling deformations).

Laboratory tests on Abadia marls were performed to characterize its hydromechanical behaviour and to supply data necessary to understand the causes of the degradation phenomena observed. Simple paraffin tests to determine the density of marl samples with different water contents were performed. It was found that the void ratio depended on water content (Fig. 25). The swelling behaviour of Abadia marl was confirmed in swelling tests where rock samples were confined under oedometric conditions and subjected to different vertical stresses and different suction changes (suction paths are presented in Fig. 26 in a vertical stress–suction space). Measured swelling strains (Fig. 26) indicate that the driest rock samples show higher volumetric deformations (volume increment) as a result of wetting for all the vertical stress applied. As expected, the applied vertical stress reduces swelling strains. The recorded swelling strains of the driest specimens initially equilibrated with

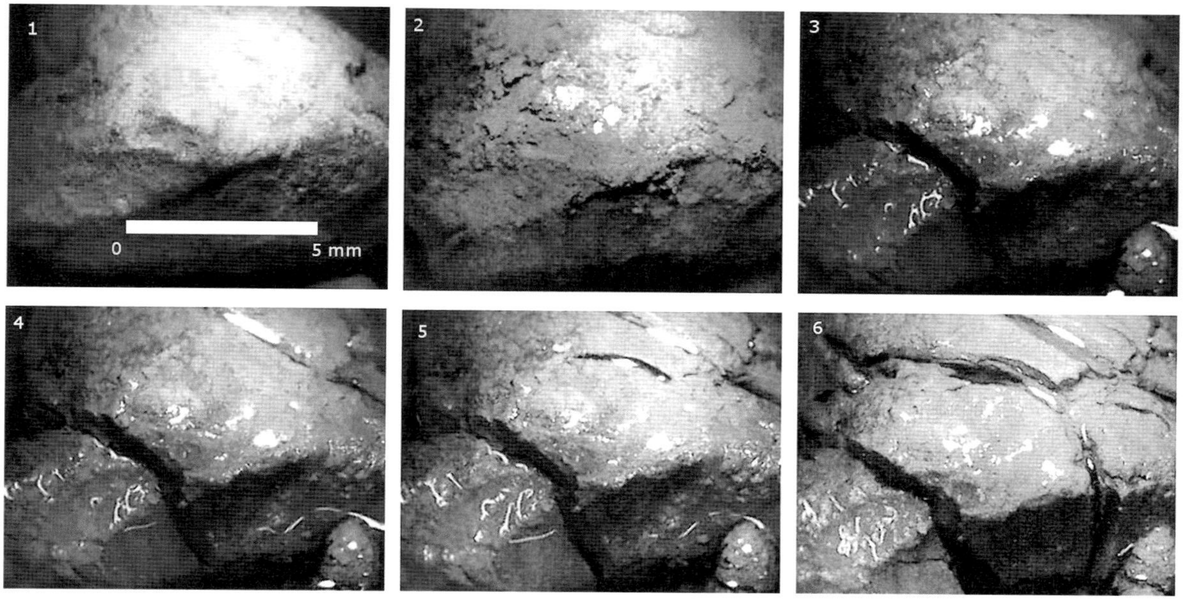

Fig. 23. Crack opening caused by wetting. 1, fragment before wetting; 2, full saturation by total submersion; 3, first crack (*c.* 3 s after submersion); 4, following crack opening; 5, crack stabilization; 6, fragment drying, showing the cracks.

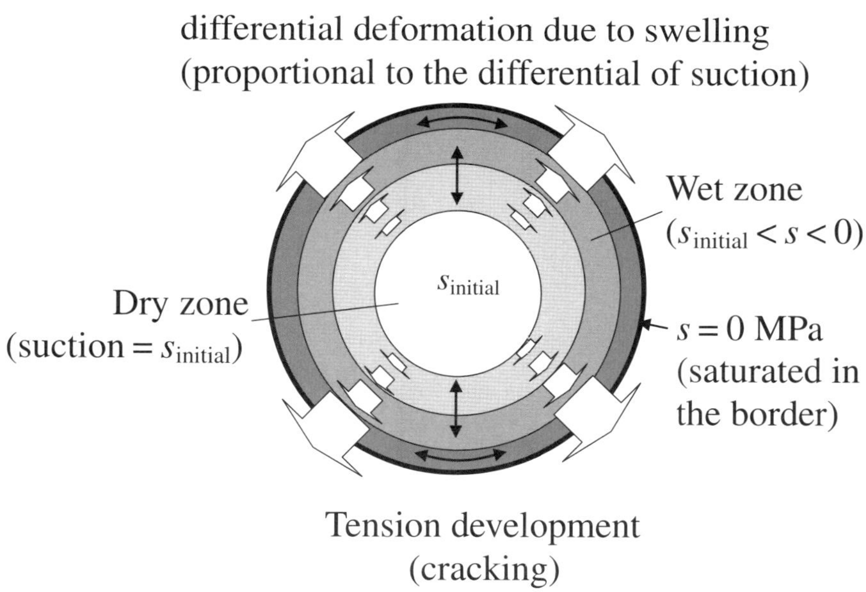

Fig. 24. Degradation mechanism explained by differential swelling displacements caused by suction gradients inside the rock.

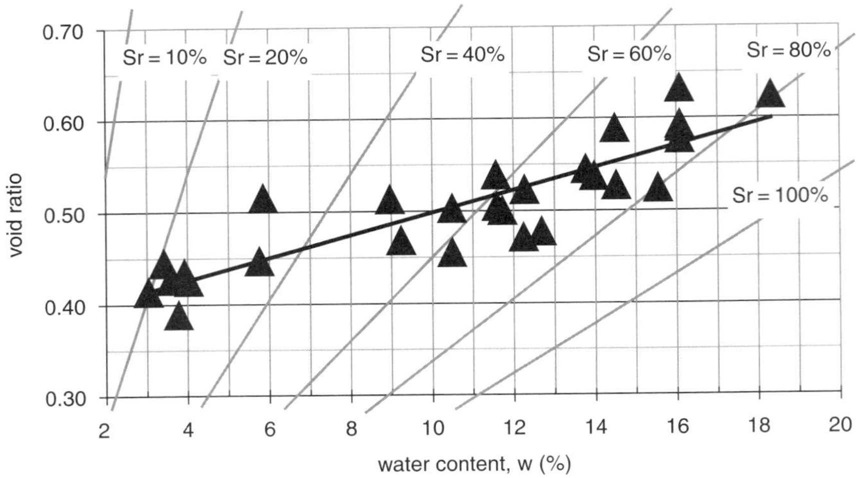

Fig. 25. Void ratio dependence on water content (saturation degree).

high suctions (a suction $s = 135.9$ MPa corresponds at 20 °C to RH of 37%; if $s = 11.0$ MPa, RH = 92%; and if $s = 4.8$ MPa, RH = 97%) are significant. They occur in the real world, as the RH values indicated are common in many field environments (in Lisbon, e.g. average RH varies between 48 and 85%). This indicates that suction gradients can lead to important swelling displacements and, therefore, to disaggregation.

The water retention curve (WRC) was determined for the marl. It was also determined for the marl after three wetting and drying cycles and for the remoulded marl ($w = 1.35w_L$; Burland 1990). The procedure followed consists in measuring suction with WP4 water dewpoint equipment (the equipment and calibration have been described by Cardoso *et al.* 2007). Suction was measured for controlled decreasing water contents (the sample was dried in the laboratory environment) and then for increasing values (wetting with controlled addition of distilled water) to full saturation. The measurements were made allowing 24 h for equilibrium of the soil with the current suction applied.

All the curves were fitted by a modified form of the Van Genuchten (1980) expression proposed by Romero & Vaunat (2000) and are presented in Figure 27. In this figure it can be confirmed that cyclic wetting and drying causes loss of structure, as the curve obtained after three cycles approaches that of the remoulded sample.

Suction also affects the marl compressibility, as confirmed by the results of the oedometer tests performed on marl samples under constant suction ($s = 124$ MPa and $s = 0$ MPa). The results of these tests are presented in Figure 28 [(e, log σ_v) plane]. It can be seen that the compressibility of the saturated sample (original compressibility $C_c = 0.057$) is higher than the compressibility of the dry sample ($C_c = 0.040$). Swelling occurred when the sample was saturated (at $\sigma_v = 100$ kPa). By comparing the unloading path of the saturated and the unsaturated samples it can be seen that there is a significant decrease in the unloading stiffness as a result of saturation, which can be explained by the loss of structure of the marl caused by suction decrease and, also, by the lower prevailing suction. This behaviour can be explained by loss of bonding between rock constituents and the release of swelling potential, a behaviour observed in some soft rocks (e.g. Calabresi & Scarpelli 1985; Leroueil & Vaughan 1990; Picarelli 1991). During unloading, the behaviour of soft rocks is not necessarily elastic because of destructuring mechanisms that affect irreversibly the mechanical properties of the material. However, the definition 'elastic' for stiffness in unloading paths is usually adopted for simplification (values in Table 7). Figure 29 compares the results of oedometer tests performed on a saturated marl sample and on a remoulded sample (fine fraction from crushed marl passing sieve ASTM No. 20, prepared at $w = 1.35w_L$) which represents the intrinsic properties of the completely disaggregated material (Burland 1990). As expected, the compressibility of the saturated rock ($C_c = 0.057$) is much lower than that of the remoulded sample ($C_c = 0.325$).

Figure 29 also presents the compressibility curve for a remoulded sample consolidated to 1000 kPa, unloaded and dried to $s = 100$ MPa, reloaded to 50 kPa and then fully saturated, loaded and unloaded (red curve) (consolidation was carried out in a oedometric cell and drying was in unconfined conditions in the laboratory environment; then the dry sample was placed in another oedometric cell used for testing small rock specimens). This curve was included to show the effect of drying on structure. For the same stress levels, the void ratios of the remoulded sample, saturated after being dried, are greater than those of the remoulded

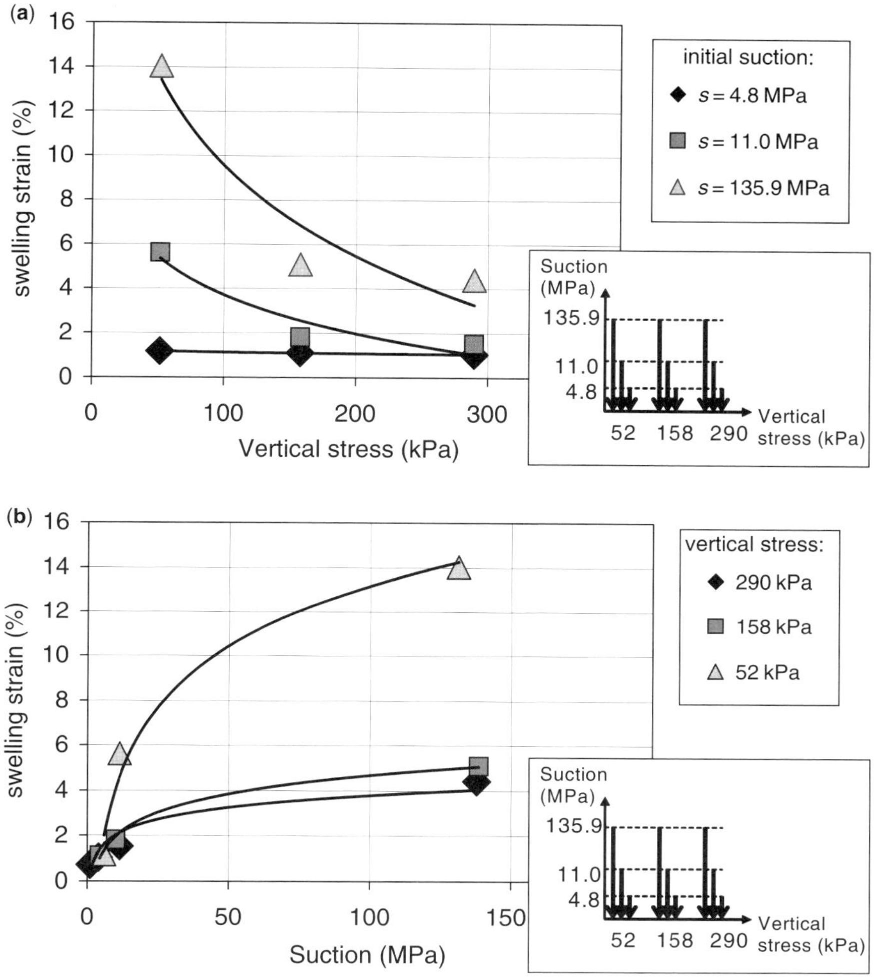

Fig. 26. Result of the swelling tests on Abadia marls. (**a**) Same initial suction, different vertical stress on wetting. (**b**) Same initial vertical stress, different initial suction on wetting.

material from slurry. It is inferred that the suction increment provided some structure to the material (Leroueil & Vaughan 1990). The structure induced by drying allows the compression curve to enter 'impossible states' to the right of the original saturated compression line for the remoulded material. The compressibility of this structured sample is high for stresses in excess of 2000 kPa ($C_c = 0.660$) because it shows the progressive breakdown of the open structure induced by drying. The compression curves for the structured and remoulded specimens approach each other for high vertical stress, when the structure associated with the drying process is completely lost. Finally, the curve for the rock is below those for the remoulded samples because the void ratio of the rock is much lower. This result was expected (Fig. 21), as loss of structure causes a porosity increase. There is no information about the intersection between the curves found for the rock and for the remoulded sample because of the limitation of the maximum vertical stress applied with the equipment.

A numerical simulation of the degradation of a fragment of marl was performed with the purpose of understanding the degradation mechanism related to wetting, illustrated in Figure 30. A more detailed description of the results of the analysis performed has been given by Cardoso & Alonso (2009). The finite element program CODE_BRIGHT (Olivella *et al.* 1996; UPC-DLT 2002) was used to perform a coupled hydromechanical analysis. Wetting was assumed to start in a uniform manner at the boundary

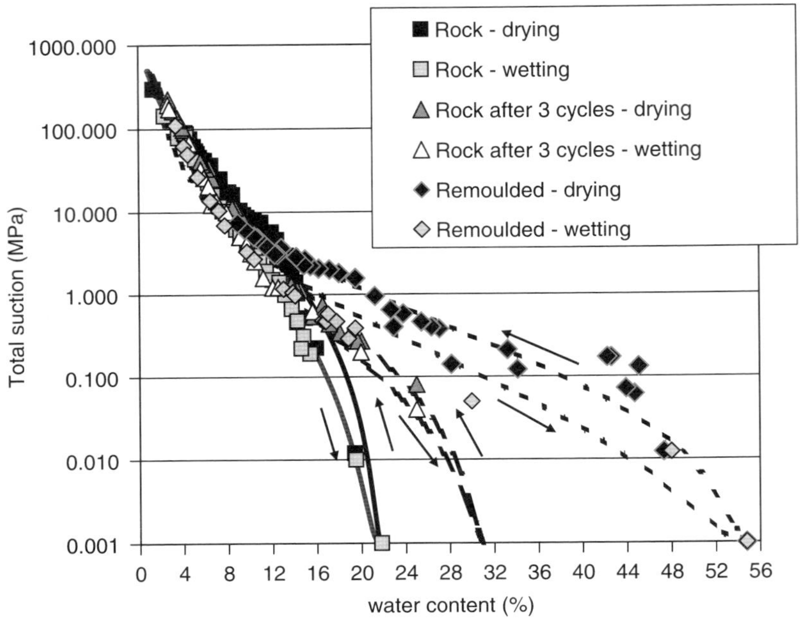

Fig. 27. Water retention curves for Abadia marls.

of a circular fragment (9 mm diameter) under plane strain conditions. The anisotropy introduced by stratification was not considered.

The hydraulic constitutive equations consider the balance of water in the liquid and gas phases, adopting Darcy's law in the calculation of the conductive flux. The intrinsic permeability was calculated by Kozeny's model. Concerning the hydraulic parameters, the intrinsic permeability coefficient of the marl was determined in permeability tests on saturated species and it was found to be 8×10^{-21} m^2, which corresponds to a very low Darcy coefficient ($k_{sat} = 8 \times 10^{-14}$ m s^{-1}). In the calculation the WRC of

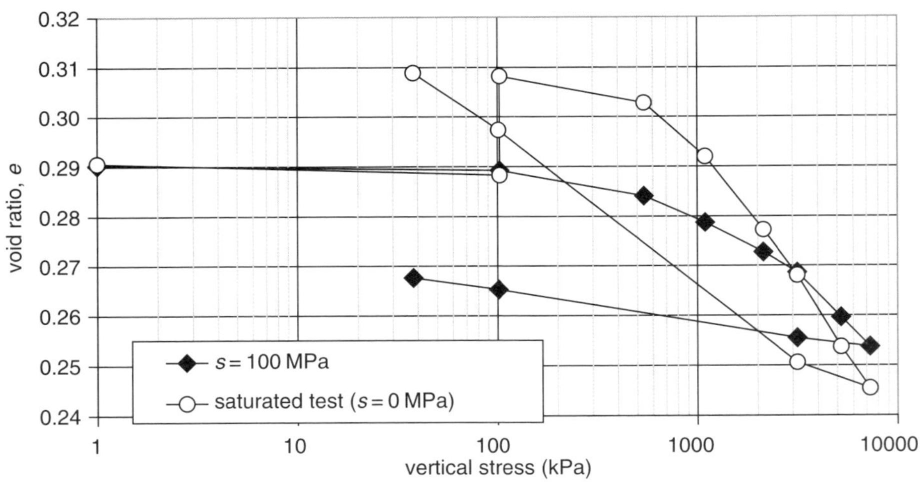

Fig. 28. Oedometer tests of the marl under different suctions.

Table 7. *Parameters necessary for BBM calibration*

Parameter	*Description*	*Value*
$\lambda(0)$	Stiffness parameter for changes in net mean stress for original states of the soil in saturated conditions	0.027
$\lambda(s)$	Stiffness parameter for changes in net mean stress for original states of the soil	Defined by equation (4)
r	Parameter defining maximum soil stiffness	0.65
β	Parameter controlling the rate of stiffness increase with suction	0.05 MPa^{-1}
κ	Elastic stiffness parameter for changes in net mean stress	0.007
κ_s	Elastic stiffness parameter for changes in suction	0.020
p_0^*	Pre-consolidation stress for saturated conditions	1.00 MPa
p_c	Reference stress	0.23 MPa
M	Slope of the critical state lines	1.0
k	Parameter describing the increase in cohesion with suction	0.01
c_0	Tensile strength for saturated conditions	0.4 MPa
ν	Poisson coefficient, used in the calculation of shear modulus, G	0.25

the intact marl was used fitted with the Van Genuchten (1980) expression

$$S_e = \left[1 + \left(\frac{P_g - P_l}{P}\right)^{\frac{\lambda}{1-\lambda}}\right]^{-\lambda} \quad (3)$$

where S_e is the degree of saturation at the current liquid pressure, P_l, P_g is the gas pressure (assumed to be equal to the atmospheric pressure, 0.1 MPa), P is the pressure corresponding to the air entry value and λ controls the shape of the curve. P and λ are material parameters. Values of $P = 0.3$ MPa and $\lambda = 0.20$ were obtained for the drying branch, and $P = 0.9$ MPa and $\lambda = 0.20$ for the wetting branch.

The Barcelona Basic Model (BBM; Alonso *et al.* 1990) for unsaturated materials was adopted as the mechanical constitutive model. This simple model can simulate the swelling behaviour of the material (Fig. 31). The constitutive model is defined in terms of net mean stress (excess of total stress over air pressure) and suction. The parameters adopted for the calibration are presented in Table 7 and were obtained from experimental data, explained as follows.

In Table 7, the small value adopted for the preconsolidation pressure is intended to reproduce strong degradation observed when the material is hydrated under small vertical stress (shown in Fig. 23). The oedometer tests on rock samples allowed definition of: (1) the stiffness parameter for changes in net mean stress for original states of the soil in saturated conditions, $\lambda(0)$; (2) the parameters r and β necessary for the definition of soil stiffness for unsaturated conditions, $\lambda(s)$ [see equation (4) below]; (3) the elastic stiffness parameter for changes in net mean stress, κ (assumed to be the average between the values found for the saturated and the unsaturated test because only one parameter can be used in the program and the transition between the two cases must be reproduced). In the absence of data for the horizontal stresses occurring during loading in the oedometer cell it was assumed that they were half of the vertical stresses applied:

$$\lambda(s) = \lambda(0)[(1 - r)e^{-\beta s} + r]. \quad (4)$$

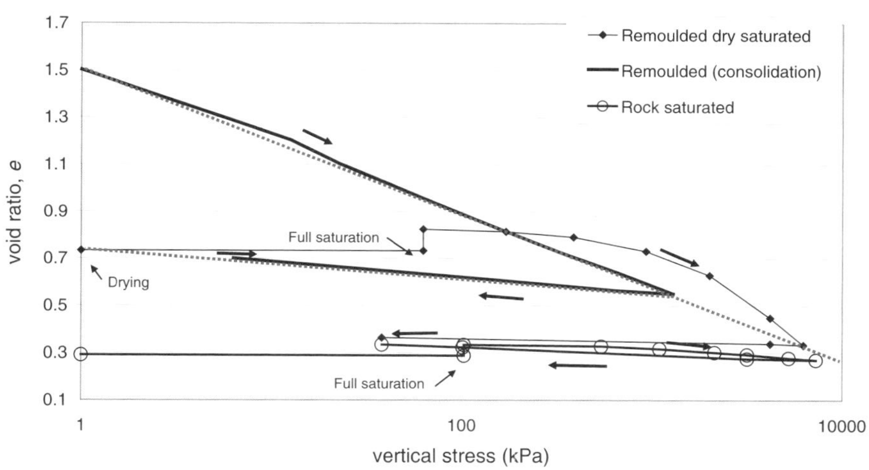

Fig. 29. Saturated oedometer tests of the rock and remoulded samples.

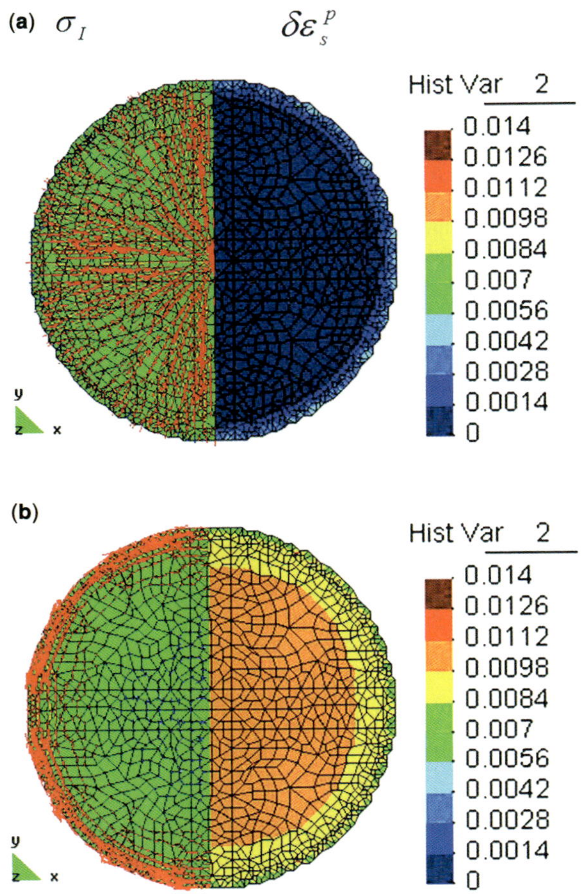

Fig. 30. Cracking pattern inside the particles: (**a**) 30 min after wetting; (**b**) 5 hours after wetting.

With the results of the paraffin tests (Fig. 25) and the WRC (Fig. 27) used for establishing a relation between water content and suction, it was possible to obtain a relation between void ratio and suction. The elastic stiffness parameter for changes in suction, κ_s, can be derived from this relationship in a (ln s, v) space ($\kappa_s = 0.020$).

Results from Brazilian tests performed on marl samples under different suctions were used for the definition of the parameter describing the increase in cohesion with suction, k ($k = 0.007$). As will be mentioned in the next section, the test results allow definition of the saturated cohesion c_0 ($c_0 = 0.4$ MPa). The Poisson's coefficient was determined with data from the unconfined compression tests performed, also presented in the next section.

Because fracture phenomena are not incorporated in the numerical model adopted, crack development is identified by the patterns of tensile stresses, shear stresses and plastic deformations (volumetric and deviatoric) developed during wetting. The analysis made it possible to identify the evolution in time of the degradation: swelling progresses from the boundary towards the centre and tensile stresses and plastic deformations develop in time while the fragment is becoming fully saturated.

The vectors corresponding to the tensile stress developed in time (Fig. 30) can help to understand breakage phenomena. Tensile stresses develop internally while the fragment outer annulus increases in volume and detaches from the core of the particle. This mechanism leads to cracking interpreted as peripheral detachment along the entire perimeter. This is the case for a time of 30 min after wetting. When the water reaches the inner core it starts to swell, pushing out the outer layers. This mechanism corresponds to radial cracking observed 5 h after wetting. The plastic shear deformations calculated (named hist_var 2), also presented in Figure 30, concentrate in the outer border and increase with increasing saturation. Therefore, patterns of cracking

Fig. 31. Expected cracking pattern from numerical analysis and a real case.

are controlled by the evolving distribution of suction inside the rock fragment.

Cracking will occur if the tensile stress developed is higher than the tensile strength. This strength reaches the minimum value when the fragment is saturated (see the next section) and this is the situation when the tensile stress reaches the maximum values. This indicates that rupture will occur near saturation and is consistent with the breakage phenomena observed when the marl is saturated.

Figure 31 presents the expected cracking pattern of the fragment and a comparison with a typical pattern found in real cases. They agree qualitatively; however, the stratification found in real cases introduces significant strength anisotropy and this has a major influence on cracking development. There is no doubt that strata are weakness planes where cracking starts. The introduction of stress anisotropy into the numerical model would not add information on the degradation mechanism, as it would be equivalent to forcing the location of the cracks into these planes. Moreover, the calibration of the anisotropy would require experimental data from tests different from those presented.

Strength

Besides compressibility, strength also depends on water content and suction. Brazilian splitting tests and unconfined compression tests were performed in marl samples under different suctions (Figs 32 & 33). Some dispersion was observed in the results, which can be explained mainly by the heterogeneity of the marl matrix in some samples and by the existence of micro-cracks not detected before the tests. Both tensile strength and stiffness under compression increase with suction as expected.

The loading cycles applied in the unconfined compression tests on marl samples under different suctions (Fig. 33a) affected the stiffness of the marl. Significant slaking (Fig. 33b) was observed and the stiffness was progressively

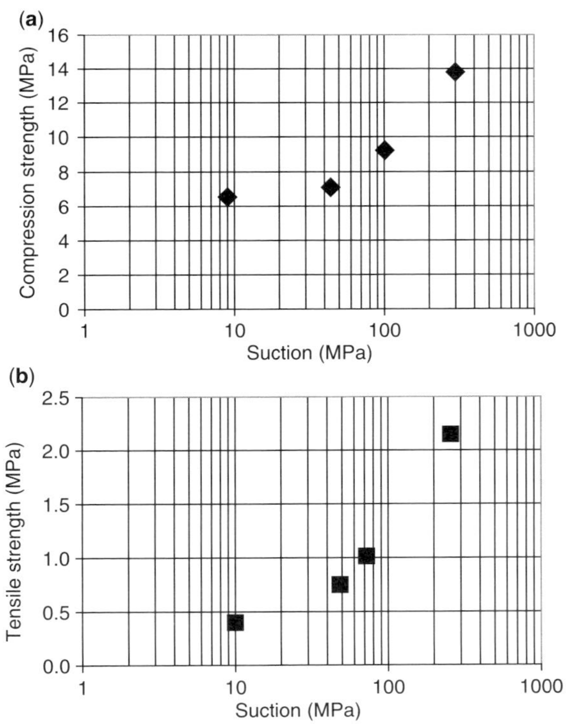

Fig. 32. Tests performed in marl samples under different suctions: (**a**) unconfined compression tests; (**b**) Brazilian splitting tests.

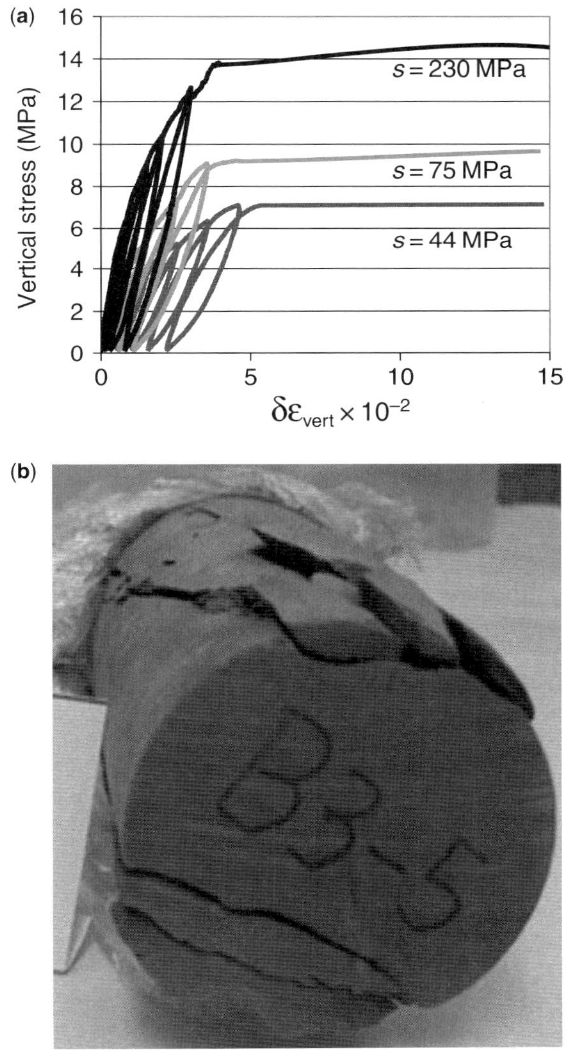

Fig. 33. Unconfined compression tests: (**a**) cyclic loading; (**b**) degradation observed.

diminishing with increasing loading cycles. The samples had shown some ductile behaviour, as high values for axial extension (larger than 15%) were reached, independently of suction. Hysteretic behaviour was also measured, being more significant for the lower suctions. This can be explained by the closure and reopening of the cracks, and indicates that the presence of cracks does not affect the marl strength for relatively high stress levels. However, cracks induced by loading cycles are different from those induced by wetting and drying cycles because the second type of cracks lead to stronger breakage and disaggregation.

Direct shear tests were performed by Oliveira (2006) on disturbed marl samples with a significant weathering degree. A friction angle of 20° and a small effective cohesion of 5 kPa (peak values) were measured. The residual friction angle was 8°, a very low value. As mentioned above, natural slopes in Abadia marls show an average slope gradient of 20°, in agreement with the peak friction angle measured in the tests. The peak friction angle measured in remoulded samples is interpreted as a measure of the 'fully softened strength' (Stark & Eid 1998). This may be the state of shallow levels of the weathered marl, which are prone to sliding, when natural reworking is intense. As a first approximation (the equilibrium of the slope is in practice controlled by several factors, including the pore-water pressure) it may be said that the sliding strength is approximated by a fully softened state but is still from residual conditions.

Data from the Brazilian tests (Fig. 32a) indicate that the tensile strength of the marl is maintained for zero suction. Its value ($c_0 = 0.4$ MPa $= 400$ kPa) can be assumed to be similar to (or of the same order of magnitude as) the effective cohesion of the marl for saturated conditions. The value expected is much larger than that measured in the shear tests for the disaggregated sample ($c' = 5$ kPa), indicating a significant loss of cohesion after disaggregation. The existence of cohesion explains the stability of slopes with steep angles maintained during excavation processes (shown in Fig. 19b). However, these slopes must be protected to avoid slope instability phenomena as a result of weathering, induced by suction changes (illustrated in Fig. 19a). Suction increase has an important role in this disaggregation as a result of internal cracking or debonding phenomena caused by internal stress development (Fig. 24), which is equivalent to a de-cohesion process. Therefore the destructuring mechanisms activated by wetting and drying cycles (atmospheric actions) are stronger and have a greater negative impact on shear strength than changes in effective stress caused by water-level oscillations.

Concluding remarks, Abadia marls

Slides on natural slopes of Abadia marls are triggered by weathering of this evolutive rock. Wetting and drying cycles causes strong particle breakage and disaggregation. The results presented led to the conclusion that weathering was mainly due to physical disaggregation, as there is no significant difference in the mineralogical composition of weathered and non-weathered marls. The results of artificial weathering tests presented confirm this idea.

Significant swelling strains were measured in swelling tests on rock samples with different initial suctions and under different vertical stresses. The suction values imposed correspond to relative humidity values common in many field environments. The occurrence of differential swelling displacements explains the cracking and loss of structure observed in real cases as a result of the development of tensile stresses and plastic deformations. This degradation mechanism was confirmed in the calculation of the effects of full saturation on a fragment of marl. In the real material cracking and disintegration is also controlled by the stratification.

The hydromechanical properties of Abadia marls are strongly dependent on suction. MIP tests and oedometer tests performed on rock samples and on reconstituted or remoulded samples (fully disaggregated material) showed that the fully disaggregated material has higher porosity. The consequences are loss of stiffness and strength as a result of disaggregation, and also changes in the water retention properties of the material. It was also shown that drying is responsible for some stiffness change in the remoulded material, indicating that it gives it some structure. However, this effect is rapidly lost when the destructured dry material is wetted and loaded. This indicates that the saturation of the disaggregated material reduces its strength, which may result in instability phenomena as soon as the equilibrium conditions are changed.

The equilibrium of slopes is in practice controlled by several factors. Pore-water pressure and sliding shear strength are the most important. The mechanism of slope instability caused by weathering is very different from that of instability caused by water-level oscillations. The second mechanism is associated with changes in the effective stresses whereas the first is associated with the degradation of the mechanical properties of the materials. For the initial sliding strength (given by the friction angle and/or cohesion) the slope may be stable for a fixed change in the pore-water pressure (in the case of intense rainfall, e.g.), but the same slope may become unstable for the same change in the pore-water pressure if the sliding strength is reduced because of weathering. Slope instability occurs frequently after rainfall because that is when the most unfavourable conditions are met. This instability process is continuous, as the superficial layers, when exposed to weathering, show reduction in their strength characteristics. If small landslides result from this process, the recently exposed lower layers will progressively become weathered. This is the mechanism for the reactivation of landslides caused by weathering.

Drying effects on the rock lead to cracking and disaggregation. These effects are different from those on the disaggregated material because it was shown that suction increment had an effect on rock structure. For the rock, the disaggregation observed can be explained by a mechanism similar to the that studied for wetting. Because the rock *in situ* is almost saturated, its exposure to the environment in a recent cut creates a suction gradient and the disaggregation of the material. This mechanism explains the erosion observed mainly in summertime on recently excavated cut slopes.

Brazilian splitting tests and unconfined compression tests on rock samples under different suctions indicated that stiffness and strength depend on the suction and are strongly reduced near saturation. This explains landslide occurrence after heavy rain periods when marls are fully saturated. Cyclic wetting and drying under constant stress leads to stronger degradation than cyclic loading under constant suction because of the physical degradation associated with the first case. This can be explained by the removal of bonding or cementation of the rock constituents owing to suction reduction.

The usual provisions of slope cover and protection adopted in the design of artificial slopes in this material are justified, as weathering and its negative consequences are inevitable. On a geological time scale, sliding of the exposed material occurs until a stable configuration is reached. This corresponds to almost flat slopes (slope gradient equal to the residual friction angle, 8°) typical of the Abadia region and justifies the name 'basin' given to the region of the Abadia geological formation.

Conclusions

This paper discusses the weathering behaviour of two marls: Pamplona marls in Spain and Abadia marls in Portugal. Both marls show similar responses as a consequence of physical weathering processes (suction changes and stress changes). Unloading changes effective stresses and leads to swelling of rocks if active minerals are present. On the other hand, water content changes (suction changes) induce plastic deformations producing fissuring of the rock. The induced cracks are an easy way for the water to enter into the rock, leading to loss of stiffness and strength. These aspects are discussed in this paper.

The reported experimental study on Pamplona marls concentrated on its strength characteristics as determined in direct shear tests. The study was associated with the analysis of the shear strength available at the foundation–rock interface of a gravity dam. The dam stability conditions depend on the available strength parameters of the marl. Most of the shear tests reported were performed in a large shear box (30 × 30 cm). Block specimens were recovered immediately after excavation in a level of unweathered marl.

Pamplona marls have a high calcite content (around 50%). Clay minerals include illite and chlorite. Mica was also identified. If undisturbed the marl has a relatively high unconfined compression strength (10–30 MPa). Anisotropy is provided by the sedimentation planes and other fracture systems that are less developed. The unweathered marl, at the site investigated, is not saturated. Measured degrees of saturation range between 0.73 and 0.95. Given the low average void ratio (0.20) and the presence of a significant clay fraction, relatively high suction values (a few megapascals) are expected. Wetting (suction reduction) leads to a rapid degradation of the material. Effective friction and cohesion are strongly reduced when the material is fully saturated. The material brittleness is also reduced. Tests performed on concrete–marl contacts also show a strong reduction in strength if compared with values measured on rock specimens. Two reasons may be given for this reduction: an increase in the marl water content, probably during the process of setting of the concrete pad against the exposed surface of the marl, and the shearing process itself of two surfaces having different stiffness.

Pamplona marls are strongly sensitive to the first wetting, an unavoidable situation in exposed cuts. The dam foundation described here is also a case of forced wetting. The reported results provide data on the expected strength reduction, but it should be accepted that selecting the correct strength parameters remains difficult.

Abadia marls have a higher *in situ* void ratio (0.20–0.35) than Pamplona marls and a substantial proportion of clay minerals (more than 50%; the main minerals are illite, kaolinite and mixed layer clay minerals). This clay component results in higher plasticity (the liquid limit reaches 50%) and in a low residual friction angle (8° has been reported, compared with 16° for Pamplona marls). The natural undisturbed marl remains essentially saturated. Unconfined compression strength of the saturated undisturbed material is close to 6 MPa. When exposed to the atmosphere degradation is intense. This study concentrated on the effect of suction and suction cycles on the mechanical and hydraulic evolution of the marl. In swelling tests under constant confining stress the marl exhibits an expansive behaviour. A numerical simulation of the effect of boundary wetting on an isolated circular fragment of rock suggests that the weathering process is associated with the development of internal surfaces where tensile stresses are developed. They are a consequence of the differential swelling inside the rock. To a certain extent this explanation must be valid also for Pamplona marls.

Anisotropy is common for both materials studied and in marls in general. The existence of weakness planes affects the strength properties of these materials irrespective of their weathering degree because they are preferential sliding surfaces. As exemplified in the study on the foundation of Monreal dam on Pamplona marls, the existence of joints must be considered in the design of foundations in these materials and also for other applications (tunnels and slopes, e.g.). Lack of confinement and the application of mechanical actions induced by construction processes contribute to the reduction of the strength and to the modification of other hydromechanical properties of this geological material. Weathering effects, however, must be considered also in the evaluation of the mechanical properties of the material, as they make an important contribution to this reduction and are inevitable in practice.

Finally, the mechanism of slope instability linked to weathering is very different from that for instability caused by water-level oscillations. In the first case there is a reduction of the sliding shear strength of the evolutive materials such as the marls studied. In the second case, instability is caused by changes in the effective stresses. The reduction of the sliding shear strength explains the occurrence of landslides in the geological formations with higher degree of weathering.

References

ALONSO, E. E. & ALCOVERRO, J. 2002. Swelling rocks. *In*: SCHANZ, T. (ed.) *Proceedings of the 4th Workshop Schriftenreihe Geotechnik*. Bauhaus-Universität Weimar, 11–21.

ALONSO, E. E. & BERDUGO, I. R. 2005. Expansive behaviour of sulphate-bearing clays. *In*: *Proceedings of International Conference on Problematic Soils, Eastern Mediterranean University, Famagusta, N. Cyprus*, 1–25.

ALONSO, E. E., GENS, A. & JOSA, A. 1990. A constitutive model for partially saturated soils. *Géotechnique*, **40**, 405–430.

ALONSO, E. E., PINEDA, J. A. & ROMERO, E. E. 2005. *Embalse de Monreal: Informe de la Cimentación*. Geotechnical Report, Department of Geotechnical Engineering, Technical University of Catalonia, Barcelona [in Spanish].

ANONYMOUS 2002. *Mapa Geotécnico del Área de Pamplona*. Dirección General de Obras Públicas del Gobierno de Navarra.

ASTM 1992. ASTM D4644, Test method for slake durability of shales and similar weak rocks. *In*: *1999 Annual Book of ASTM Standards. Soil and Rock; Geosynthetics, Section 4*, **04.08**.

BARTON, N. & CHOUBEY, V. 1977. The shear strength of rock joints in theory and practice. *Rock Mechanics and Rock Engineering*, **10**, 1–54.

BARTON, N., BANDIS, S. & BAKHTAR, K. 1985. Strength, deformation and conductivity coupling of rock joints. *International Journal of Rock Mechanics and Mining Sciences & Geomechanical Abstracts*, **22**, 121–140.

BERDUGO, I. R. 2007. *Tunnelling in sulphate-bearing rocks—expansive phenomena*. PhD thesis, Technical University of Catalonia, Barcelona.

BIENIAWSKI, Z. T. 1989. *Engineering Rock Mass Classifications*. Wiley, New York.

BOSSART, P., MEIER, P. M., MOERI, A., TRICK, T. & MAYOR, J. C. 2002. Geological and hydraulic characterisation of the excavation disturbed zone in Opalinus clay of the Mont Teri Rock Laboratory. *Engineering Geology*, **66**, 19–38.

BOTTS, M. E. 1998. Effects of slaking on the strength of clay shales: a critical state approach. *In*: EVANGELISTA, A. & PICARELLI, L. (eds) *The Geotechnics of Hard Soils–Soft Rocks*. Balkema, Rotterdam, 447–458.

BURLAND, J. B. 1990. On the compressibility and shear strength of natural clays. *Géotechnique*, **40**, 329–378.

CALABRESI, G. & SCARPELLI, G. 1985. Effects of swelling caused by unloading in overconsolidated clays. *In*: *Proceedings of 11th International Conference on Soil Mechanics and Foundation Engineering*, **2**, 411–414.

CARDOSO, R. & ALONSO, E. E. 2009. Degradation of compacted marls: a microstructural investigation. *Soils and Foundations*, **49**, 315–327.

CARDOSO, R., ROMERO, E., LIMA, A. & FERRARI, A. 2007. A comparative study of soil suction measurement using two different high-range psychrometers. *In*: *Proceedings of 2nd International Conference on Mechanics of Unsaturated Soils*. Bauhaus-Universität Weimar, 79–94.

FERREIRA, D. B. 1981. *Corte Geomorphologique de Portugal*. Memórias do Centro de Estudos Geográficos de Lisboa, **6**.

HARDIN, B. O. 1985. Crushing of soil particles. *Journal of Geotechnical Engineering ASCE*, **111**, 1177–1192.

HOEK, E. & BRAY, J. 1981. *Rock slope engineering*, 3rd edn. Institution of Mining and Metallurgy, London.

ISRM 1979. Suggested method for determining water content, porosity, density, adsorption and related properties and swelling and slake-durability and index properties. *International Journal of Rock Mechanics and Mining Science & Geomechanical Abstracts*, **18**, 85–110.

JEREMIAS, F. T. 2000. *Geological controls on the engineering properties of mudrocks of the north Lisbon area*. PhD thesis, University of Sheffield.

LEINFELDER, R. 1993. A sequence stratigraphic approach to the Upper Jurassic mixed carbonate–siliciclastic succession of the central Lusitanian Basin, Portugal. *Profil*, **5**, 119–140.

LEROUEIL, S. & VAUGHAN, P. 1990. The general and congruent effects of the structure in natural soils and weak rocks. *Géotechnique*, **40**, 467–488.

OLIVEIRA, S. 2006. In situ *characterization of soft marls and embankments. Contribution to modelling*. MSc thesis, Faculdade de Ciências e Tecnologia da Universidade Nova de Lisboa.

OLIVELLA, S., GENS, A., CARRERA, J. & ALONSO, E. E. 1996. Numerical formulation for simulator (CODE_BRIGHT) for coupled analysis of saline media. *Engineering Computations*, **13**, 87–112.

OLIVIER, H. J. 1987. Some aspects of the influence of mineralogy and moisture redistribution on the weathering behaviour of mudrocks. *In*: *Proceedings of 4th Congress, International Society of Rock Mechanics, Montreux*, 467–474.

PICARELLI, L. 1991. Discussion on the paper 'The general and congruent effects of the structure in natural soils and weak rocks' by S. LEROUEIL & P. R. VAUGHAN. *Géotechnique*, **41**, 281–284.

PINYOL, N., VAUNAT, J. & ALONSO, E. E. 2007. A constitutive model for soft clayey rocks which includes weathering effects. *Géotechnique*, **57**, 137–151.

ROMERO, E. & VAUNAT, J. 2000. Retention curves of deformable clays. Experimental evidence and theoretical approaches in unsaturated soils. *In*: TARANTINO, A. & MANCUSO, C. (eds) *Proceedings of International Workshop on Unsaturated Soils, Trento, Italy*. Balkema, Rotterdam, 91–108.

Romero, E. E. 2005. *Ensayos de corte directo sobre contacto hormigón–marga embalse de Monreal*. Laboratory Report, Department of Geotechnical Engineering, Technical University of Catalonia, Barcelona.

Stark, T. D. & Eid, H. T. 1994. Slope stability analysis in stiff fissured clays. *Journal of Geotechnical and Geoenvironmental Engineering*, **123**, 335–343.

UPC-DLT 2002. *CODE_BRIGHT—User's Guide*. Departamento de Ingeniería del Terreno, ETS Ingenieros de Caminos, Canales y Puertos de Barcelona, Universidad Politécnica de Cataluña, Barcelona.

Van Genuchten, M. Th. 1980. A closed-form equation for predicting the hydraulic conductivity of unsaturated soils. *Soil Science Society of America Journal*, **49**, 143–159.

Zêzere, J. L. 2002. Landslide susceptibility assessment considering landslide typology. A case study in the area north of Lisbon (Portugal). *Natural Hazards and Earth System Sciences*, **2**, 73–82.

Zêzere, J. L., Ferreira, A. B. & Rodrigues, M. L. 1999. Landslides in the north of Lisbon region (Portugal): conditioning and triggering factors. *Physics and Chemistry of the Earth (A)*, **24**, 925–934.

Landslide mechanisms in Hong Kong

S. R. Hencher[1,2,3]* & S. G. Lee[4]

[1] Halcrow China Ltd., Level 10 Millennium City 6, Kwun Tong Road, Kowloon, Hong Kong, PR China
[2] School of Earth and Environment, University of Leeds, Leeds LS2 9JT, UK
[3] Department of Earth Sciences, Hong Kong University, Hong Kong, PR China
[4] Department of Civil Engineering, University of Seoul, Seoul, South Korea

*Corresponding author (e-mail: stevehencher@btinternet.com)

Abstract: This paper reviews the nature and mechanics of landslides in the weathered terrain of Hong Kong. The vast majority of landslides are very shallow (a few metres depth) and occur during intense rainstorms. Deeper-seated landslides, in contrast, may occur days or weeks after intense rainstorms. The time of occurrence of landslides can be linked to hydrological and hydrogeological factors, and a hydrogeological grouping of landslide mechanisms is introduced related to timing in a storm. A relationship is presented that links intensity of landsliding to 24 h rainfall. The gradual deterioration and internal erosion of slopes prior to detachment is discussed and allows some realistic opportunity for identifying progressive major landslides. In particular, the growth of natural piping systems and infilling of dilated fracture networks are recommended as important indicators of landslide development. The conclusions are supported by case examples of slope failures, the study of some of which has been taken to a forensic level.

Landslides are a common hazard in Hong Kong and this hazard has been addressed over many years by the concerted efforts of the Hong Kong Government, its consultants and contractors through a specialized office, previously known as the Geotechnical Control Office (GCO), and now as the Geotechnical Engineering Office (GEO). This office vets the designs of all new slopes and is going through a process of upgrading old slopes through the implementation of extensive landslide preventive works, wherever necessary. Over the last few years, it has also begun to tackle, in a managed way, the risks from landslides issuing from the steep and rugged natural terrain that is typical of most of the territory. The early history of the Hong Kong Government on dealing with the risk of landslides has been discussed by Brand (1984), and the situation has been updated by Malone (1997) and Wong (2005).

Ground conditions

Geology

Current understanding of the geology of Hong Kong is summarized in two government publications (Fyfe *et al.* 2000; Sewell *et al.* 2000). The major urban areas are underlain by Jurassic and Early Cretaceous volcanic rocks and by granitic rocks of similar age but intrusive into the volcanic rocks. The volcanic rocks, which in their fresh state can have unconfined compressive strength well in excess of 200 MPa, largely make up the higher mountainous areas surrounding Hong Kong harbour and peaks on Lantau Island, as illustrated in Figures 1 and 2. The central harbour basin and other low-lying areas and more flat-lying islands comprise granitic rocks that include granite, granodiorite and quartz monzonite (Figs 3–5). Differences in petrology can be reflected to a degree in engineering behaviour. For example, quartz monzonite has been linked to particular ground problems such as preferentially weathered zones encountered in construction of the Aberdeen tunnel, which necessitated extensive dewatering and grouting (Cochrane 1984) and low shear strength rock joints in the Yip Kan Street landslide (Hencher 1981, 2000).

Weathering

Distribution of weathered rock

Hong Kong rocks are locally deeply weathered to depths as great as 60 m and sometimes more. Typical profiles are illustrated schematically in Figure 6. Active weathering processes and the implications for geotechnical engineering have been discussed by Hencher (2006). The weathered rocks range from completely restructured and fairly dense residual soil to soil-like weathered rocks that retain the original rock texture and fabric but have dry densities as low as $1.2\ Mg\ m^{-3}$, which is less than half the density of rock in its unweathered state and indicative of a very porous, open and sensitive condition. At a mass scale the weathering front is sometimes very sharp, with a zone of uniform soil-like weathered rock overlying almost fresh rock (Fig. 7). Elsewhere, the change is more gradual, with mixtures of rock

From: Calcaterra, D. & Parise, M. (eds) *Weathering as a Predisposing Factor to Slope Movements*.
Geological Society, London, Engineering Geology Special Publications, **23**, 77–103.
DOI: 10.1144/EGSP23.6 0267-9914/10/$15.00

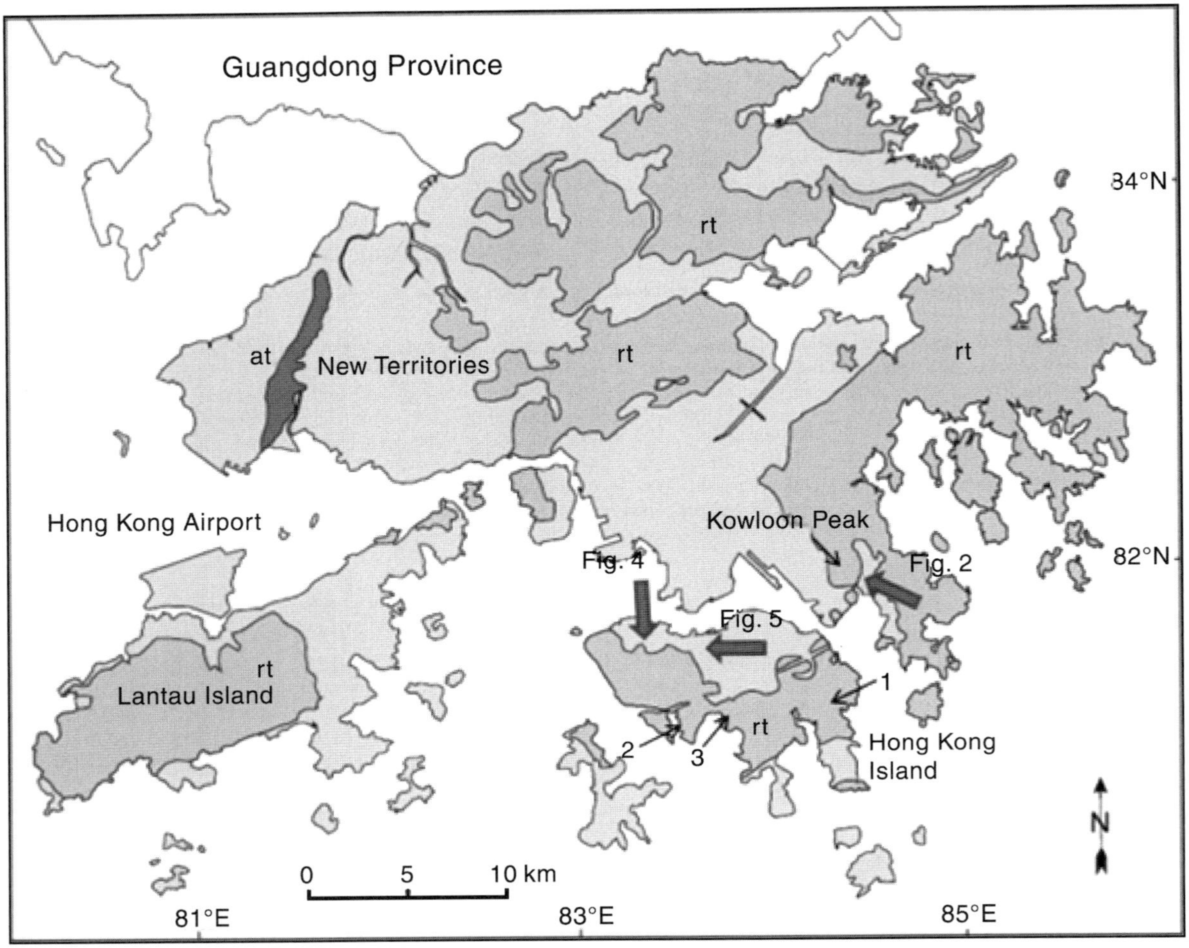

Fig. 1. Distribution of major volcanic outcrop in Hong Kong (adapted from Fletcher 2004). rt, predominantly rhyolite tuff with some lava and sedimentary rock; at, predominantly andesitic lava with some tuff and sedimentary rock; 1, Chai Wan Road; 2, Shum Wan Road, Aberdeen; 3, South Bay Close, Repulse Bay (numbers refer to locations mentioned in the text).

and soil occurring between zones that are fully soil or fully rock (Figs 8 & 9), or the weathering profile can be completely haphazard (see Fig. 10).

The distribution of weathering can be difficult to predict from current topography. Sometimes hills are underlain by thick weathered profiles but at other locations the hills comprise 100% rock from the ground surface. Valleys may be the location of deep weathering but elsewhere rivers run directly on exposed rock even where the valley reflects the presence of a fault. The difficulties for investigation and characterization are obvious. Surface mapping ('regolith mapping') is advocated by some researchers as a first stage in identifying susceptibility to natural terrain landslides (e.g. Fletcher *et al.* 2002) but there remain great uncertainties in interpreting the underlying ground conditions from surface expression alone and surprises are common.

Weathering terminology

The terminology for weathering used in this paper is that advocated in *Geoguide 3* (GCO 1988) and essentially the same as that recommended by Anonymous (1995) and adopted in BS5930 (BSI 1999). The scheme is illustrated in Figure 11. It is generally found useful in Hong Kong to classify hand-size (material) samples at the scale of core logging and laboratory testing according to the left-hand table in the figure. The classification does not cater readily for all the varieties of weathered rock that are encountered; for example, it is difficult to apply to rocks that are highly disintegrated. It does, however, often allow relatively uniform weathered rock of similar strength to be grouped using simple index tests (Hencher & Martin 1982; Martin 1986). At the larger scale, for engineering works and landslide analysis, units or zones need to be differentiated

Fig. 2. Kowloon Peak (Fei Ngo Shan) in volcanic rock with weathered granite making up the foothills. (Note large colluvial lobes on the side slopes.) Lion Rock (resistant granite) in distant background.

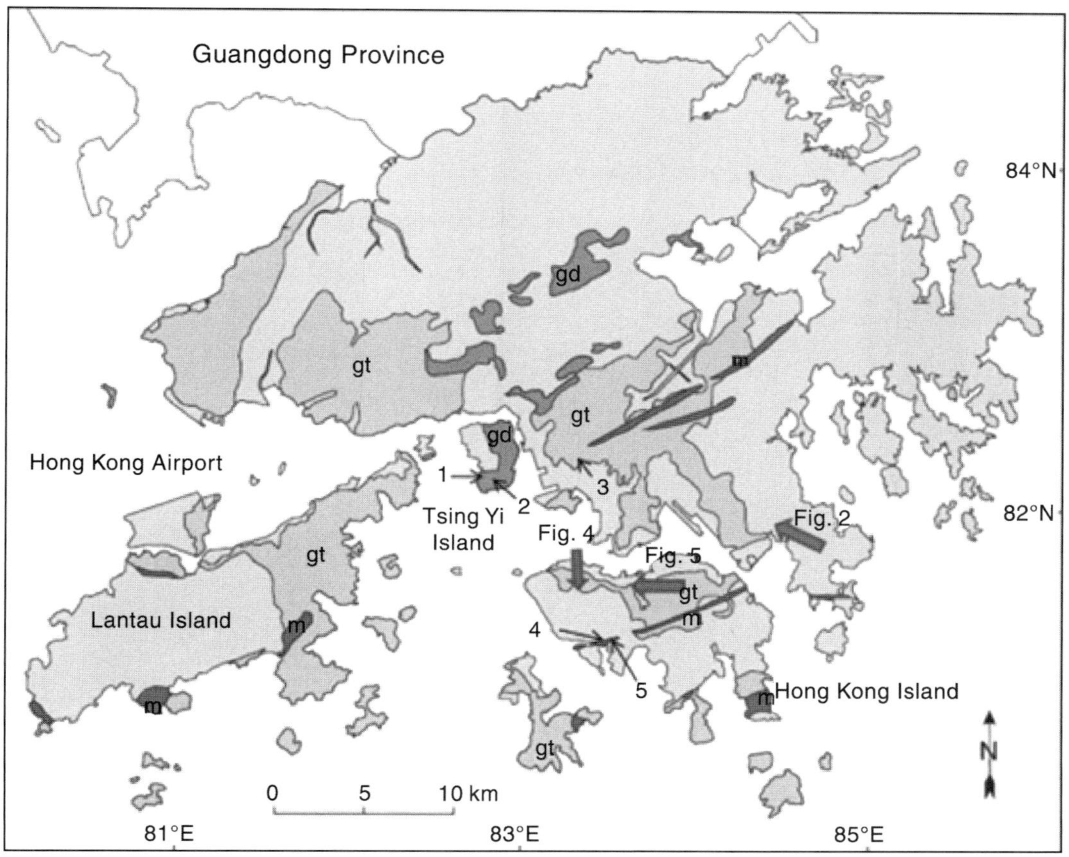

Fig. 3. Distribution of main granitic rock bodies in Hong Kong (adapted from Fletcher 2004). gt, granite; gd, granodiorite; m, quartz monzonite; 1, Tsing Yi 1; 2, Tsing Yi 2; 3, Ching Cheung Road; 4, Yip Kan Street; 5, Aberdeen Tunnel Portal (numbers refer to locations mentioned in the text).

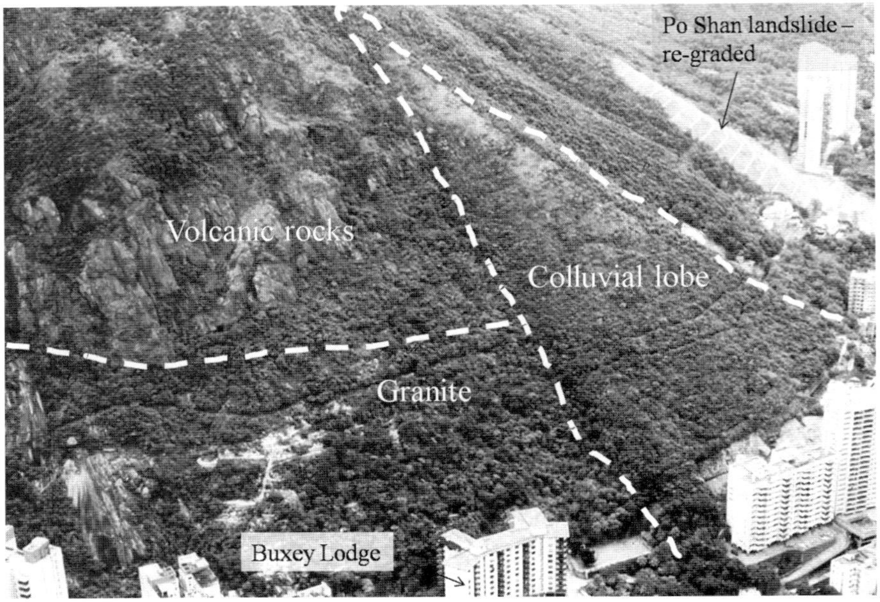

Fig. 4. Aerial view (IR photograph) of part of Mid Levels District above Hong Kong Harbour.

Fig. 5. View westwards across northern part of Hong Kong Island showing change in gradient from steep volcanic rhyolite tuffs to underlying more weathered granite.

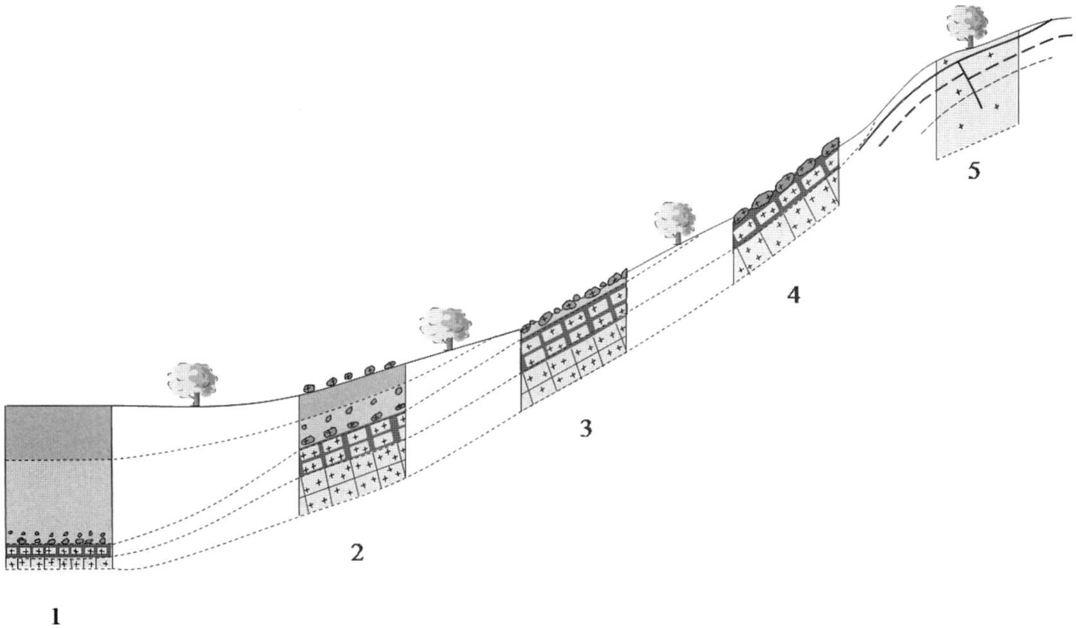

Fig. 6. Variety of weathering profiles to be anticipated in different parts of slope profile (schematic and not to be relied upon). Sections 1–4 redrawn from the Ruxton & Berry (1957) representation of an old age weathering profile. Sheeting joint section (5) is area where erosion rate exceeds weathering.

Fig. 7. Sharp weathering front in weathered granite, Tai Tam Reservoir, Hong Kong.

Fig. 8. Corestone development in granite. Tai Tam Reservoir, Hong Kong. Height of exposure is approximately 4 m.

Fig. 9. Corestone development in volcanic rocks, Victoria Peak, Hong Kong.

according to variations in mass properties. Prescriptive zonal classifications are sometimes used in Hong Kong following the principles discussed by Martin & Hencher (1986). The schemes of *Geoguide 3* (GCO 1988) and Anonymous (1995) for heterogeneous weathered profiles are illustrated on the right-hand side of Figure 11. In the author's experience such zonal schemes for weathered rock masses are actually of little use other than for the broad communication of rock mass conditions, because they are inflexible and fail to account for important factors such as geological structure and fracture spacing, which are often of overriding importance to stability and properties at the mass scale. No one, to the author's knowledge, has every attempted to link ranges of geotechnical parameters such as E or ϕ values to particular prescriptive weathering zones, which is very wise because it would be very difficult to generalize meaningfully, given the often extremely heterogeneous nature of zones of weathered rock. In the author's opinion, classification is no substitute for description at the mass scale. Incidentally, it is to be noted that the weathering classification schemes illustrated in Figure 11 are not used world-wide and, confusingly, many of the terms defined in that figure are used elsewhere in different ways. This problem was addressed in detail by Anonymous (1995) and Martin & Hencher (1986), but despite the inevitable confusion, committees continue to invent new schemes, redefining the same terms, often apparently without any research or reasoning. Hencher (2008) has provided a critique of the new scheme adopted for Eurocode 7, where a new mass classification has been prescribed that conflicts with all other such classifications world-wide yet fails to provide any guidance on material-scale weathering classification, which experience shows is the most useful.

Fig. 10. Highly heterogeneous profile with single very large corestone. Right-hand photograph shows detail with very rapid transition from Grade V, completely decomposed rock, to almost fresh rock within the corestone. Tai Po, Hong Kong. Width of right-hand photograph approximately 1 m.

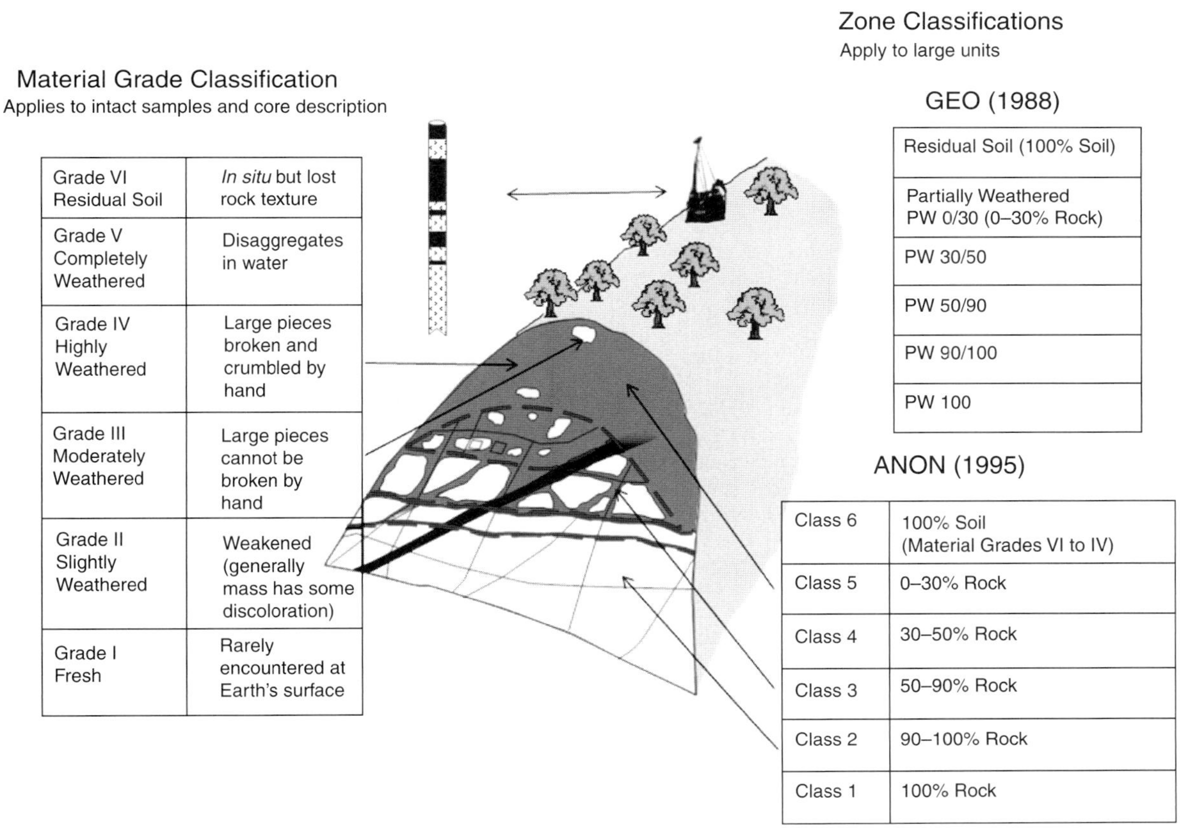

Fig. 11. Weathering classification.

Fig. 12. Relict joint through Grade IV, highly weathered granite. Near Shek O, Hong Kong.

Engineering properties

Properties of weathered rocks can be difficult to ascertain because sampling and testing of very weak and sensitive weathered materials without disturbance can be impossible, and because of the heterogeneous nature of weathered rock masses. The difficulty of conducting high-quality laboratory tests that provide realistic properties is one of the main reasons for employing a material weathering classification linked to simple index tests. Parameters can be transferred from one location to another. At the mass scale one may wish to account for contributions from corestones within the profile. These problems have been addressed partly by Hencher & McNicoll (1995) and Hencher (2006).

Geological structure, fabric and texture survive at least in part right through to the final stage of weathering, residual soil, where the soil collapses and becomes reworked, so relict joints (Fig. 12) and joints that develop during weathering and unloading (Hencher 2006; Hencher & Knipe 2007) dominate stability in many slopes. Larger structures such as dykes, sills and faults often result in differential weathering within the rock mass. Variation in soil particle grading, texture and clay content results in permeability contrasts that control groundwater flow and partitioning, as discussed below with respect to landsliding. Groundwater flow through the rock mass can cause erosion pipes to form, typically at discontinuity intersections but also commonly as part of the overall degradation and progressive failure of slopes (Nash & Dale 1984; Hencher 2006). These erosion pipes can lead to relatively rapid infiltration and throughflow, and are a common feature associated

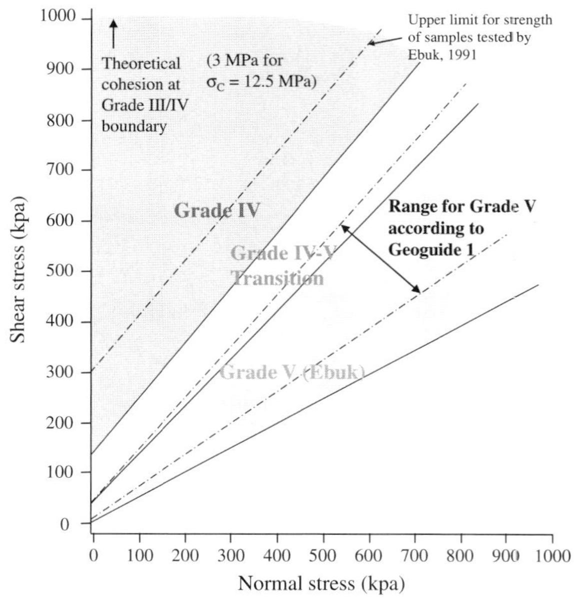

Fig. 13. Range of shear strength envelopes measured by Ebuk (1991) as reported by Ebuk *et al.* (1993) and presented by Hencher (2006). *Geoguide 1* is GEO (1993).

with large, deep-seated landslides, as discussed in more detail below.

Properties of typical weathered rock materials in Hong Kong have been discussed by Irfan (1996, 1999) and Hencher (2006). In granite saprolite (predominantly Grades IV and V materials), friction angles are typically $\geq 30^\circ$ (Martin 1986). El-Ramly *et al.* (2005) reported little difference between friction values for Grades IV and V granite from Hong Kong, with an average ϕ value of 37.7°. Hencher (1983*a*, 2006) reported tests on samples of Grade IV granite of various dry densities, corrected for dilation angle at peak strength, which gave non-dilatant, basic ϕ values in the range of 38–40°, which is the same as the corrected non-dilational friction angle for rough granite joints (Hencher & Richards 1982) and actually the same as for many other silicate rocks (Papaliangas *et al.* 1996). Volcanic rock joints, however, sometimes give lower dilation-corrected shear strengths, down to 32° for

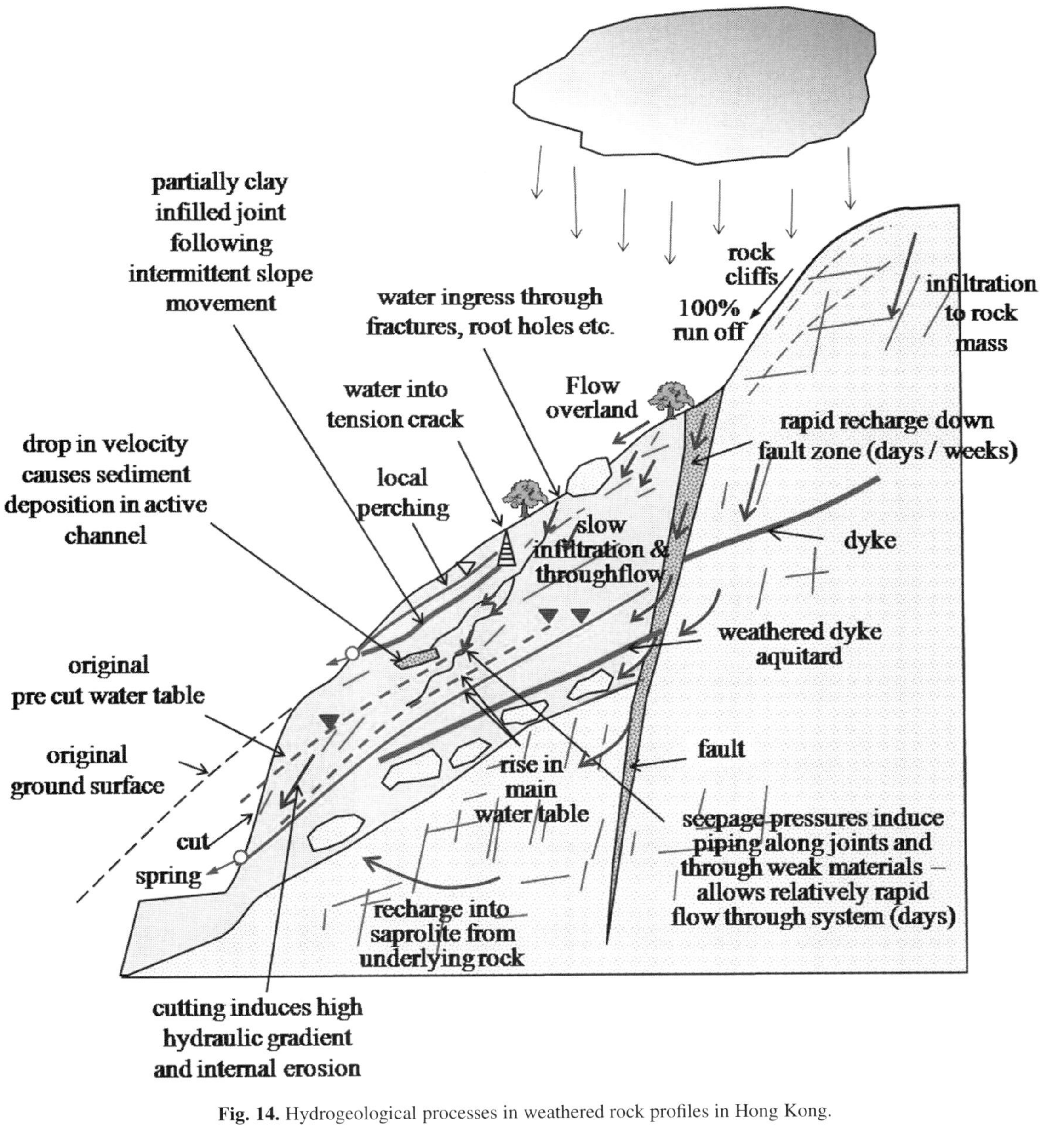

Fig. 14. Hydrogeological processes in weathered rock profiles in Hong Kong.

fine-grained varieties. Coatings on joints can lead to considerably reduced shear strength of 17° at low stresses for chlorite (Hencher 1981; Brand *et al.* 1983), and values for joints infilled with clay can be even lower, of the order of 20–10° (Koor *et al.* 2000). In the case of clay-infilled discontinuities, much depends upon the continuity of the clay infill and its thickness relative to the wall rock roughness. If it occurs only as accumulations in hollows along an intermittently dilating rock joint, as in examples discussed below, then the shear strength characteristics of the infill may be irrelevant although its contribution to reduced permeability may be significant. Proper geological characterization of such structures is necessary. One slightly undulating and very persistent (>60 m laterally) clay-infilled discontinuity of unknown geological origin contributed to the Shek Kip Mei landslide of 1999 (FMSWJV 2000) but such features are fortunately rare.

True cohesion in the sense of adhesional bonding between grains is present in soil-like, weathered rock material (Grades VI–IV) as a result of relict bonding from the parent rock and sometimes because of secondary cementation but reported values from Hong Kong are usually very low. El-Ramly *et al.* (2005), in a review of selected data, gave a range of 0–25 kPa. Ebuk (1991) measured cohesion up to 300 kPa for Grade IV granite from Hong Kong (Fig. 13), but it is considered that even higher values probably exist *in situ* for intact material, given that the upper bound unconfined compressive strength of Grade IV is about 10–15 MPa (break by hand), as discussed by Hencher (2006). It is suspected that the lack of higher reported values is due to a lack of systematic study.

Suction (negative water pressure in partially saturated soil) clearly provides some temporary increase in effective stress and therefore strength in many slopes (Fredlund

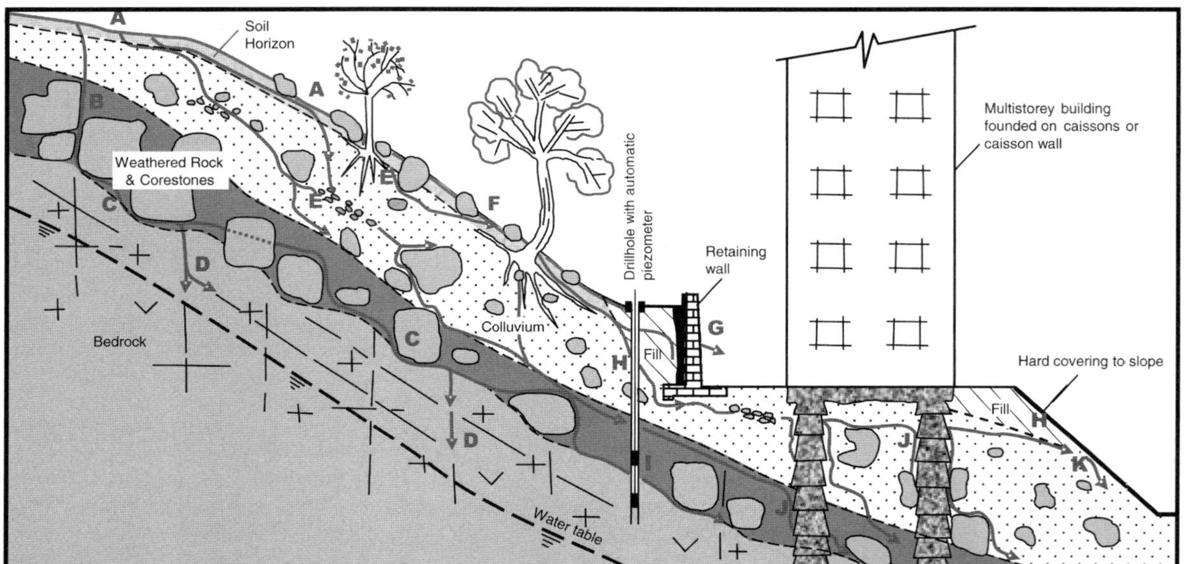

NATURAL SLOPE	DEVELOPED SLOPE
A Groundwater inlet zones in small depressions, via cracks, animal burrows, root systems and around boulders.	**G** Pipe flow through poorly compacted fill. Retaining walls (with or without drainage filters) may cause damming.
B Movement of water along relict joints sometimes forming conduit flow.	**H** Pipe flow beneath fill and redirected by retaining structure.
C Movement of water around corestones, also influenced by relict joints and other 'permeability' boundaries.	**I** Piezometer locally intercepting pipe. Other pipes not intercepted.
D Fissure flow along joints both vertically and horizontally which may feed the more permanent water table.	**J** Voids caused by caisson construction redirect pipe flow.
E Pipe flow along gravel bands and around boulders	**K** Damming of pipe flow by hard cover to slope
F Efflux points on slope	

Fig. 15. Schematic representation of hydrogeological conditions close to developments (after Nash & Dale 1984).

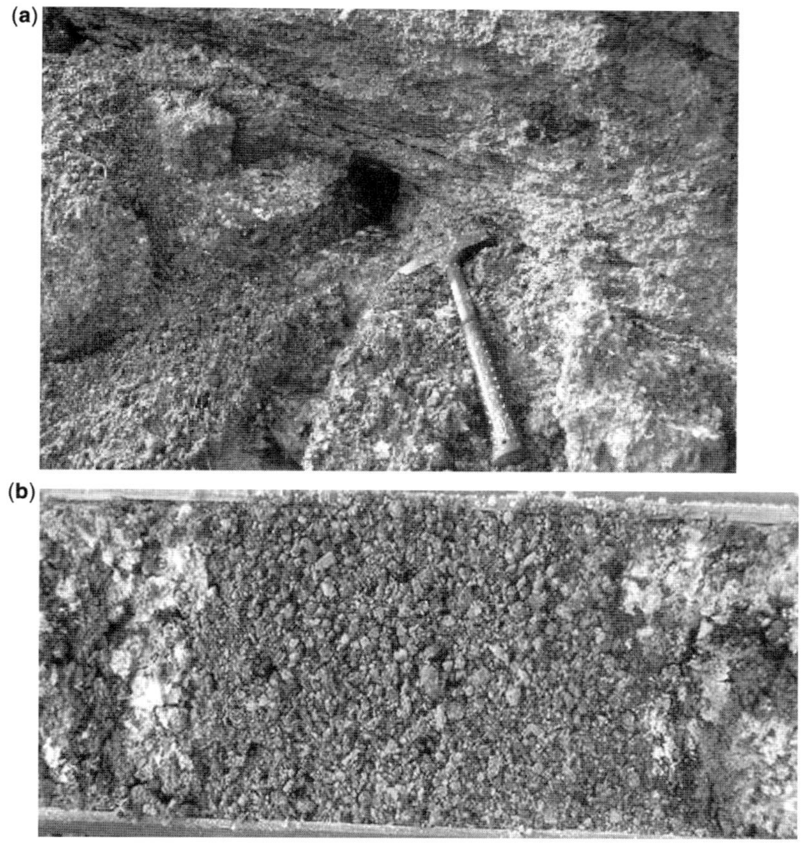

Fig. 16. (**a**) Natural pipe through decomposed granite exposed in landslide at Tsing Yi (2); (**b**) alluvial sand sandwiched between decomposed granite within continuous Mazier sample from depth at location of Ching Cheung Road landslide, 1997.

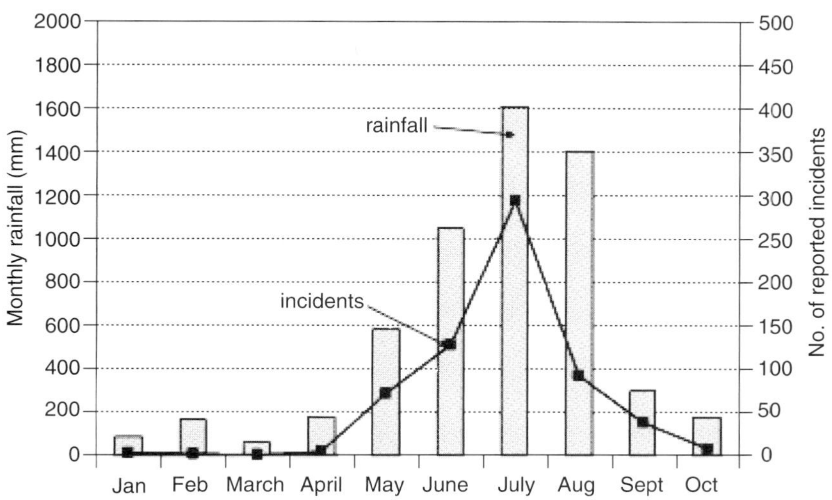

Fig. 17. Landslide incidents reported to GEO v. monthly rainfall, 1997.

1981) but cannot be relied on for design purposes because it can be lost rapidly during a rainstorm (GCO 1982; Rodin *et al.* 1982). Nevertheless, its loss on wetting is probably a factor in many landslides.

Hydrogeology

As illustrated below, almost all landslides in Hong Kong are triggered by rainfall in some way and hydrogeological controls are all-important. The heterogeneity and activity of the weathered rock profiles, often mantled by thick colluvial deposits, lead to a wide variety of hydrogeological conditions, as illustrated schematically in Figure 14. Many of these conditions were identified previously in the insightful paper by Deere & Patton (1971). In Hong Kong these conditions are strongly influenced by anthropogenic activities such as deep bored pile foundations and site formations including very deep cuttings and retaining walls (Pope & Ho 1982; Nash & Dale 1984; Blake *et al.* 2003), as illustrated in Figure 15.

Hydraulic gradients are often very steep, running parallel to natural hillsides, and throughflow is enhanced by the development of natural pipe systems, particularly in colluvium and saprolite (Fig. 16). At some locations large-diameter underground stream systems develop, leading to karstic-like collapse, even in weathered granite terrain (Halcrow China Ltd. 2003; Hencher *et al.* 2008). Channel flow and piping is also very important in less weathered rock, as evidenced by direct observation and preferential discoloration.

Generally infiltration is slow but locally will be faster down tension cracks or exploiting root and pipe systems. Rises in the deeper, permanent water table in the saprolite can be dramatic, with perhaps 10 m rise in head, and there is evidence that some of the water is fed by channel flow through the underlying rock and then upwards, linked back hydraulically to much higher terrain and consequent artesian conditions (GCO 1982; Leach & Herbert 1982; Jiao *et al.*

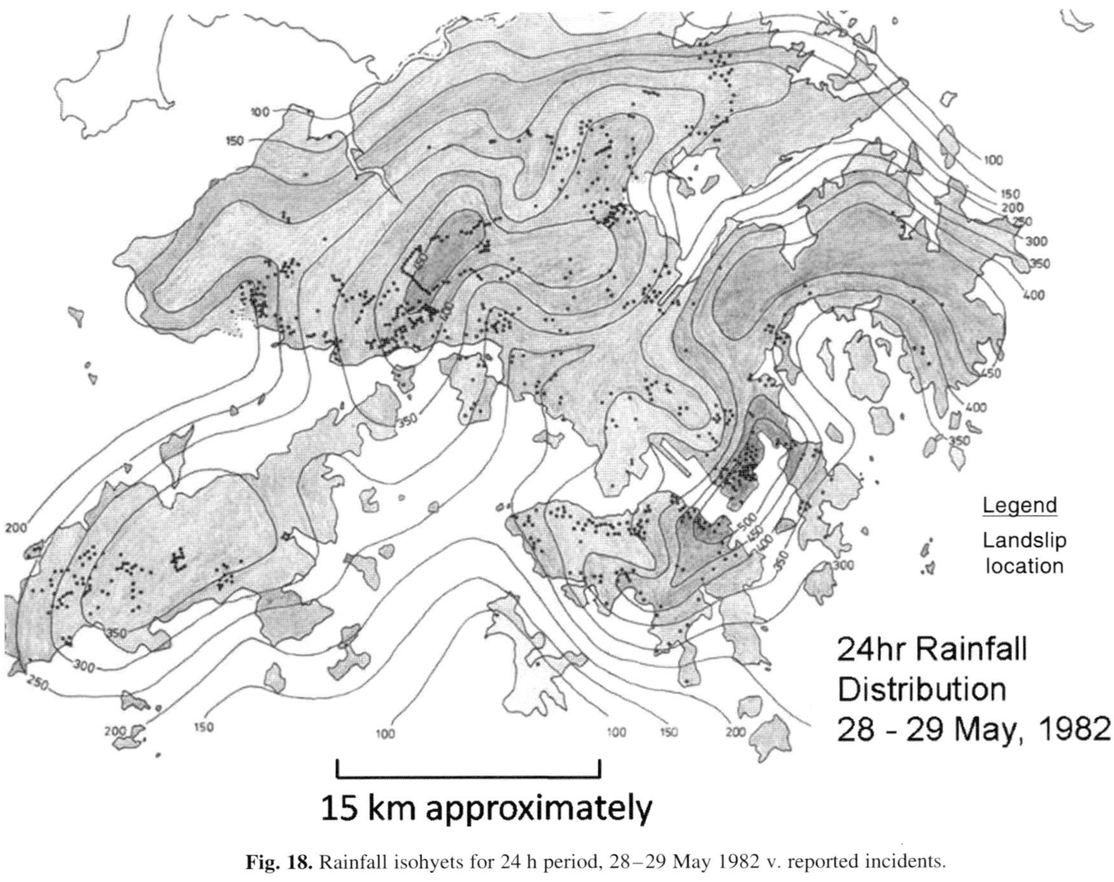

Fig. 18. Rainfall isohyets for 24 h period, 28–29 May 1982 v. reported incidents.

2005, 2006). An important and common condition, contributing to shallow landslides, is perching of local water above aquicludes at geological discontinuities and bands of clay concentration (Au 1986; Kirk *et al.* 1997; Irfan 1998). Such factors in the activity of weathered profiles as leaching, washing out and redeposition of fines, formation of pipes followed by their collapse and self re-establishment together with gradual intermittent slope movements are all very important contributors to the hydrogeological conditions and a general 'ripening' of slopes as the factor of safety diminishes over many years (de Mello 1972; Hencher 2006; Malone 1998).

Landslides and rainfall

Almost all landslides in Hong Kong are associated with periods of high rainfall, which is highly seasonal. Intense storms are to be anticipated from about April to September, although intense storms sometimes occur outside that period and similarly cause extensive failures (e.g. Wong & Ho 1995). The strong link between landsliding and rainfall may be inferred from Figure 17, which gives data for 1997, a record high yearly rainfall for Hong Kong. Lumb (1975) related the occurrence of landslides to rainfall over the preceding 15 day period and rainfall on the day of the majority of landslides. Analysis of calls to emergency services later identified the importance of short-period rainfall intensity of about 70 mm h^{-1} or more, which often coincides with the onset of numerous failures (Brand *et al.* 1984), although this criterion alone does not correlate particularly well with landslide occurrence: according to Malone

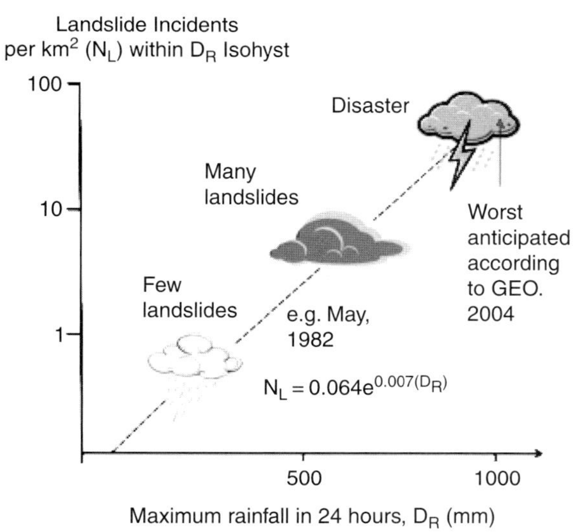

Fig. 19. Twenty-four hour rainfall v. reported incidents (after Hencher *et al.* 2006).

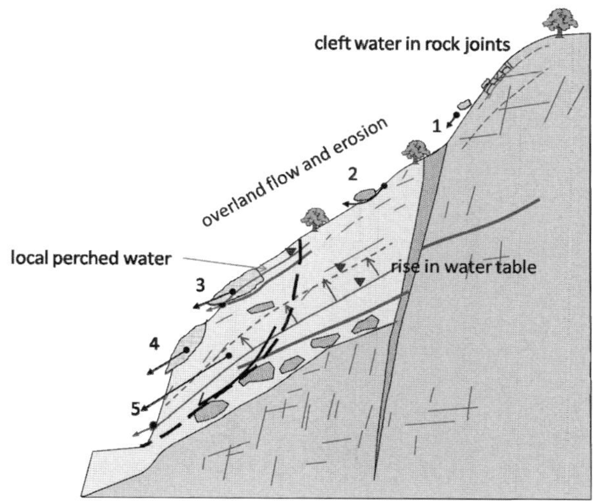

Fig. 20. Typical modes of failure: 1, rock fall; 2, washouts and boulder falls; 3, translational slide triggered by perched water; sometimes above aquiclude, sometimes above rock head or on sheeting joint; 4, shallow failure at crest or in face of cut slope; 5, deep-seated failure.

(1996), over one-third of such rainstorms between 1979 to 1995 did not trigger landslides.

Two of the intense storms listed by Malone (1996) occurred in the same year, 1982, and caused large numbers of landslides, including many in man-made slopes. Eleven of the most significant ones were investigated in some detail (Hencher 1983*b*). One of the things that was most striking about the landslides caused by those storms was the way in which they were geographically discrete. Areas devastated by the first storm in May 1982 were relatively untouched by the second storm in August and vice versa. The same was observed for landslides triggered by Typhoon Rusa in Korea in 2002 (Lee & Hencher 2007): the area devastated by landsliding was very discrete. Following the 1982 storms it was noted that in some natural terrain, sides of valleys facing in one direction were scarred by many landslides whereas the opposing valley sides were unscarred, which was probably an orographic effect. It is recalled that along one road, in Tsing Yi Island, almost every cut slope showed some signs of distress. An interpretation is that the intense storm had 'tested' each slope and found at least one section with an insufficient reserve of strength to resist the triggering action. Following those storms in 1982, landslide incidents reported to the Geotechnical Control Office were plotted against maps showing the maximum 24 h rainfall as in Figure 18. Analysis of these and other data by counting the number of incidents within each isohyet and dividing by area allowed a plot to be developed showing intensity of landslides per square kilometre v. maximum 24 h rainfall (Fig. 19). For the 1982 rainstorms, in areas where the 24 h rainfall exceeded about 500 mm, the

Table 1. *Hydrogeological processes and their link to timing of landslides*

Timing	*Typical size*		*Hydrogeological process*	*Example*	*Comments*
I Early during intense storm	Small to medium	Ia	Surface flow causing erosion and undermining	Gulleying, shallow washouts, boulder fall	Convergence of flow towards topographic hollows
		Ib	Water pressure build-up in rock joint	Rock fall Rock slab failure above adversely oriented joint	May be indications of progressive, intermittent failure before final detachment
		Ic	Loss of suction or softening	Shallow face failure in steep, cut slope	Where stability relies upon apparent cohesion
		Id	Shallow perched water	Shallow slip of soil (colluvium or saprolite) over more competent rock	Very common type of failure in natural terrain (typically less than 2 m depth)
II Medium term late in storm or delayed	Medium to large	IIa	Deeper perched water table	Perching above aquiclude within saprolite	Aquiclude may be clay-infilled discontinuity or geological unit of lower permeability such as a weathered sill or infilled fault
		IIb	Rapid infiltration and throughflow via piping or along other high-permeability channels at depth	Through natural pipes at base of thick colluvial deposits Along permeable fault or shear zone	Where linked to higher catchment or source of recharge, can result in large failures with pore pressures maintained until final detachment despite displacement and dilation of mass
III Delayed by days or weeks	Large	IIIa	Deep rise in water table, possibly by recharge from underlying bedrock	Fractured rock zone at top of rock head channelling water from high topography to location of landslide	May be caused by delayed flow along channels
		IIIb	May be apparently triggered by only minor rainfall	Progressive failure leads to increased vulnerability	Delay can be because of complex mechanism that takes time to develop

density of reported incidents was about 5–10. Extrapolating to the worst rainfall anticipated for Hong Kong (GEO 2004) indicates a density of incidents in the areas of extreme rainfall of perhaps 20–50 km^{-2}. Brand *et al.* (1984) reported that for 24 h rainfall less than 100 mm in Hong Kong only a few minor landslide 'events' would be anticipated.

It is acknowledged that the above analysis is crude not least because there is no definition of reported 'incident', which might range from a minor rock fall to a deep-seated landslide. Also, the incidents on which Figure 18 is based are only those reported to GCO/GEO and therefore biased to those affecting urban areas and infrastructure. No doubt the concentration would be higher if one were to count all natural terrain landslides and partial detachments. By comparison, a recent study of Malaysian cut slopes in weathered granite that appear superficially very similar to those of Hong Kong identified that large, serious landslides can be triggered in that country at much lower rainfall intensities than would be associated with similar incidents in Hong Kong. The reason for this is not clear, but it does indicate that the relationship in Figure 19 is not generally applicable, probably because, in some way, terrain susceptibility is in balance with local climate (Corominas 2000). Attempts have been made to define relationships for the onset of landsliding using rainfall intensity and duration normalized for mean annual precipitation (MAP) or 'rainy-day normal' (RDN) for different climatic regimes but these have met with little success (Guzetti *et al.* 2008). Clearly, there remains research to be done on susceptibilities and the underlying mechanisms and predisposing factors that control the response of a particular terrain to rainfall intensity. Perhaps the Malaysian granite saprolite has different density profiles for equivalent weathering grades compared with Hong Kong, as is certainly true of the hydrothermal weathering profiles from granite in the UK compared with weathered granite from Hong Kong, where, for example, Grade V material is much denser than the equivalent material from Hong Kong (Hencher *et al.* 1990). It is to be anticipated that such differences in the weathered profile characteristics will have knock-on effects on permeability, wetting rates, suction and true cohesion and hence the impact of particular types of rain storms, which vary in their intensity and duration.

Landslide mechanisms and occurrence

The landslides that occur in the weathered terrain of Hong Kong are varied because of the relatively complex geology and structure for a small area and the diverse geomorphological environments, which range from steep rock cliffs to deeply weathered foothills (Fig. 20). Furthermore, there is potential for natural terrain landslides of all types and for failures in man-made cuts and embankments. Malone (2000) identified many of the settings contributing to the occurrence of landslides in Hong Kong.

One of the chief aspects of Hong Kong landslides is the link to severe storms, and therefore a classification based on the hydrogeological factors leading to the failure is an attractive concept. It is evident that there are many different ways that water can influence stability, ranging from surface erosion to wetting infiltration and loss of suction and rise in water pressure, locally above an aquiclude or more generally through a rise in water table. It is also immediately obvious that the time frame for these various influences may be very different. Erosion will probably happen during a storm but rises in a deep groundwater table will probably be delayed by hours if not days and weeks, as illustrated in data from GCO (1982) discussed by Jiao *et al.* (2006). So (1971) reported that for the severe rainstorm of June 1966, the majority of landslides occurred during the intense rainfall but landslides continued to occur for more than 2 weeks. This has since been confirmed for many rainstorms, with deep-seated landslides often occurring many days after an intense rainstorm (e.g. Hudson & Hencher 1984).

Table 1 summarizes the main hydrogeological factors that trigger landslides in Hong Kong, grouped according to timing with respect to a major storm (which may last for several days in Hong Kong).

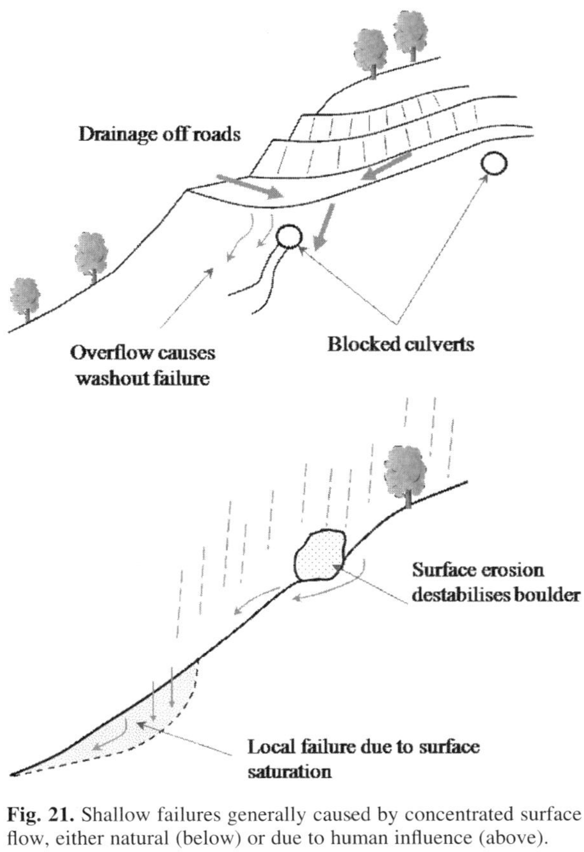

Fig. 21. Shallow failures generally caused by concentrated surface flow, either natural (below) or due to human influence (above).

Fig. 22. Washing out of newly constructed fill slope on Lantau Island following blockage of drains and flooding of the road above combined with an inability of surface channels to carry such large quantities of water.

Landslides that occur relatively early during an intense storm

Shallow washouts, erosion and rockfall (Types Ia and Ib)

Shallow landslides caused by erosion or shallow processes such as cleft water pressures behind loose rock blocks are the commonest types of failure in Hong Kong and probably the most significant in terms of risk. Even a very minor rockfall from a steep roadside slope can cause a serious incident where it affects traffic. Almost 50% of landslides reported to GEO in Hong Kong have volumes less than 5 m^3 according to Premchitt *et al.* (1994). They typically occur during or very shortly after severe rain storms. Seventy-four rock and boulder falls incidents were reported between 1978 and 1995 with nine reported fatalities since 1926. Some of the processes are illustrated in Figure 21. Figure 22 shows a recently constructed fill slope damaged by surface flooding, and Figure 23 shows the fatal Shum Wan Road landslide triggered by concentrated surface water flow channelled into the landslide area following partial failure of a small fill slope supporting the passing-bay on the road above (GEO 1996). The concentration of surface water towards topographic depressions has been discussed by Anderson *et al.* (1983).

Loss of suction (Type Ic)

Suction may provide additional apparent cohesion of the order of a few tens of kPa to a slope in its partially saturated condition, which is of the same order as true cohesion owing to relict bonding in completely weathered rocks (Fredlund 1981; Shen 1998). Suction probably plays a role in many landslides although it is generally difficult to prove as a primary candidate because of the many other factors and lack of definitive monitoring. The landslide in Figure 24 was, however, judged to be such a case, largely because there was no catchment above the slope, which made it unlikely that there was any positive pore pressure development (Hencher 1983*c*; Hencher *et al.* 1984).

Shallow perched water table (Type Id)

Landslides involving a thin soil slab overlying rock head or within a rock sheeting joint are very common forms of failure both in Hong Kong and Korea (Lee & Hencher 2007). Because of their shallow nature, infiltration and the build-up of water pressure can be rapid during a storm, so that this type of failure will be triggered early on. Figure 25 illustrates the mechanism. The case example is from a detailed study of a slope above a sheeting joint that had been evidently deteriorating progressively over many years prior to final detachment. Full details have been given by Halcrow China Ltd. (2001) and illustrated by Hencher (2006). This was a particularly illuminating case demonstrating the progressive loss of shear strength with intermittent movements, and one conclusion was that evidence of the kind of deterioration seen at that site could be

Fig. 23. Site of the Shum Wan Road landslide of 1995 that killed two people and was considered to have been triggered by flow down roads at the head of the landslide exacerbated by blocked drains (GEO 1996). Photograph following preliminary remediation works and with continuing ground investigation.

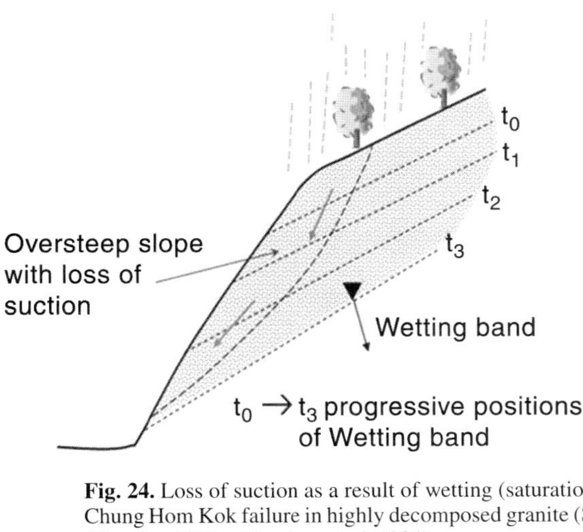

Fig. 24. Loss of suction as a result of wetting (saturation) of partially saturated material. Photographs on the right are of Chung Hom Kok failure in highly decomposed granite (Schmidt hammer rebound values up to 17) before and after failure (Hencher 1983*c*; Hencher *et al.* 1984).

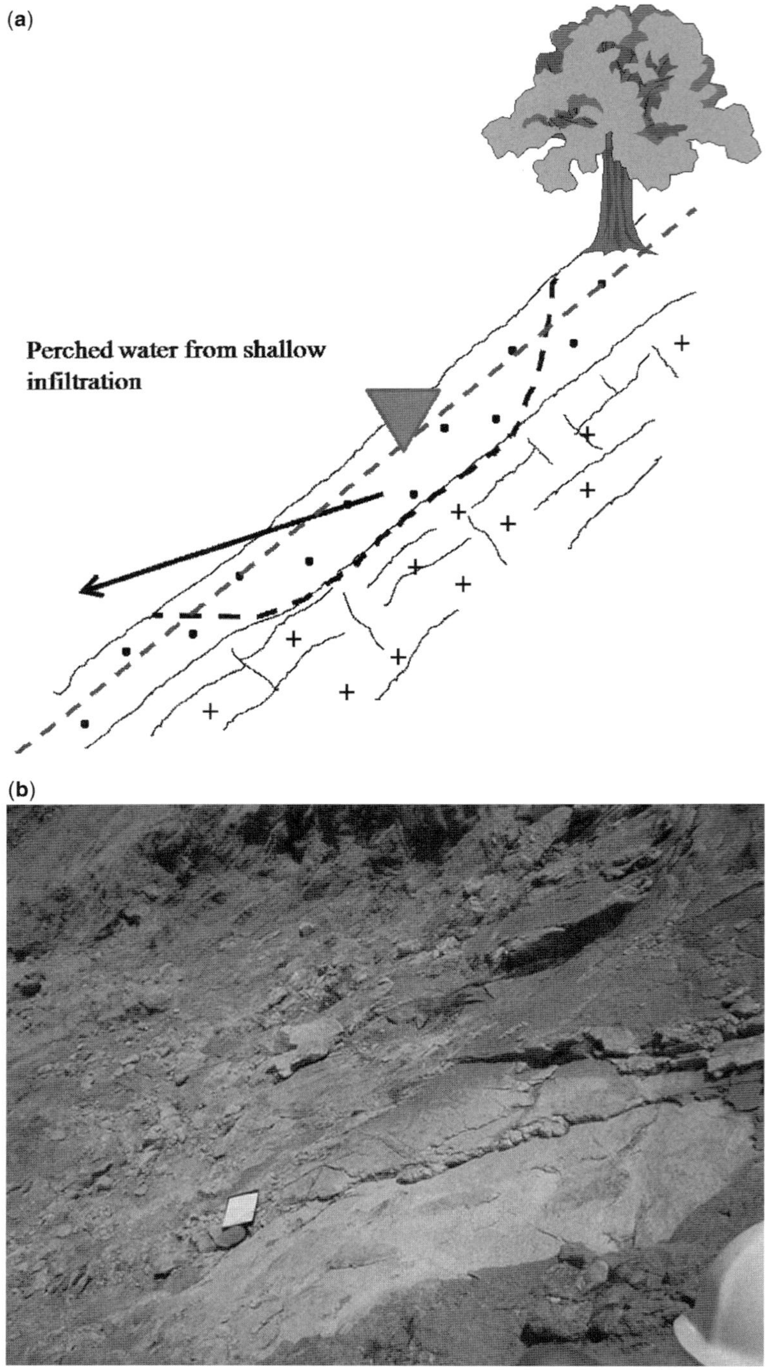

Fig. 25. (**a**) Schematic representation of perching of water above shallow rock head. (**b**) Debris trail from landslide above Leung King Estate (Halcrow China Ltd. 2001). In the foreground, in front of the clip board is exposed the pale grey, underlying wall of an extensive sheeting joint. The rock above the sheeting joint is discoloured as the result of many years of intermittent movement and deterioration prior to final detachment, as explained in detail by Halcrow China Ltd. (2003) and Hencher (2006).

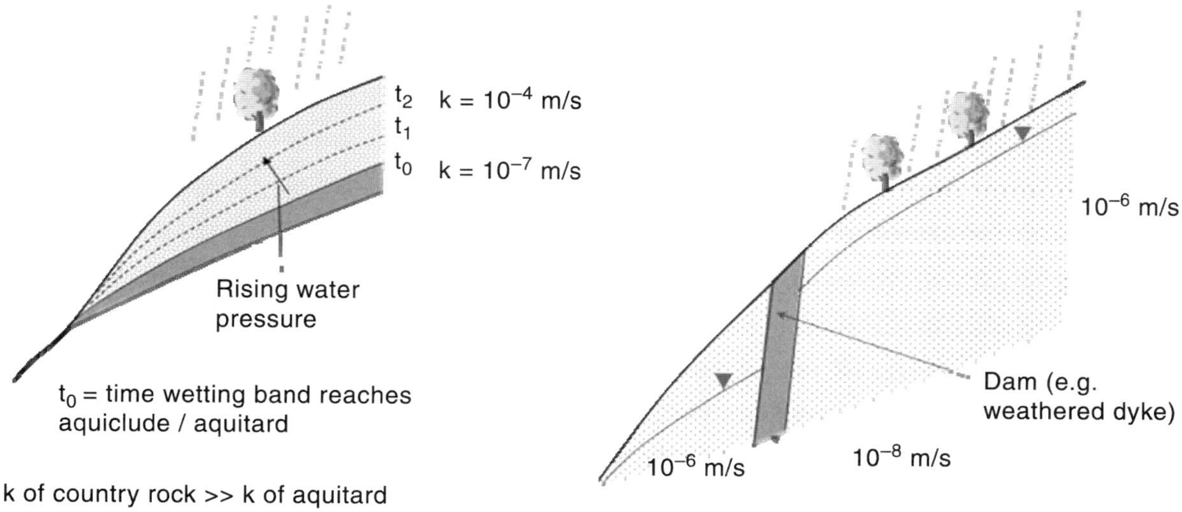

Fig. 26. (**a**) Perched water above deep aquitard. (**b**) Damming of ground water as a result of permeability contrasts.

taken, elsewhere, to indicate a slope progressively moving towards collapse.

Landslides occurring in the medium term

Perching at depth within the weathered rock profile or partitioning causing damming of ground water (Type IIa)
This group is differentiated from Type Id on the basis of scale and timing although the mechanisms are essentially the same. This type of landslide is common and is caused because infiltration is restricted by the presence of a low-permeability layer so that water pressure can build up above that layer. The concepts are illustrated in Figure 26. Such failures may be delayed because of the time it takes for infiltration to occur, although in many cases it is suspected that pre-existing tension cracks and other flaws might have facilitated relatively fast infiltration; much faster than that generally envisaged adopting normal

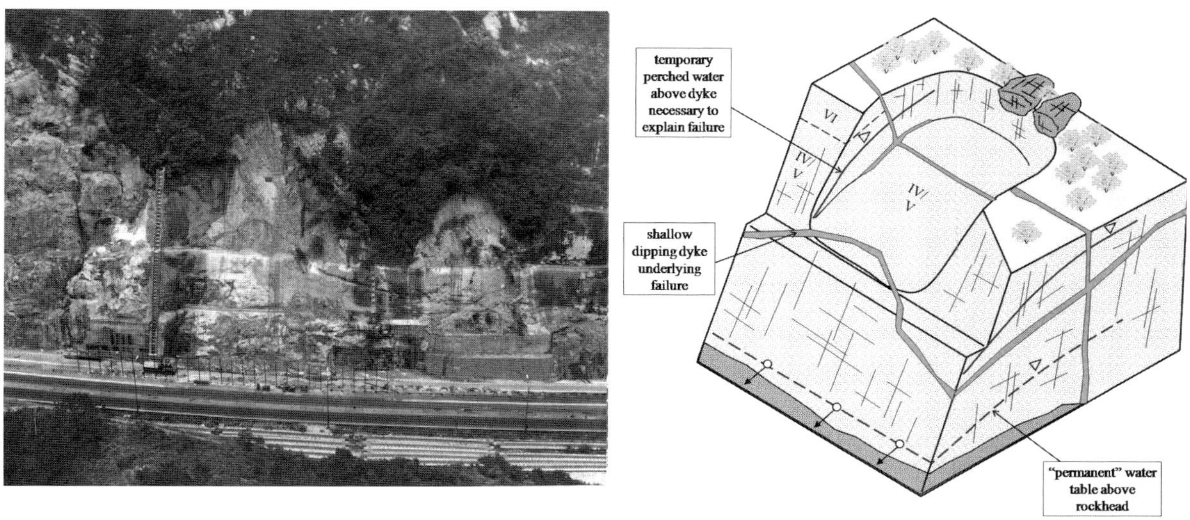

Fig. 27. Tuen Mun Highway landslide, 1982. The landslide was discussed in detail by Hencher (1983*a*, 2000) and Martin & Hencher (1984).

Fig. 28. Landslide in a section of Ching Cheung Road, 1982, that failed 4 days after heavy rain. A steeply dipping, thick decomposed dolerite dyke was mapped across the failure scarp with water seepage from above, after the failure.

permeability values, for example as discussed by Ruxton (1980) and Iverson (2000). Weathered rock profiles have many such features: persistent discontinuities can become filled with clay, derived from weathering of the rock above and around (Parry *et al.* 2000). Dykes and sills of fine-grained rock such as basalt and dolerite will weather differently compared with the host parent rock, say granite, with a resultant permeability contrast at the interfaces.

Examples of this type of failure have been reported by Hencher (1983*a*, *d*), Hudson & Hencher (1984), Martin & Hencher (1984) and Au (1986). The slope illustrated in Figure 27 is on the Tuen Mun Highway in Hong Kong. Post-failure investigation and analysis established that positive pore-water pressure was necessary to explain the landslide yet a piezometer through the landslide scar and field observation established that the main groundwater level a few days post-failure was much deeper than the failure surface. A series of decomposed dolerite dykes with lower permeability than the granitic country rock were mapped through the slope with a disposition such that the development of perched water was likely during high rainfall, and it was deduced that this was the triggering factor for the failure (Hencher 1983*a*). An example of groundwater being partitioned as in Figure 26b was identified at Ching Cheung Road (Fig. 28) (Hencher 1983*e*; Hudson & Hencher 1984). A thick dyke with water seeping above it was mapped within the failure scar, and its presence was used to explain various unusual aspects of this failure, which occurred on a dry day, 4 days after an intense storm.

High throughflow along permeable channels (Type IIb)

Some landslides are associated with rapid throughflow of water along highly conductive layers as illustrated in Figure 29. These will affect the pore pressure distribution within a slope and the timing of water pressure development, as demonstrated through careful instrumentation and monitoring along a sheeting joint by Richards & Cowland (1986). Such zones can comprise preferentially fractured layers such as faults or shear zones, or pipes, which can coalesce to become underground stream systems (Hencher *et al.* 2008). Examples of failures associated with such highly transmissive zones include the Chai Wan 1982 landslide (Hencher 1983*f*; Hencher *et al.* 1984), and the landslides of South Bay Close (Hencher 1983*g*) and Lai Ping Road (Sun & Campbell 1999; Koor & Campbell 2005). In many such cases the transmissive feature leading into the back of the landslide can be seen issuing large volumes of water long after the landslide has occurred and rainstorm has passed, which illustrates an important component of the landslide mechanism. In the case of a shallow failure (e.g. Type I), initial dilation of the sliding mass can result in a reduction of water pressures leading to cessation of the movement. In the case of these deeper-seated failures, fed by underground stream systems, there is a much larger

(a)

(b)

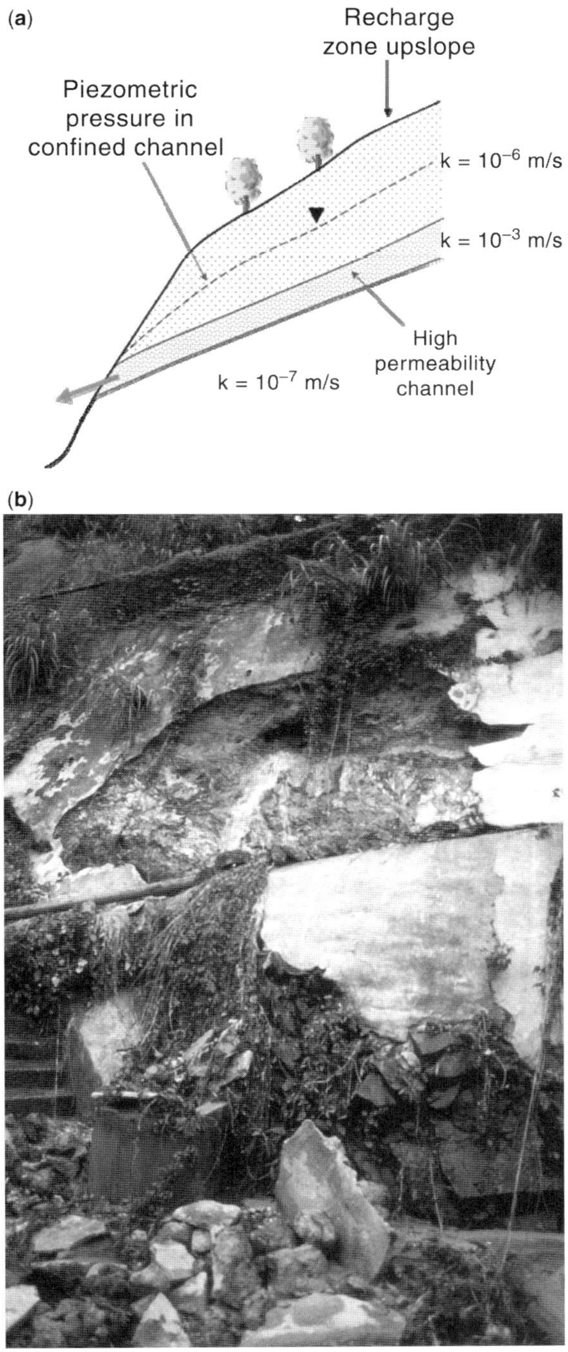

Fig. 29. Above, high-permeability zone allowing rapid through flow and build-up of potentially artesian pore pressure. Below, photograph of slope on Chai Wan Road, 1982 with large pipes at boundary between colluvium and underlying volcanic rock.

source of water available that can drive the failure to a full collapse.

Delayed failures

General rise in deep water table (Type IIIa)

Perhaps the simplest concept of hydrogeological control of landslides involves infiltration leading to a general wetting and an eventual rise in the main groundwater table with increased water pressure and then failure. Lumb (1975) discussed infiltration rates in detail and calculated that only the upper few metres could become saturated following continuous heavy rainfall for more than half a day in granite and rather longer in volcanic rocks. Rises in deeper-seated groundwater table will take even longer (Fig. 14) and a delayed response may therefore be expected for deep-seated landslides. Well-documented cases are remarkably rare for Hong Kong, despite this being identified long ago as an area of knowledge that might be improved considerably through systematic study (Sweeney & Robertson 1979; Hencher *et al.* 1984; Brand 1985). Jiao *et al.* (2005, 2006) have recently revisited some of the data from the early 1980s as well as more recent cases in an attempt to analyse the driving factors for large-scale, deep-seated landslides. One example of a deep-seated failure illustrating most of the key characteristics (evidence of deep recharge through a transient and continuously changing natural pipe system and delayed response of piezometric head to rainfall and delayed failure) is illustrated in Figure 30. This failure was described in detail by Halcrow Asia Partnership (1998) and Hencher (2000, 2006).

The failure illustrated in Figure 31 comprised the dilation and partial detachment of a section of a cut slope in 1982 (Choot 1983). Water was seen issuing from the toe of the distressed area several days after rainfall, and an area to the left and above the failure was observed to be very wet and vegetated. A few days later water issue ceased from the zone of the failure. It was deduced that heavy rainfall on the natural hillside above the slope had flowed towards an old river valley, choked with boulders, which emerged in the upper left side of the cut slope (the wet and vegetated zone). During formation of the slope the valley cross-section had been dammed by a hard surface cover. Consequently, as water flowed down the old river, water pressure built up behind the concrete cover and this in turn led to a rise in water pressure within the relict joints in the adjacent zone of cut slope where the failure took place. The reduced effective stress and consequent decrease in shear strength combined with active cleft water pressure led to some initial translational movement on relict joints (about 0.5 m at the toe). The translation movement along the adverse joint sets involved dilation of the rock mass and increased permeability, and this in turn resulted in a drop in driving water pressure and cessation of movement as illustrated in Figure 31.

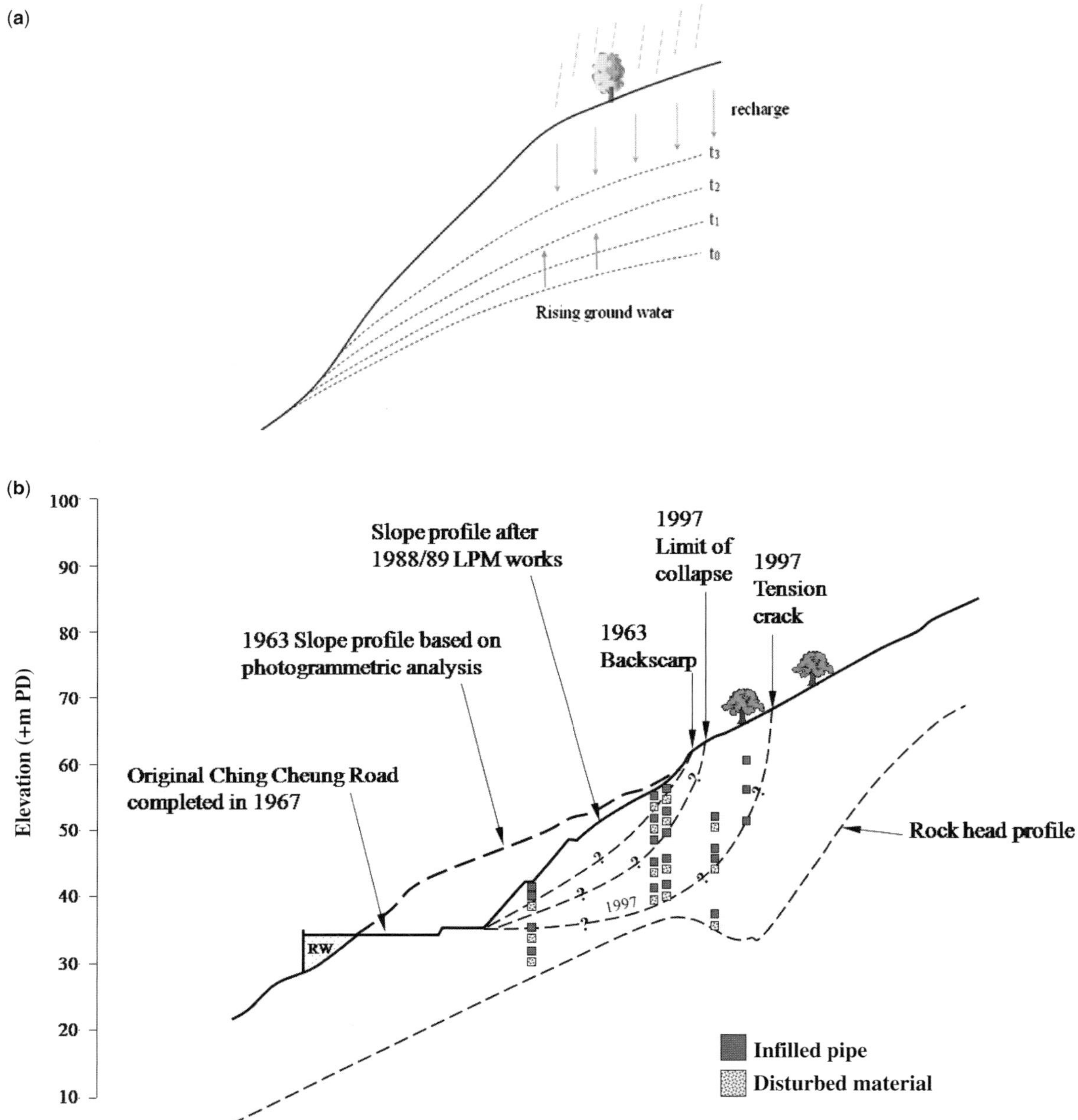

Fig. 30. (**a**) Rise in groundwater table from time t_0 to t_3. (**b**) Cross-section of the Ching Cheung Road landslide of 1997. (Note the number of choked natural pipes sampled in each borehole and that they extend to considerable depth (even below the 1997 slip surface), which indicates the length of recharge path to the groundwater table.)

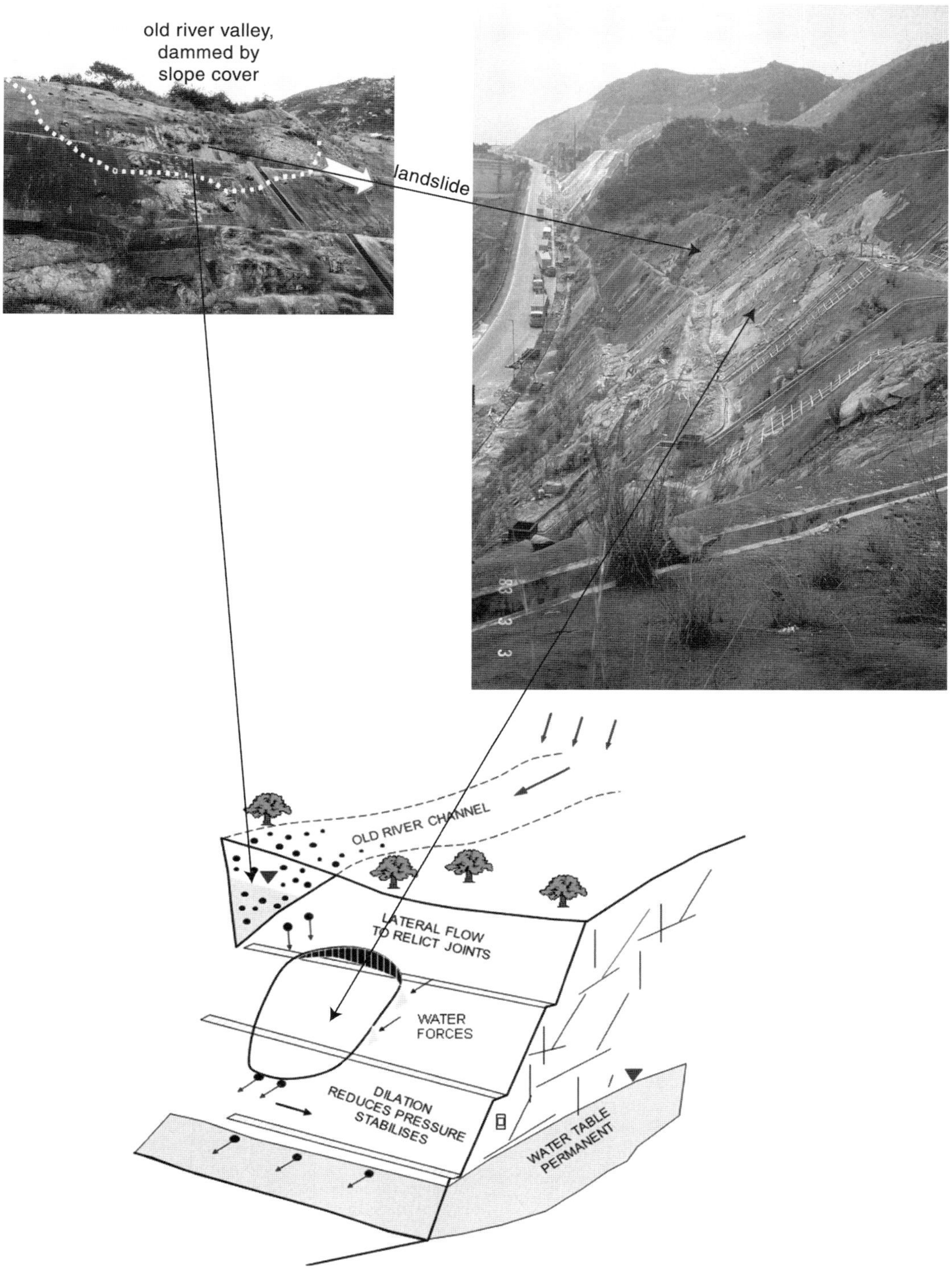

Fig. 31. Tsing Yi (2) landslide.

Other examples of delayed rises in deep groundwater level have been given by Insley & McNicholl (1982) and Jiao (2006).

Progressive deterioration (Type IIIb)
Finally, very large and disastrous failures can occur late in a storm or triggered by some later event, sometimes triggered by relatively minor rainfall. This can happen because the slope has deteriorated progressively over a period, making it susceptible to a final trigger. In the case of the Ching Cheung landslide of 1997 (Halcrow Asia Partnership 1998; Hencher 2000, 2006), the first major movement occurred as a delayed failure, several days after a rainstorm, a month before the final detachment that blocked both sides of a dual carriageway. Fortunately there were no injuries. The famous Po Shan landslide of 1972 occurred after 3 days of heavy rainfall and numerous collapses and settlement giving some warning of the final landslide that destroyed a block of flats and killed 67 people (Vail & Attewill 1976). Careful mapping and investigation of the Lai Ping landslide of 1997 demonstrated that it had apparently been moving intermittently for 20 years (Sun & Campbell 1999; Koor & Campbell 2005).

Conclusions

Landslides occur every year in Hong Kong and these are almost always associated with intense rainstorms. Shallow failures, including minor rock falls, are the most common and generally occur during the storm when the intensity of rainfall is highest, and pose the greatest risk because of their frequency. Deeper-seated landslides are often delayed, even many days after the rainstorm that is subsequently identified as the triggering factor. It is suggested that the timing of landslides directly relates to the hydrogeological factors triggering each landslide. Recent investigations of landslides in Hong Kong have identified the importance of precursory movements and continuing deterioration prior to the final failure, which lends support to the 'ripening' concept of de Mello (1972). The deterioration generally involves dilation and enhanced permeability with the development of underground stream systems and sedimentation. These features can be sought in ground investigations, giving a potential tool for identifying imminent landsliding. Large, deep-seated landslides pose a major threat and many of these are delayed. One category involves rising deep groundwater, sometimes linked to hydraulic systems much higher up a hill slope. A second category of large landslides involves a period of deterioration and movement that need to be monitored and managed. The situation may be irretrievable through normal landslide preventive measures but should allow the risk to the public to be minimized.

Acknowledgements. The author is grateful for comments and suggestions by reviewers A. Malone and M. Parise, which have improved the ideas expressed here. D. Campbell was very helpful in discussing the development of a 'hydrogeological classification' of landslides. Much of the work reported here was carried out in association with the Hong Kong Government, either as an employee of the Geotechnical Control Office in the 1980s or more recently through various consultancies and general discussions with colleagues from the Geotechnical Engineering Office as it now is. This research was supported in part by a grant (NEMA-06-NH-05) from the Natural Hazard Mitigation Research Group, National Emergency Management Agency (NEMA), Ministry of Public Administration and Security, South Korea.

References

ANDERSON, M. G., MCNICHOLL, D. P. & SHEN, J. M. 1983. On the effect of topography in controlling soil water conditions, with specific regard to cut slope piezometric levels. *Hong Kong Engineer*, **11**, 35–41.

ANONYMOUS 1995. The description and classification of weathered rocks for engineering purposes. Geological Society Engineering Group Working Party Report. *Quarterly Journal of Engineering Geology*, **28**, 207–242.

AU, S. W. C. 1986. Decomposed dolerite dykes as a cause of slope failure. *Hong Kong Engineer*, **14**, 33–36.

BLAKE, J. R., RENAUD, J.-P., ANDERSON, M. G. & HENCHER, S. R. 2003. Prediction of rainfall-induced transient water pressure head behind a retaining wall using a high-resolution finite element model. *Computers and Geotechnics*, **30**, 431–442.

BRAND, E. W. 1984. Landslides in Southeast Asia: A State-of-the-Art Report. *In*: *Proceedings, 4th International Symposium on Landslides, Toronto*, Vol. 1, 17–59.

BRAND, E. W. 1985. Predicting the performance of residual soil slopes. *In*: *Proceedings of the Eleventh International Conference on Soil Mechanics and Foundation Engineering, San Francisco*, Vol. 5, 2541–2548.

BRAND, E. W., HENCHER, S. R. & YOUDAN, D. G. 1983. Rock slope engineering in Hong Kong. *In*: *Proceedings of the Fifth International Rock Mechanics Congress, Melbourne*, Vol. 1, C17–C24.

BRAND, E. W., PREMCHITT, J. & PHILLIPSON, H. B. 1984. Relationship between rainfall and landslides in Hong Kong. *In*: *Proceedings of the Fourth International Symposium on Landslides, Toronto*, Vol. 1, 377–384.

BSI 1999. *BS5930, Code of Practice for Site Investigation*. British Standards Institution, London.

CHOOT, E. B. 1983. *Landslide Studies 1982 Case Study No. 9 Tsing Yi (2)*. Geotechnical Control Office, Special Project Report **SPR 10/83**.

COCHRANE, R. A. 1984. Ground problems at Hong Kong's Aberdeen Tunnel. (Report by O. Bevan of British Tunnelling Society Meeting, 24 November 1983). *Tunnels and Tunnelling*, **16**, 33–34.

COROMINAS, J. 2000. Landslides and climate. *In*: BROMHEAD, E. N. (ed.) *VIII International Symposium on Landslides, Cardiff, UK (2000) Keynote Lectures* [CD-ROM].

DEERE, D. U. & PATTON, F. D. 1971. Slope stability in residual soils. *In*: *Proceedings 4th Pan-American Conference Soil Mechanics and Foundation Engineering*, Vol. 1, 87–170.

DE MELLO, V. F. B. 1972. Thoughts on soil engineering applicable to residual soils. *In*: *Proceedings 3rd Southeast Asian Conference on Soil Engineering*, 5–34.

EBUK, E. J. 1991. *The influence of fabric on the shear strength characteristics of weathered granites.* PhD thesis, University of Leeds.

EBUK, E. J., HENCHER, S. R. & LUMSDEN, A. C. 1993. The influence of structure on the shearing mechanism of weakly bonded soils. *In*: *Proceedings 26th Annual Conference of the Engineering Group of the Geological Society, the Engineering Geology of Weak Rock, Leeds*, 207–215.

EL-RAMLY, H., MORGENSTERN, N. R. & CRUDEN, D. M. 2005. Probabilistic assessment of a cut slope in residual soil. *Géotechnique*, **55**, 77–84.

FLETCHER, C. J. N. 2004. *Geology of Site Investigation Boreholes from Hong Kong*. AGC, Hong Kong.

FLETCHER, C. J. N., MASSEY, C. I., WILLIAMSON, S. J. & PARRY, S. 2002. Importance of bedrock and regolith mapping for natural terrain hazard studies: an example from the Tsing Shan area, Hong Kong. *In*: *Natural Terrain—A Constraint to Development? Proceedings of Conference of the Institution of Mining & Metallurgy, Hong Kong Branch*, 61–76.

FMSWJV (FUGRO MAUNSELL SCOTT WILSON JOINT VENTURE) 2000. *Report on the Shek Kip Mei Landslide of 25 August 1999. Vol 1—Findings of the Landslide Investigation.* Geotechnical Engineering Office, Hong Kong.

FREDLUND, D. G. 1981. The shear strength of unsaturated soil and its relationship to slope problems in Hong Kong. *Hong Kong Engineer*, **9**, 37–45.

FYFE, J. A., SHAW, R., CAMPBELL, S. D. G., LAI, K. W. & KIRK, P. A. 2000. *The Quaternary Geology of Hong Kong*. Geotechnical Engineering Office, Hong Kong.

GCO 1982. *Mid-Levels Study: Report on Geology, Hydrology and Soil Properties*. Geotechnical Control Office, Hong Kong.

GCO 1988. *Guide Rock and Soil descriptions (Geoguide 3)*. Geotechnical Control Office, Hong Kong.

GEO 1993. *Guide to Retaining Wall Design (Geoguide 1)*, 2nd edn. Geotechnical Engineering Office, Hong Kong.

GEO. 1996. *Report on the Shum Wan Road Landslide of 13 August 1995*. Geotechnical Engineering Office, Hong Kong.

GEO 2004. *Rainstorm Severity and Landslide Potential.* Geotechnical Engineering Office, Information Note **30/2004**.

GUZZETTI, F., PERUCCACCI, S., ROSSI, M. & STARK, C. 2008. The rainfall intensity–duration control of shallow landslides and debris flows: an update. *Landslides*, **5**, 3–17.

HALCROW ASIA PARTNERSHIP 1998. *Report on the Ching Cheung Road Landslide of 3 August 1997*. Geotechnical Engineering Office Report **78**.

HALCROW CHINA LTD 2001. *Detailed Study of Selected Landslides above Leung King Estate of 14 April 2000*. Geotechnical Engineering Office, Landslide Study Report **LSR 9/2001**.

HALCROW CHINA LTD 2003. *Interim Report on Detailed Hydrogeological Study of the Hillside near Yee King Road, Tai Hang*. Geotechnical Engineering Office, Landslide Study Report **LSR 4/2003**.

HENCHER, S. R. 1981. *Report on Slope Failure at Yip Kan Street (11SW-D/C86) Aberdeen on 12th July 1981*. Geotechnical Control Office Report **GCO 16/81**.

HENCHER, S. R. 1983*a*. *Landslide Studies 1982 Case Study No. 5 Tuen Mun Highway*. Geotechnical Control Office, Special Project Report **SPR 6/83**.

HENCHER, S. R. 1983*b*. *Summary Report on Ten Major Landslides in 1982*. Geotechnical Engineering Office, Special Project Report **SPR 1/83**.

HENCHER, S. R. 1983*c*. *Landslide Studies 1982 Case Study No. 3 Chung Hom Kok*. Geotechnical Control Office, Special Project Report **SPR 4/83**.

HENCHER, S. R. 1983*d*. *Landslide Studies 1982 Case Study No. 2 Junk Bay Road*. Geotechnical Control Office, Special Project Report **SPR 3/83**.

HENCHER, S. R. 1983*e*. *Landslide Studies 1982 Case Study No. 10 Ching Cheung Road*. Geotechnical Control Office, Special Project Report **SPR 11/83**.

HENCHER, S. R. 1983*f*. *Landslide Studies 1982 Case Study No. 1 Chai Wan Road*. Geotechnical Control Office, Special Project Report **SPR 2/83**.

HENCHER, S. R. 1983*g*. *Landslide Studies 1982 Case Study No. 4 South Bay Close*. Geotechnical Control Office, Special Project Report **SPR 5/83**.

HENCHER, S. R. 2000. Engineering geological aspects of landslides. Keynote paper. *In*: *Proceedings Conference on Engineering Geology HK 2000*. Institution of Mining and Metallurgy, Hong Kong, 93–116.

HENCHER, S. R. 2006. Weathering and erosion processes in rock – implications for geotechnical engineering. Invited Paper. *In*: *Proceedings Symposium on Hong Kong Soils and Rocks, March 2004*. Institution of Mining, Metallurgy and Materials and Geological Society, London, 29–79.

HENCHER, S. R. 2008. The 'new' British and European standard guidance on rock description. A critique by Steve Hencher. *Ground Engineering*, **41**, 17–21.

HENCHER, S. R., ANDERSON, M. G. & MARTIN, R. P. 2006. Hydrogeology of landslides. *In*: *Proceedings of International Conference on Slopes, Malaysia*, 463–474.

HENCHER, S. R. & KNIPE, R. J. 2007. Development of rock joints with time and consequences for engineering. *In*: *Proceedings of the 11th Congress of the International Society for Rock Mechanics, Lisbon*, Vol. 1, 223–226.

HENCHER, S. R. & MARTIN, R. P. 1982. Description and classification of weathered rocks in Hong Kong for engineering purposes. *In*: *Proceedings of the 7th Southeast Asian Geotechnical Conference, Hong Kong*, Vol. 1, 125–142.

HENCHER, S. R. & MCNICHOLL, D. P. 1995. Engineering in weathered rock. *Quarterly Journal of Engineering Geology*, **28**, 253–266.

HENCHER, S. R. & RICHARDS, L. R. 1982. The basic frictional resistance of sheeting joints in Hong Kong Granite. *Hong Kong Engineer*, **11**, 21–25.

HENCHER, S. R., MASSEY, J. B. & BRAND, E. W. 1984. Application of back analysis to some Hong Kong landslides. *In*: *Proceedings of the 4th International Symposium on Landslides, Toronto*, Vol. 1, 631–638.

HENCHER, S. R., EBUK, E. J., ABRAMS, J. H. & LUMSDEN, A. C. 1990. Field description and classification of hydrothermally altered granite from SW England. *In*: *Proceedings 10th Southeast Asian Geotechnical Conference, Taipei*, 303–308.

HENCHER, S. R., SUN, H. W. & HO, K. K. S. 2008. The investigation of underground streams in a weathered granite terrain in Hong Kong. *In*: HUANG, A-B. & MAYNE, P. W. (eds)

Geotechnical and Geophysical Site Characterization. Taylor & Francis, London, 601–607.

HUDSON, R. R. & HENCHER, S. R. 1984. The delayed failure of a large cutting in Hong Kong. *In*: *Proceedings of the International Conference on Case Histories in Geotechnical Engineering, St. Louis, MO*, Vol. 2, 679–682.

INSLEY, H. T. M. & MCNICHOLL, D. 1982. Groundwater monitoring of a soil slope in Hong Kong. *In*: *Proceedings of the 7th Southeast Asian Geotechnical Conference, Hong Kong*, Vol. 1, 63–75.

IRFAN, T. Y. 1996. Mineralogy, fabric properties and classification of weathered granites in Hong Kong. *Quarterly Journal of Engineering Geology*, **29**, 5–35.

IRFAN, T. Y. 1998. Structurally controlled landslides in saprolitic soils in Hong Kong. *Geotechnical and Geological Engineering*, **16**, 215–238.

IRFAN, T. Y. 1999. Characterization of weathered volcanic rocks in Hong Kong. *Quarterly Journal of Engineering Geology*, **32**, 317–348.

IVERSON, R. M. 2000. Landslide triggering by rain infiltration. *Water Resources Research*, **36**, 1897–1910.

JIAO, J. J., WANG, X.-S. & NANDY, S. 2005. Confined groundwater zone and slope instability in weathered igneous rocks in Hong Kong. *Engineering Geology*, **80**, 71–92.

JIAO, J. J., DING, G. P. & LEUNG, C. M. 2006. Confined groundwater near the rockhead in igneous rocks in the Mid-Levels area, Hong Kong, China. *Engineering Geology*, **84**, 207–219.

KIRK, P. A., CAMPBELL, S. D. G., FLETCHER, C. J. N. & MERRIMAN, R. J. 1997. The significance of primary volcanic fabrics and clay distribution in landslides in Hong Kong. *Journal of the Geological Society, London*, **154**, 1009–1019.

KOOR, N. P. & CAMPBELL, S. D. G. 2005. *Geological characterisation of the Lai Ping Road Landslide*. Geotechnical Engineering Office, Report **166**.

KOOR, N. P., PARRY, S. & YIN, J. H. 2000. The shear strength of infilled and slickensided discontinuities of some Hong Kong saprolites. *In*: *Proceedings of the Conference on Engineering Geology HK 2000*. Institution of Mining and Metallurgy, Hong Kong Branch, 187–192.

LEACH, B. & HERBERT, R. 1982. The genesis of a numerical model for the study of the hydrogeology of a steep hillside in Hong Kong. *Quarterly Journal of Engineering Geology*, **15**, 243–259.

LEE, S. G. & HENCHER, S. R. 2007. Slope safety and landslide risk management in Korea. *In*: *Proceedings of the 2007 International Forum on Landslide Disaster Management, Hong Kong*, Vol. 1, 125–168.

LUMB, P. 1975. Slope failures in Hong Kong. *Quarterly Journal of Engineering Geology*, **8**, 31–65.

MALONE, A. W. 1996. *Areal extent of intense rainfall in Hong Kong 1979 to 1995*. Geotechnical Engineering Office, Report **62**.

MALONE, A. W. 1997. Risk management and slope safety in Hong Kong. *Transactions of the Hong Kong Institute of Engineers*, **4**, 12–21.

MALONE, A. W. 1998. Slope movement and failure: evidence from field observations of landslides associated with hillside cuttings in saprolites in Hong Kong. *In*: *Proceedings 13th Southeast Asian Geotechnical Conference, Taipei*, Vol. 2, 81–90.

MALONE, A. W. 2000. Panel Report: Slope movements associated with hillside cuttings in saprolites in Hong Kong. *In*: EVANGELISTA, A. & PICARELLI, L. (eds) *The Geotechnics of Hard Soils–Soft Rocks*, Vol. 3. Balkema, Rotterdam, 1593–1596.

MARTIN, R. P. 1986. Use of index tests for engineering assessment of weathered rocks. *In*: *Proceedings of the Fifth International Congress of the International Association of Engineering Geology, Buenos Aires*, Vol. 1, 433–450.

MARTIN, R. P. & HENCHER, S. R. 1984. The failure of a cut slope on the Tuen Mun Road in Hong Kong. *In*: *Proceedings of the International Conference on Case Histories in Geotechnical Engineering, St. Louis, MO*. Vol. 2, 683–688.

MARTIN, R. P. & HENCHER, S. R. 1986. Principles for description and classification of weathered rocks for engineering purposes. *In*: HAWKINS, A. B. (ed.) *Site Investigation Practice: Assessing BS5930*. Geological Society, London, Engineering Geology Special Publication, **2**, 299–308.

NASH, J. M. & DALE, M. J. 1984. Geology and hydrogeology of natural tunnel erosion in superficial deposits in Hong Kong. *Geological Society of Hong Kong Bulletin*, **1**, 61–72.

PAPALIANGAS, T., HENCHER, S. R. & LUMSDEN, A. C. 1996. A comprehensive peak shear strength criterion for rock joints. *In*: *Proceedings 8th International Congress ISRM, Tokyo*, Vol. 1, 359–366.

PARRY, S., CAMPBELL, S. D. G. & FLETCHER, C. J. N. 2000. Typical kaolin occurrences in Hong Kong – a model for their origin and implications for landslide development. Landslides in research, theory and practice. *In*: *Proceedings 8th International Symposium on Landslides, Cardiff*, Vol. 3, 1177–1181.

POPE, R. G. & HO, C. S. 1982. The effect of piles and caissons on groundwater flow. *Hong Kong Engineer*, **10**, 25–27.

PREMCHITT, J., BRAND, E. W. & CHEN, P. Y. M. 1994. Rain-induced landslides in Hong Kong, 1972–1992. *Asia Engineer*, June, 43–51.

RICHARDS, L. R. & COWLAND, J. W. 1986. Stability evaluation of urban rock slopes in a transient groundwater regime. *In*: *First International Conference on Rock Engineering and Excavation in an Urban Environment*. Institution of Mining and Metallurgy, Hong Kong, 357–363.

RODIN, S., HENKEL, D. J. & BROWN, R. L. 1982. Geotechnical study of a large hillside area in Hong Kong. *Hong Kong Engineer*, **10**, 37–45.

RUXTON, B. P. 1980. Slope problems in Hong Kong: a geological appraisal. *Hong Kong Engineer*, **8**, 29–38.

RUXTON, B. P. & BERRY, L. 1957. The weathering of granite and associated erosional features in Hong Kong. *Geological Society of America Bulletin*, **68**, 1263–1292.

SEWELL, R. J., CAMPBELL, S. D. G., FLETCHER, C. J. N., LAI, K. W. & KIRK, P. A. 2000. *The Pre-Quaternary Geology of Hong Kong*. Geotechnical Engineering Office, Hong Kong.

SHEN, J. M. 1998. Soil suction in relation to slope stability: a summary of research carried out in Hong Kong in 1978–1997. *In*: *Proceedings of the Geotechnical Division Seminar on Slope Engineering in Hong Kong*. Hong Kong Institution of Engineers, 93–99.

SO, C. L. 1971. Mass movements associated with the rainstorm of June 1966 in Hong Kong. *Transactions of the Institute of British Geographers*, **53**, 55–65.

SUN, H. W. & CAMPBELL, S. D. G. 1999. *The Lai Ping Road Landslide of 2nd July 1997*. Geotechnical Engineering Office, Report **95**.

SWEENEY, D. J. & ROBERTSON, P. K. 1979. A fundamental approach to slope stability problems in Hong Kong. *Hong Kong Engineer*, **7**, 35–44.

VAIL, A. J. & ATTEWILL, L. J. S. 1976. The remedial works at Po Shan Road. *Hong Kong Engineer*, **4**, 18–30.

WONG, H. N. 2005. Landslide risk assessment for individual facilities. *In*: HUNG, O., FELL, R., COUTURE, R. & EBERHARDT, E. (eds) *Landslide Risk Management*. Taylor & Francis, London, 237–296.

WONG, H. N. & HO, K. K. S. 1995. *General report on landslips on 5 November 1993 at man-made features in Lantau*. Geotechnical Engineering Office, Report **144**.

Weathering in the crystalline rocks of Calabria, Italy, and relationships to landslides

Domenico Calcaterra[1]* & Mario Parise[2]

[1] Department of Hydraulic, Geotechnical and Environmental Engineering—Section of Engineering Geology, Federico II University of Naples, Piazzale Tecchio 80, 80125 Naples, Italy
[2] National Research Council, IRPI, Via Amendola 122-I, 70126 Bari, Italy

**Corresponding author (e-mail: domenico.calcaterra@unina.it)*

Abstract: This paper reviews approximately 40 years of studies focused on landslides in weathered crystalline rocks of Calabria, southern Italy. In these studies geological, geomorphological, mineralogical, petrographical, engineering-geological and geotechnical aspects of the Palaeozoic granitoids and the high-grade metamorphic rocks cropping out in the Calabrian massifs have been considered. This in turn has allowed depiction of the regional landsliding scenario, where almost all the main typologies of mass movements have been recognized, ranging from shallow soil slips and extremely rapid debris flows to slow-moving deep-seated gravitational slope deformations. From north to south, the Sila Massif, Coastal Chain, and Poro, Serre and Aspromonte Massifs are described, with reference to the studies dealing with weathering and slope movements in the different geological and morphological settings. The regional extent of the weathering processes gave researchers the opportunity to adopt and calibrate mapping methods, specifically devoted to analysis and mitigation of the landslide hazard in weathered materials, which have also been applied to important civil engineering works. A further distinctive character of the Calabrian weathering is its age, which has been ascribed to Tertiary time. Such an ancient initiation could be regarded as one of the main reasons for the depth of the weathered mantle, which, despite the rugged topography, in some districts exceeds 150–200 m.

Weathering processes are responsible for the chemical and mechanical transformation of several types of rock under a variety of climatic conditions all over the world. Practically all the most important and widespread lithotypes (magmatic, metamorphic, siliciclastic and carbonate rocks) tend to alter and to decompose when they interact with the atmosphere, the hydrosphere and the biosphere. These latter, in turn, can result in the onset of adverse geomorphological consequences. When silicic rocks become deeply weathered, whatever their geological origin is, they are subject to several kinds of slope instabilities, usually rapidly moving, which may be superficial or deep-seated depending on the rock masses or soil-like materials. The occurrence of adverse effects in weathered regions and the related engineering-geological problems have induced the scientific community to make efforts to better comprehend both the mechanisms and the features of the weathering processes. These started to be studied using modern approaches in the 1950s, and contributions such as those by Moye (1955) and Ruxton & Berry (1957) were pioneer works in recognizing the presence and the importance of weathered materials resting on a relatively intact parent rock.

However, in those regions of the world where weathered rock masses occur, descriptions, even if qualitative, of the weathering features had been provided well before that time. This is the case for Calabria, the southernmost region of the Italian peninsula, whose backbone is essentially made up of Palaeozoic plutonic and metamorphic rocks. The importance of the weathered rocks in the regional geological setting was first recognized at the end of the 19th and start of the 20th centuries, in some monographs on Calabrian geology.

Cortese (1895), describing the geological setting of Calabria, was the first to underline that, despite what was known in cold and tropical regions of the world, in Calabria as in other mid-latitude regions rainfall, freeze–thaw cycles and high temperatures all contribute to decompose the crystalline rocks. Cortese identified the deep chloritization of micas as one of the distinctive features of the Calabrian granites, along with the diffuse presence of a grey–yellowish coarse sand, which resembled a lacustrine deposit: in this latter respect, dykes and pegmatites testify to the *in situ* alteration of the parent rock, without any transportation. Cortese (1895) also noted some additional features of the weathering in Calabria: in contrast to the situation in tropical regions, in Calabria only occasionally is a weathered granite transformed into residual clayey deposits; corestones are widespread, and they can also be found as blocks within Pliocene conglomerates; sudden thermal changes, common during winter, are responsible for the detachment of thin crusts; and the interaction between brackish water and freshwater makes weathering effective also along the coastal zones.

A few years later, De Lorenzo (1904) mentioned the difficulties in carrying out studies on the age of the Calabrian

From: CALCATERRA, D. & PARISE, M. (eds) *Weathering as a Predisposing Factor to Slope Movements*.
Geological Society, London, Engineering Geology Special Publications, **23**, 105–130.
DOI: 10.1144/EGSP23.7 0267-9914/10/$15.00

granites, because of their 'deep state of denudation and decomposition' and related the weathering of the granitoid rocks to 'the continuous movements occurred within them and to the modifications developed over them, since their formation up to now'.

When the geographer Roberto Almagià published the first detailed inventory of landslides in southern Italy (Almagià 1910), he indicated weathering as the most important causal factor in the instability of several areas, mentioning the severe hazard related to slope instability in the crystalline rocks. After some decades of apparent absence of interest on the topic, Montanari (1939, 1940, 1941) recognized a number of mass movements in the weathered rocks of all the Calabrian provinces. Nicotera carried out a comprehensive, regional-based study devoted to the analysis of the ground conditions predisposing erosion and landslides (Nicotera 1959*a*; Fig. 1). He found that some catchments in the Serre and Aspromonte Massifs, widely characterized by crystalline rocks, were among those most affected by slope instabilities.

In the 1960s and 1970s a few researchers focused their attention on the effect of weathering on urban settlements and man-made structures. Nicotera *et al.* (1968) and Ietto (1969, 1976), based on some important designed and/or constructed engineering works (dams, tunnels, railways, highways), considered the behaviour, geomechanical properties and technical problems of weathered rocks in various parts of the region.

However, the key paper that showed the importance of weathering in Calabria to an international audience is the short note by Guzzetta (1974). In that paper Guzzetta, comparing the weathering profile commonly observed in the Calabrian granites with that in present-day and ancient rainforest tropical areas, suggested that, under the actual Mediterranean climatic conditions, an intense erosion of an exposed, fossil, deeply weathered mantle occurs (Guzzetta 1974). The existence of the latter can be explained only by several controlling factors not fulfilled in Calabria during the recent geological past: tropical conditions, peneplained landscape, prolonged tectonic quiescence, and a slight amount of erosion.

From the 1970s onward, weathering of Calabrian rocks and the related implications became a major topic for several researchers involved in the fields of geomorphology, engineering geology and geotechnical engineering. A great impulse to these studies was given by the IRPI (Institute of Research for the Hydrogeological Protection) at Cosenza, a branch of the National Research Council created with the aim of deepening the knowledge on geohazards such as landslides and floods, and to mitigate the related risk. Thanks to the IRPI researchers, as well as experts from other southern Italian scientific institutions, Calabria has become one of the most studied regions in the world as regards weathering-related issues.

For the above reasons, after some decades of modern studies, a review of the present-day knowledge on weathering in Calabria represents an interesting contribution not only for those directly engaged in this region, but also for a wider audience, which may find here useful issues for comparison with similar topics in other parts of the world.

The main aim of this paper is to illustrate the distinctive features of weathering in Calabria, and their possible relationships to slope instabilities, focusing on the main crystalline massifs of the region. In each geographical area, the bedrock geology, associated weathering, engineering properties and slope failure types are described. For this purpose, the available scientific literature has been used, although it should be noted that not all the Calabrian districts are evenly covered. However, although partial, the description of the less studied areas can also be regarded as interesting, providing useful hints for comparison with other settings.

Geological setting

The geological structure of Calabria is based on two distinct chain segments, the Calabrian–Lucanian Apennines to the north and the Calabrian Arc to the south. The former in Calabria include the Pollino Mountains and the northern part of the Coastal Chain, and are made up of several tectonic units, mainly of sedimentary origin and Mesozoic–Cenozoic in age. The Calabrian Arc, in turn, is represented by Sila, Serre, Poro and Aspromonte Massifs, and also includes the southern portion of the Coastal Chain. It consists of a series of Palaeozoic plutonic–metamorphic nappes, locally overlain by a Mesozoic–Cenozoic sedimentary cover. The Calabrian Arc is generally interpreted as a fragment of the Alpine Chain (Fig. 2), which formed during the Palaeogene, and was transported and overthrust during the Miocene over the Apennine Chain, which at that time was still in its formation phase.

The definition of the structural highs and depressions started during the Late Pliocene. Uplift of the arc is still active, as demonstrated by the intense regional seismicity (Tortorici *et al.* 1995, and references therein). The uplift rate has reached maximum values of the order of about 1.5 mm a^{-1}.

As regards the nature of the terrains forming the Calabrian massifs, sedimentary formations prevail in the north, whereas crystalline rocks dominate in the central–southern part of the region. The Pollino Mountains mark the border between Calabria and Basilicata, and define an almost continuous mountain system from the Ionian Sea to the Tyrrhenian Sea. They are the southernmost portions of the Apennines, and consist of highly fractured and karstified Mesozoic dolostones and limestones. The Coastal Chain is probably the most complex Calabrian range, from a geological standpoint. In its northern sector, sedimentary facies similar to those cropping out in the Pollino Mountains can be found. The southern part of the chain, in contrast, is characterized by various metamorphic rocks (gneiss, micaschists, calcic schists, phyllites) tectonically lying over the

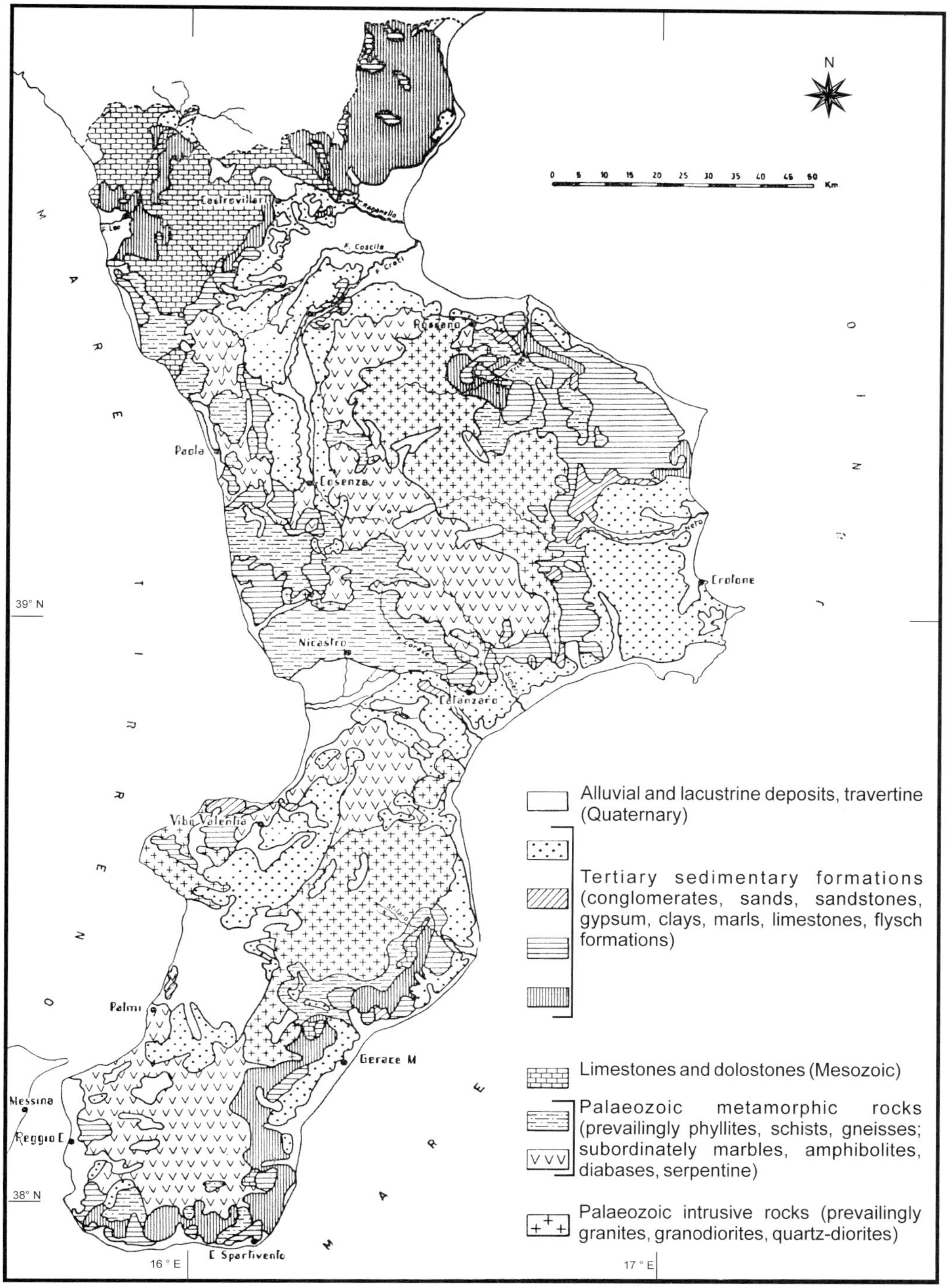

Fig. 1. Geological sketch map of Calabria (modified after Nicotera 1959*a*).

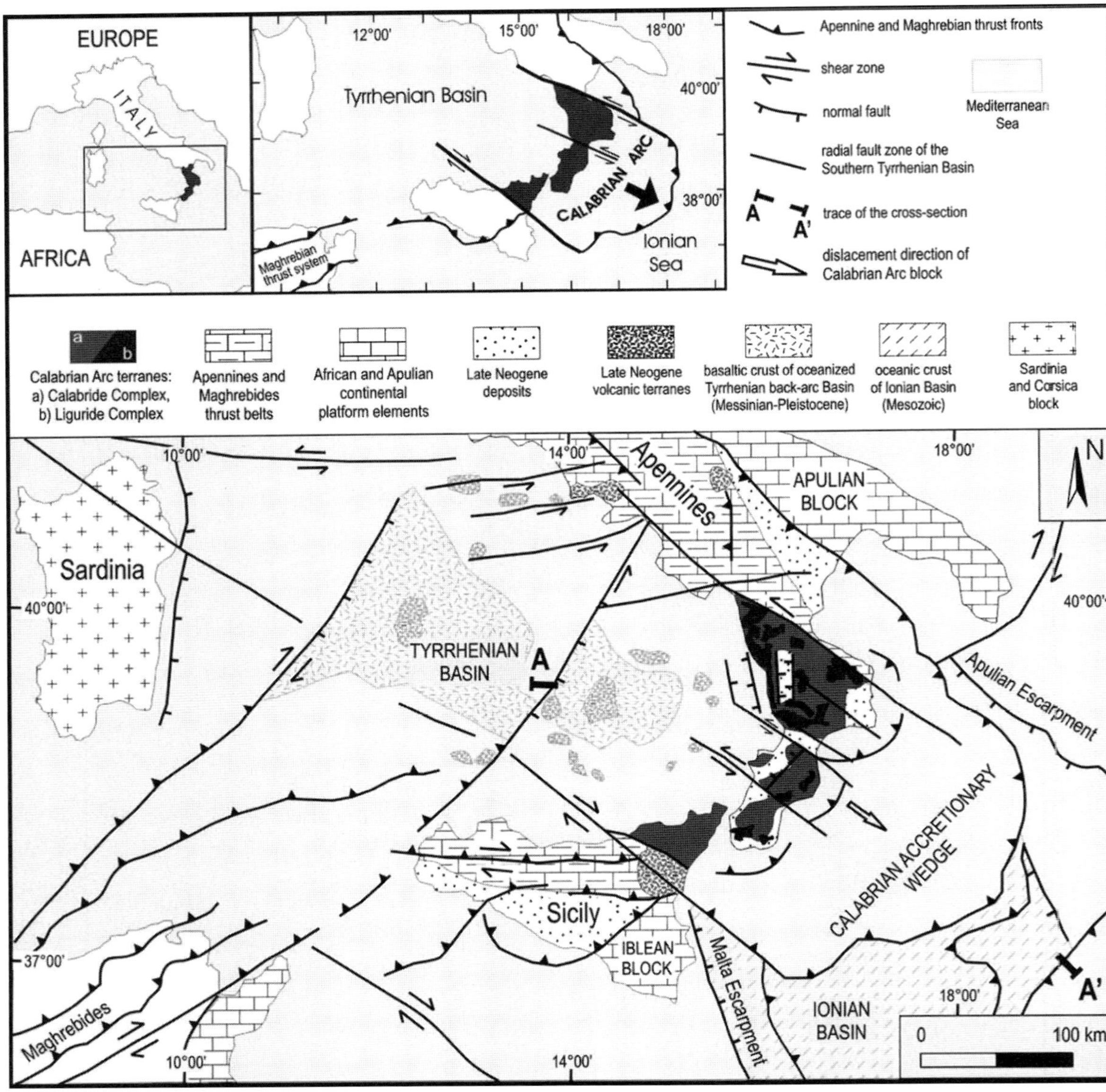

Fig. 2. Geological sketch map of the Central Mediterranean area (modified after Tansi *et al.* 2007). At the top is shown the location of the study area, and a simplified sketch of the Calabrian Arc tectonics.

sedimentary formations. The backbone of the Sila Massif consists of granitoid and gneissic rocks, whereas in the marginal areas low-grade metamorphic and sedimentary formations are present. The Poro Massif shows a geological structure formed by granitoid and high-grade metamorphic terrains, overlain by Cenozoic–Quaternary sediments. Granitoid rocks and subordinate gneiss and phyllites prevail in the Serre Massif. Gneissic rocks extensively crop out in the Aspromonte range, with subordinate amounts of phyllites and granitoids. Along the Ionian side of Aspromonte and Serre, sedimentary formations rest on the crystalline terrains.

The complexity of the geological and tectonic history of Calabria is proved by the fact that several units cropping out in the region underwent at least two metamorphic events. In addition, the crystalline units, as well as most of the Apennine units, have been intensely deformed. At the same time, the neotectonics has to be taken into account

(Ergenzinger *et al.* 1978): extremely high uplift rates characterize Calabria, with particular regard to sectors such as the Coastal Chain and the Aspromonte Massif, especially during Quaternary time (Fig. 3).

The geomorphology of the region is largely dominated by the presence, spatial distribution and activity of different types of landslides. However, some geomorphological features other than those produced by slope movements are worth mentioning. In the inner highlands, remnants of morphological surfaces can be identified: little dismantled by the erosional and gravity-related processes induced by the high uplift rates in Quaternary time, they date back probably to the end of the Early Pleistocene (Sorriso Valvo 1993).

Quaternary normal faults (developed since the Middle Pleistocene with an overall ESE–WNW trend) extend almost continuously along the inner side of the arc; they have fault escarpments showing a very young morphology that defines the fronts of the major mountain ranges of the arc (Tortorici *et al.* 1995). These normal faults are likely to be seismically active.

Tectonic uplift in the last million years intensified all of the geomorphological processes in Calabria, with particular regard to mass wasting and erosion (Sorriso Valvo 1997). Intense dismantling of the landforms predating the onset of tectonic uplift (Late Pliocene–Early Pleistocene) left remnants of the original landscape only in the inner portions of the Calabrian massifs (the so-called 'palaeosurfaces', generally recognizable at the hilltops), and active downcutting of the rivers led to the typical drainage system characterizing the steep Calabrian watersheds (Sorriso Valvo 1993).

The Sila Massif

Bedrock geology

The Sila Massif (Fig. 3) represents the northernmost crystalline massif in Calabria. It is a sector of the Alpine orogenic belt of western Europe, where allochthonous crystalline rocks are exposed at present in the highest tectonic unit of the Apennine Chain (Amodio Morelli *et al.* 1976). This massif consists of Palaeozoic intrusive and metamorphic terranes (gneisses, amphibolites, schists and phyllites), locally overlain by a transgressive and unmetamorphosed Mesozoic sedimentary cover. The relief rises to about 2000 m above sea level (a.s.l.), reaching the maximum elevation at Mt. Botte Donato. The central part of the massif is characterized by flat landscapes, where an upland plateau dominates. Several studies on weathering have been carried out here, because in this area weathering exceeds the rate at which slope processes can remove detritus, and thick profiles of weathered terranes can be observed below the soil cover (Le Pera *et al.* 2001). Gently rolling hills, often showing a marked control by structural discontinuities in the rock mass, are the most typical landforms, with the widespread presence of exfoliation boulders (Ollier 1971; Twidale 1990); the presence of the latter at the ground surface probably derives from selective erosion that caused their exhumation from the different depths at which the boulders originally were formed (Le Pera & Sorriso Valvo 2000*a*). Toward the margins of the Sila Massif, the topography abruptly changes, with a sharp increase of slope gradient along the valleys dissecting the massif rims.

Weathering

In Sila the development of weathering appears to be strongly related to topography, and the bedrock is less affected by alteration processes. This is partly due to the prevalence of mechanical erosion, which in this sector is far more pronounced than chemical weathering, with slopes being highly prone to erosion and different types of mass movements, both shallow and deep-seated.

Granite, granodiorite and gneiss in the Sila Massif are affected by weathering processes that have resulted in the presence of moderately to completely weathered rock (Cascini & Versace 1988; Calcaterra *et al.* 1998; Le Pera & Sorriso Valvo 2000*a*; Bruno 2005). The overall weathering profile is commonly complex, with abrupt changes both vertically and laterally, which are often controlled by the occurrence of joints and/or pegmatitic dykes and quartz veins. The first evidence of rock alteration is the sericitization of plagioclase along cleavage traces, together with neoformation of clay minerals within both plagioclase and K-feldspar, and partial chloritization of biotite, whereas quartz remains unaffected by chemical alteration, except for some intragranular microcracking (Le Pera & Sorriso Valvo 2000*a*). Quartz can be used to indicate the extent of chemical weathering in the Sila Massif granitoids, as transformation of the parent rock into saprolite is characterized by enrichment of quartz at the expense of feldspar and heavy minerals, so that the quartz content of the soil may be twice as high as that of the original bedrock. Those parts of rock where weathering is less developed (slightly weathered to intact rock) are, on the other hand, characterized by the confinement of mineral alteration to fractures, with the formation of ferric iron staining from the ferrous iron component of mafic minerals.

Le Pera *et al.* (2001), from field observations and mineralogical analysis of bedrock (gneiss) and saprolite, have identified two major stages in the weathering process in Sila, showing two distinct rock microfabrics: an early stage involving primarily biotite and feldspar decomposition, with increase in porosity and microfracture, and the production of a granular-framework microfabric; and a later stage of prolonged weathering characterized by mineral dissolution and the formation of authigenic clays, and resulting in a clay-matrix microfabric (Le Pera *et al.* 2001). These features are consistent with those previously observed by other researchers in granitic rocks of England (Baynes & Dearman 1978; Irfan & Dearman 1978). The processes acting in these two stages of weathering may be

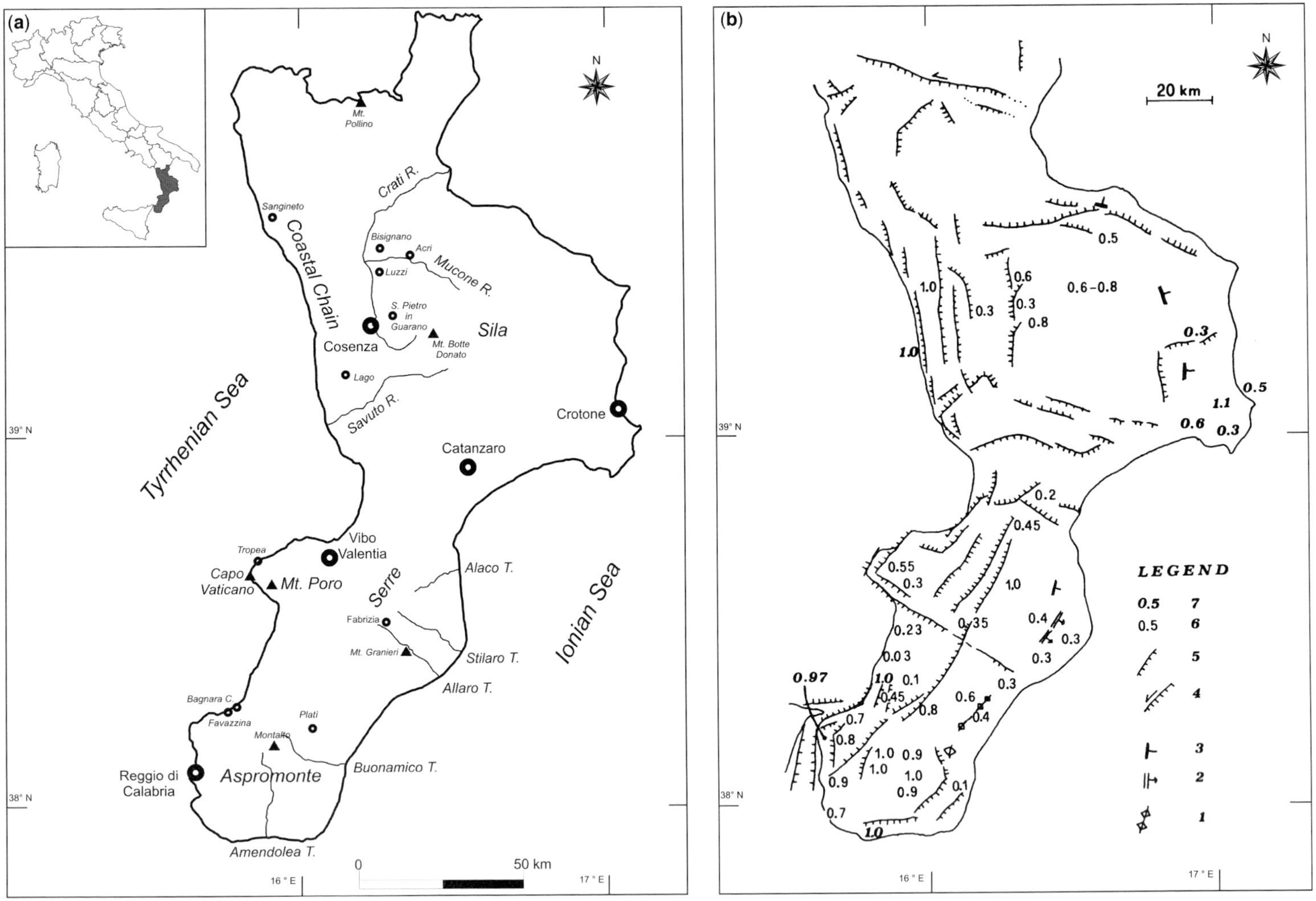

Fig. 3. (**a**) Location of the main localities mentioned in the text. (**b**) Main neotectonic morphostructures and Quaternary uplift rates in Calabria (after Sorriso Valvo 1993). Key to legend: 1, eroded anticlinal depression; 2, hogback; 3, cuesta; 4, strike-slip fault scarp; 5, normal fault scarp; 6 and 7, average uplift rates over a period of 1 Ma and 40–125 ka, respectively.

highly conducive to the occurrence of slope movements in the weathered terranes.

In S. Pietro in Guarano (Fig. 3) Critelli *et al.* (1991*a*) and Cascini *et al.* (1992*a*, *b*) adopted the six-grade classification introduced by the Geotechnical Control Office of Hong Kong (GCO 1984, 1988) to survey the weathered gneissic unit locally cropping out. Four decomposition grades were recognized over an area of about 7.5 km^2, ranging from residual and colluvial soils (grade VI) to moderately weathered gneisses (grade III), whereas fresher rock masses were not present. Some quantitative mineral–petrographic indices (X_d, Lumb 1962; I_p, Irfan & Dearman 1978) confirmed the field evidence (Critelli *et al.* 1991*a*). From I_p (% unaltered constituents/% altered constituents), in particular, an increasing alteration was found, from grade III ($I_p = 3.76$) to grade VI ($I_p = 7.77$). In the same paper, sericite, halloysite, kaolinite and Fe-oxides were identified as the main neoformation minerals, related to the decomposition of either feldspar or biotite.

The Acri area was also chosen to further validate specific procedures for surveying and characterizing the weathered horizon. In this respect, Calcaterra *et al.* (1998) showed a weathering map obtained by applying the GCO classification and, integrating the field observations with *in situ* and laboratory tests, defined the main geomechanical properties of grade III–V gneisses and grade V granitoids. Later, Borrelli *et al.* (2007) applied to the same area a method of reconnaissance surveys of the weathering grade, which was in turn related to slope instability. The weathering grade was determined by quantitative and semi-quantitative measurements and the reconnaissance mapping, carried out at 1:50 000 scale, was checked by detailed mapping at 1:10 000 scale.

Weathering-related slope failures

Most of the studies on the Sila weathered rocks and related slope instabilities focused on a few case-studies, all located on the western slope of the massif (Fig. 3): S. Pietro in Guarano, Altavilla, Acri and Luzzi.

S. Pietro in Guarano was probably the first Calabrian case-study on the interaction between weathering and landslides that was put before an international audience. Cascini & Versace (1988) introduced the main themes, among which weathering surveying, mineral–petrographical and geotechnical laboratory characterization and landslide hazard assessment were analysed in detail in a number of subsequent papers (e.g. Critelli *et al.*1991*a*, *b*; Cascini *et al.* 1992*a*, *b*; Gullà & Matano 1997).

The geomorphological analysis revealed more than 260 landslides, 54% of which involved the highly weathered materials of grade IV, covering an overall extent of about 29% of the study area (Cascini *et al.* 1992*b*). Slides, which sometimes evolved into flows, affect the most weathered materials (grades VI and V), whereas mass movements controlled by pre-existing discontinuities are typical of the rock-like horizons (grades III and IV). Eighty-nine per cent of the surveyed landslides have been related to geostructural elements (faults, joints) and/or to stream downcutting. Cascini *et al.* (1992*b*) also illustrated a scheme that accounts for the evolution of the studied slopes from the pre-Würm III stage to the present-day situation. In this respect, the San Francesco (S. Pietro in Guarano) and the Altavilla landslides are considered as representative of the proposed evolutionary scheme. The Altavilla landslide (Critelli *et al.* 1991*b*; Cascini *et al.* 1992*b*) is a slide-flow known since the beginning of the 20th century, which involved about 475 000 m^3 of completely to highly weathered gneisses. It is bounded by a detachment surface whose depth has been inferred at about 27 m below the ground surface. The San Francesco landslide, occurring within the urban area of S. Pietro in Guarano, has been long investigated using site monitoring instruments. Buried, concave-shaped palaeo-landforms were recognized, filled with landslide debris and grade VI and V materials, and impulsive, intermittent movements were related to the presence of two superimposed water tables, the uppermost of which depended upon the seasonal rainfall (Cascini & Versace 1988).

The San Francesco landslide was again considered by Cascini *et al.* (2006) as regards its response to groundwater regime variations. Those researchers proposed a saturated–unsaturated flow model, which allowed the role played by the hydraulic heterogeneity of the gneissic soils to be defined and the critical values attained during landslide reactivations to be predicted.

Acri has been the focus for several contributions by various researchers, among whom Almagià (1910) showed the importance of mass movements in weathered terrains in the Mucone river basin (Fig. 4). Here, Palaeozoic granitoids, and biotitic and kinzigitic gneisses are present, with migmatites cropping out between the intrusive and metamorphic rocks. Calcaterra & Parise (2001) found references to 71 landslide and flood events in the Acri–Bisignano area, the oldest of which dates back to the 16th century (1595). Most of such events were recognized to have been rainfall-induced, whereas a minor role was played by deforestation and earthquakes; an intermittent activity was recognized in the time-span 1955–1996. However, some reactivations of the Serra di Buda landslide between December 1998 and March 1999 drew attention to Acri when a state road was interrupted, causing serious problems to the local population. The Serra di Buda landslide, considered part of a sackung, involved gneissic and granitic rocks, for which the weathering ranged from residual soils to moderately weathered (Gullà *et al.* 2004*a*, *b*). As a result of about 3 years of surfical and deep displacement measurements, a strongly deformed zone has been identified, to a depth of 50–60 m (Fig. 5), with an overall maximum horizontal displacement of 6–19 cm; the slope sectors with larger strains coincide with the most weathered deposits (grades IV–VI of GCO 1988).

The presence of several debris-flow related fans along the main course of the Mucone river provided the subject for in-depth studies on some geomorphological and vegetational

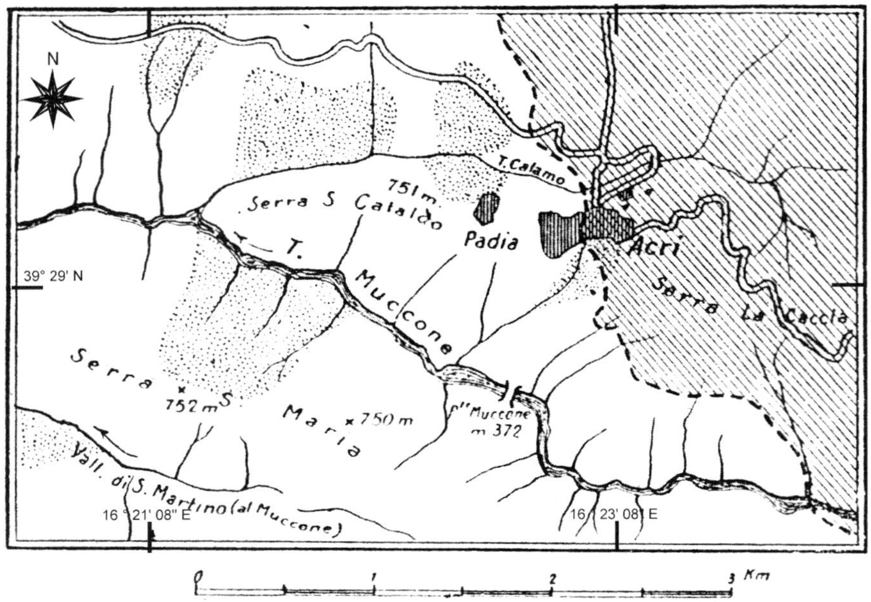

Fig. 4. Main landslides (dotted areas) around Acri (Almagià 1910). White areas, metamorphic rocks; diagonally shaded areas, shaded granitoid rocks.

topics. Parise & Calcaterra (2000) conducted a study on 10 fans, trying to infer information on the related hazard by using a method already adopted in the Canadian Rocky Mountains and in the Italian Eastern Alps (Jackson *et al.* 1987; Marchi *et al.* 1993), based upon geomorphological, stratigraphic and sedimentological evidence (Fig. 6). Using this method, alluvial, debris-flow related and mixed fans were tentatively discriminated, although the proposed method did not prove to be fully satisfactory. The dynamics of the same set of fans was analysed by Garfì *et al.* (2007), together with an assessment of the slope evolution in the period 1955–1996 (Fig. 7). Also using complementary information drawn from the analysis of land-use changes, the aim of those researchers was to obtain a more comprehensive understanding of both the present slope stability status and the future evolutionary trends. For some of the fans studied, the exposures revealed alternating layers of high-viscosity (DF) and low-viscosity (df) debris-flow

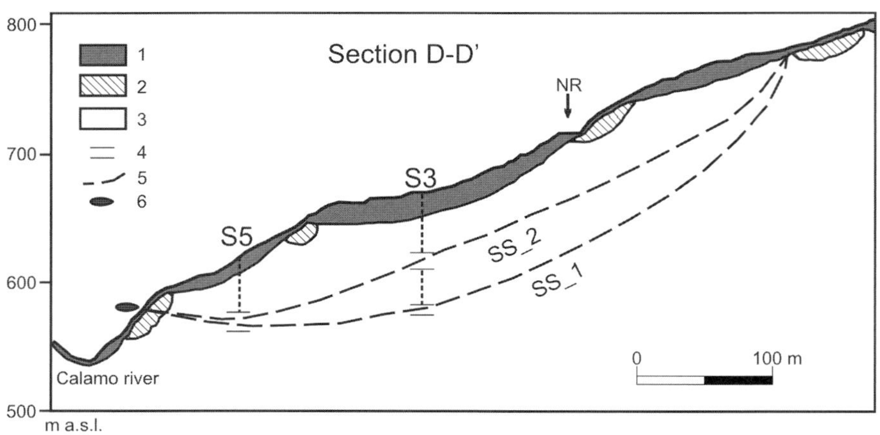

Fig. 5. Cross-section of the Serra di Buda landslide (after Gullà *et al.* 2004*b*, redrawn). 1, Rocks of classes V and VI (cover); 2, rocks of class IV; 3, rocks of classes I and II; 4, completely degraded rocks; 5, assumed slip surface; 6, spring.

Fig. 6. Debris-flow related fan deposits cropping out along the Mucone River.

deposits, whose feeding materials are usually colluvium, residual soils and completely weathered rocks. Water plays the discriminating role, because in DF it initiates the event and entrains the sediments, whereas in df it additionally makes a material selection. In the period studied, the maximum extent of areas of instability was recorded in 1955, probably resulting from the regional flood event that occurred in 1951, whereas in the following years the activity of fans tended to decline progressively. This finding was further confirmed by the land-use changes in 1955–1996, which involved a considerable expansion of natural vegetation cover at the expense of cultivated land (Garfì *et al.* 2007).

The town of Luzzi partly rests on deeply weathered Palaeozoic gneisses, garnet gneisses and biotite schists. Several landslides are present, the most important being the Santo Miele landslide, a complex mass movement that caused serious damage to infrastructure after rainfall-induced reactivations in 1996 and 1997 (Fig. 8); in this area, the oldest phenomenon was recorded during the 16th century (Calcaterra *et al.* 2004).

Geomorphological investigations allowed the reconstruction of the long-term slope evolution (from 1955 to 2006) of an area including the Santo Miele locality (Bruno *et al.* 2008). Interpretation of the 1955 aerial photographs highlighted that all of the studied sector was characterized by old instability and erosion phenomena. Since 1979 an increase in the unstable areas has been recorded; between 2000 and 2004 active landslides and reactivations of pre-existing movements were observed. On the basis of the available data (geological and geomorphological surveys, boreholes, inclinometers, piezometers, seismic refraction profiles), the active 1996 slope movement has been interpreted as the shallow part of a deeper movement, where the presence of residual soils and completely weathered materials has played an important role in predisposing the slope movement.

The Coastal Chain

Bedrock geology

The Coastal Chain shows a very complex geological setting, where sedimentary facies characterize the northern sector, whereas in the southern part of the chain various metamorphic rocks (gneiss, micaschists, calcic schists, phyllites) tectonically overlie the sedimentary formations. In addition, the uplift rates of this area are, together with those in the Aspromonte Massif, the highest in the region (Fig. 3).

Weathering and weathering-related slope failures

Although these characteristics make large portions of the chain highly susceptible to mass movements, practically no

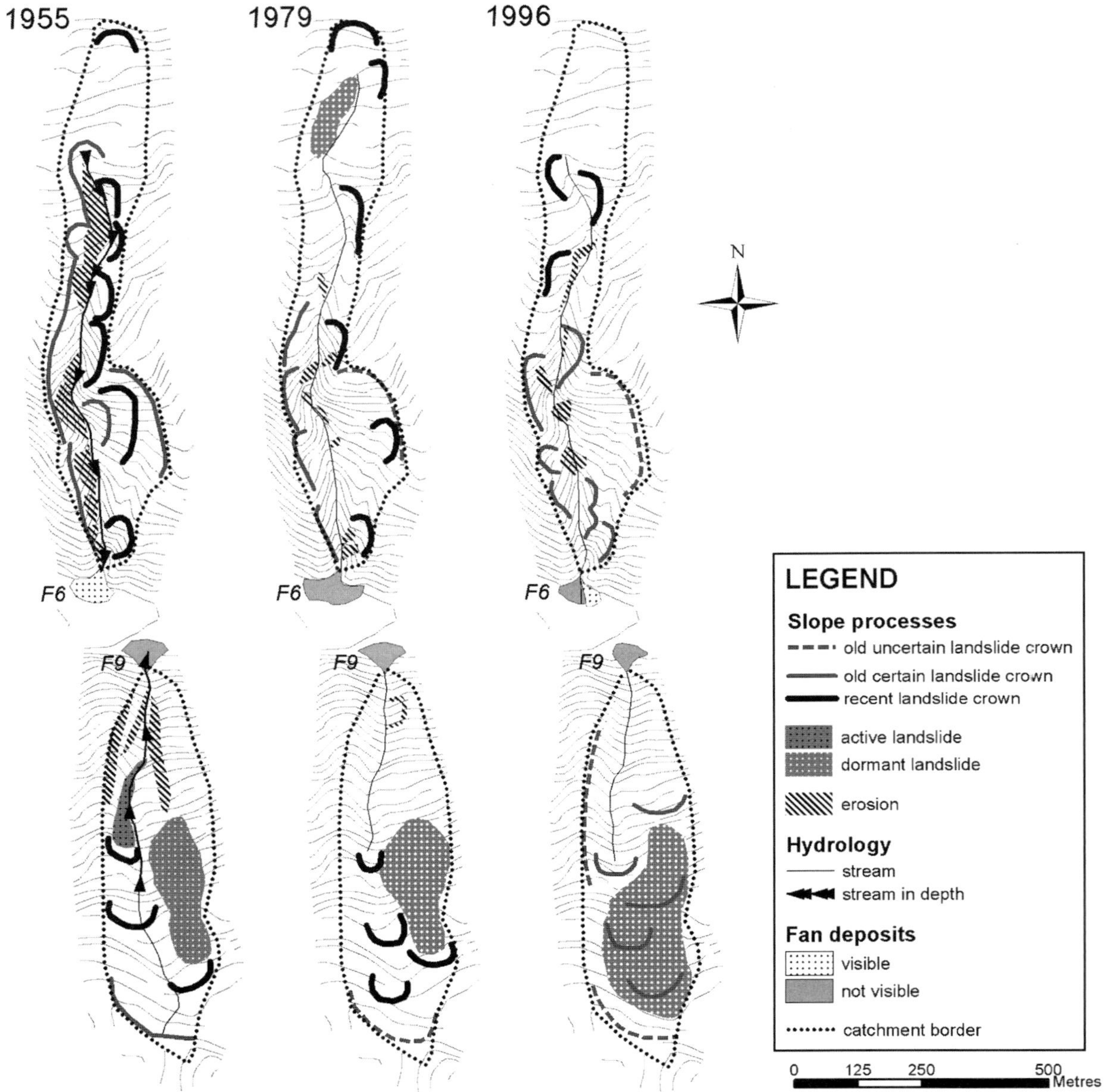

Fig. 7. Maps of geomorphological evidence in two sample catchments of the Mucone River in the period 1955–1996 (after Garfì *et al.* 2007).

studies have directly dealt with the relationships between weathering and landslides in the Coastal Chain. More commonly, weathering is cited or briefly described in papers mostly devoted to other issues, such as geomorphological evolution and tectonics. This is the case, for example, of the work by Sorriso Valvo & Sylvester (1993), where the geomorphology of the Coastal Chain was examined in relation to tectonics and geology. Among the rock units in the area, the Mesozoic sequence of phyllite and slate is the most prone to mass movements, the distribution and activity of which is claimed to be strongly controlled by recent faulting and uplift rates, but also favoured by weakness of the rock mass.

Greater detail about weathering was provided by the analysis of the Pizzotto–Greci slope, south of Mt. Cocuzzo and

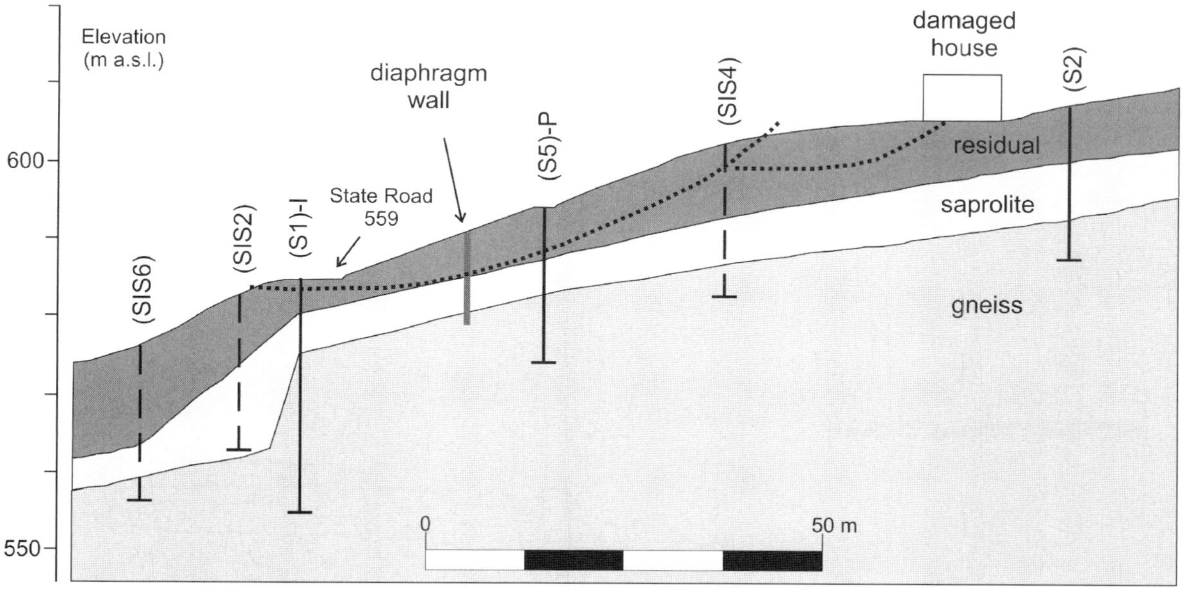

Fig. 8. Section of the S. Miele landslide (after Calcaterra *et al.* 2004, redrawn), showing the main horizons, the critical sliding surface, the location of site investigations and the stabilization measures. I, Inclinometer; P, piezometer.

near the village of Lago, by Sorriso Valvo *et al.* (1999). There, several slope movements are present, including a sackūng-type deformation and debris-flow activity that feeds a wide fan at the mouth of a creek. Within the stratigraphic sequence, several members are affected by weathering processes: the phyllites, most of which are completely weathered, with layers ranging from moderately weathered to fresh; and the metabasites, which are intensely jointed and weathered. It was observed, both at the surface and in borehole cores, that the weathering degree tends to be higher in proximity to tectonic contacts. The physical properties of the rock units, strongly affected by weathering, which also determines differing resistance in the various rock types (and, within the same rock type, differently weathered portions), were interpreted as being among the main causal factors for the development of the slope movements in the area (Sorriso Valvo *et al.* 1999).

The Poro Massif

Bedrock geology

The local bedrock is formed of a Palaeozoic granitoid suite, locally passing to schists and gneisses, overlain by a transgressive series of Miocene age, which is represented by basal conglomerate and evaporitic limestones followed by sands and weak sandstones, and by Pliocene clays and sands. The overall morphology of the massif is characterized by a flat summit plateau, culminating at about 660 m a.s.l., bordered by narrow, steep-sided valleys whose outlet often coincides with the coastline.

Weathering and weathering-related slope failures

Little attention has been paid up to now to the slope instabilities of the Poro Massif (Fig. 3), which dominates the Tyrrhenian side of the Vibo Valentia province. However, Almagià (1910) and, later, Nicotera (1959*b*) recognized the high susceptibility of the Poro weathered materials to landsliding. This propensity was tragically confirmed by an event that occurred in 2006. On 3 July, 2006, 16 municipalities in the Vibo Valentia province, including the main town, were hit by an extreme rainfall event (about 200 mm of rainfall in 2 h, with a peak hourly value of 130.2 mm). Some towns were flooded, and a number of landslides were triggered or reactivated, causing huge damage to infrastructure and urban areas. The mass movements mainly affected the weathered cover, where slumps, soil slips and debris flows prevailed. In some instances, the complete removal of the weathered horizon also favoured the detachment of blocks from the jointed crystalline bedrock, which in some cases led to the temporary closure of some important roads.

Ietto & Calcaterra (1988) recognized in the northwestern sector of the Poro Massif some deep-seated gravitational slope deformations controlled by neotectonics and

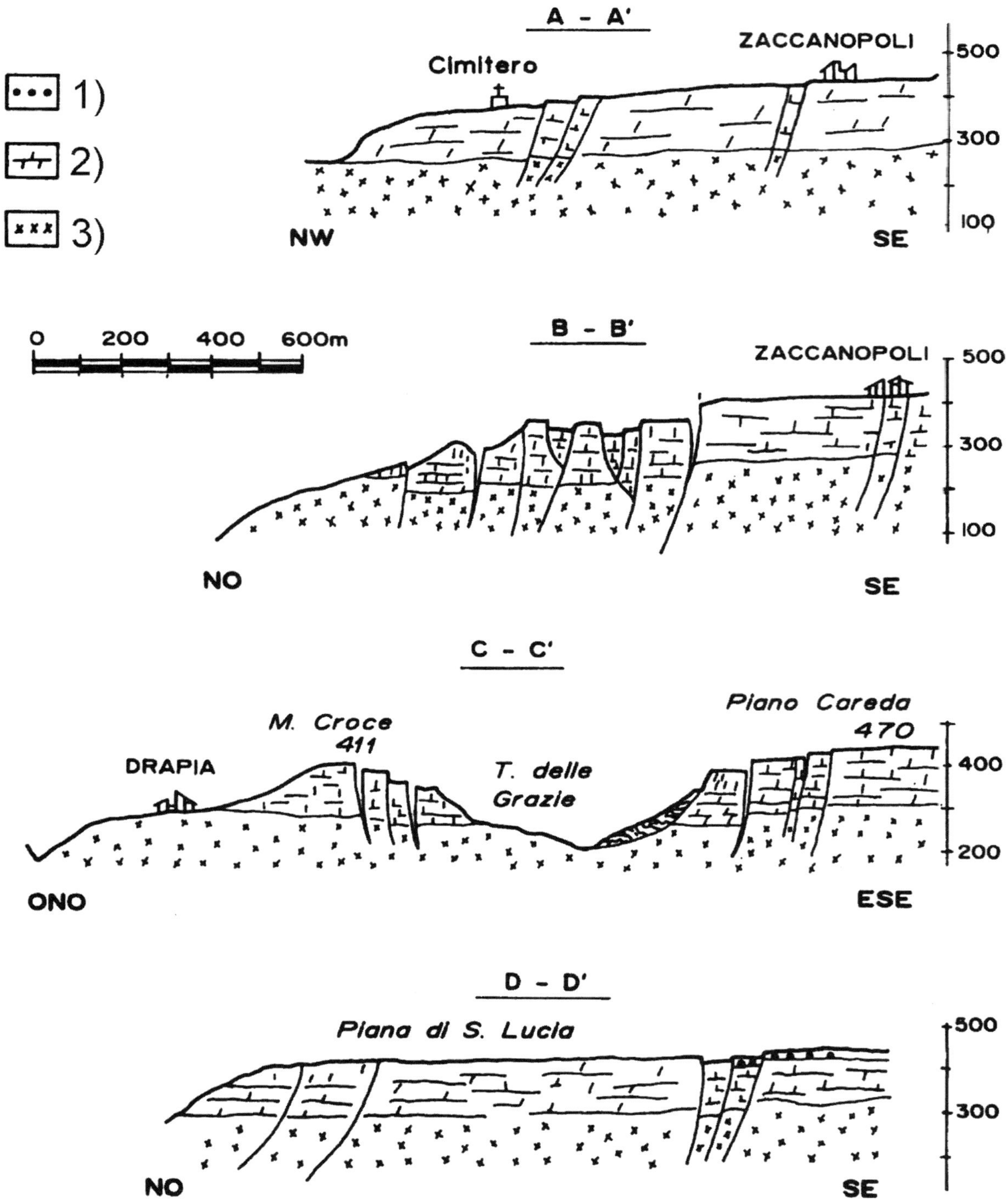

Fig. 9. Geological cross-sections through the deep-seated gravitational slope deformations near Tropea (modified after Ietto & Calcaterra 1988). 1, Pyroclastic level within topsoil deposits; 2, transgressive Miocene sequence (sands and sandstones); 3, granitoid rocks.

(a)
(b)
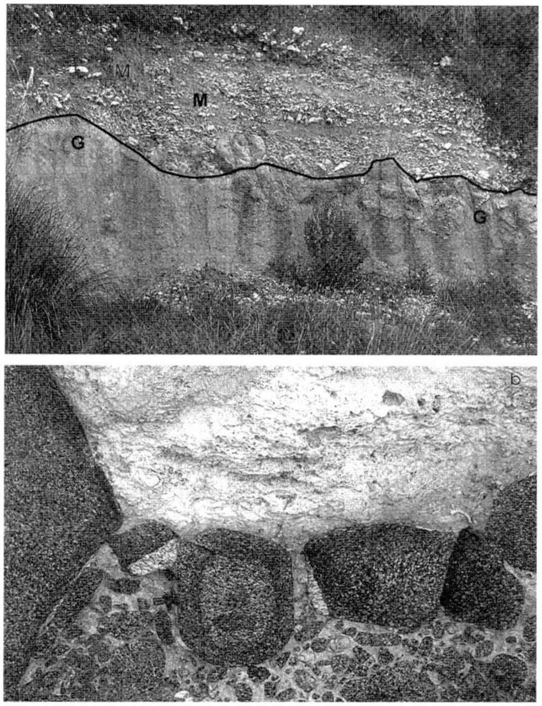

Fig. 10. Evidence of ancient weathering in the Mt. Poro granitoid rocks (modified after Ietto & Ietto 2004). (**a**) Miocene transgressive deposits (M) resting on deeply weathered granitoid (G); (**b**) weathered granitoid corestones included in the Miocene transgressive conglomerates.

associated with shallow soil slides or debris flows (Fig. 9). The slope instabilities involve deeply weathered granitoids, overlain by a Miocene sedimentary cover of sands and weak sandstones. Such a stratigraphic setting provided the basis for Ietto & Ietto (2004) to recognize evidence of an ancient weathering (Fig. 10; see also Serre Massif). In the same area, Gullà *et al.* (2004*a*) tested the surveying procedure discussed above for the Sila Massif. In the case of the Poro Massif, over an area of about 500 km^2 surveyed at a 1:50 000 scale, two macro-classes were identified in the igneous and metamorphic rocks: class A, characterized by a soil-like mechanical behaviour, and class B, with a weak-rock behaviour. In contrast, hard rock masses, belonging to class C, did not crop out. Thirty-two per cent of the surveyed area was affected by landslides. At a detailed scale (1:10 000) a sample area of about 30 km^2 centred around the town of Tropea was considered, where granitoid rocks are mainly present. Adopting the six GCO decomposition grades, rock masses belonging to grades II–VI were surveyed. Of these, grades IV and VI were found to be more extensively involved in landsliding, at 47% and 39%, respectively.

The Serre Massif

Bedrock geology

The Serre Massif (Fig. 3), south of the Catanzaro Strait, is characterized geologically by prevailing granitoid rocks with subordinate gneiss and phyllites. Along the Ionian side, sedimentary formations rest on the crystalline terrains. The massif is drained by subparallel basins, extending west–east and NW–SE from the main central divide to the Ionian coastline. Extended planations are widespread in the higher parts of the massif (at elevations between 850 and 1200 m a.s.l.), and represent the most striking geomorphological feature of the area. These high plains show evidence of continental abrasion in the underlying crystalline basement and are related to Quaternary uplift. They were initially dated as older than 1.5 Ma (Ghisetti 1980), although Ibbeken & Schleyer (1991), on different assumptions, ascribed to these continental surfaces an age of about 1 Ma. Erosion rates determine the amount of weathering debris delivered to the fluvial basins. In the Ibbeken & Schleyer study, the maximum values of erosion rates within the Serre province were obtained from the Stilaro and Assi catchment basins, at 620 and 660 mm ka^{-1}, respectively. Erosion and landsliding activity is evident in many catchments, and particularly in the main river valleys, where a number of fans are continuously fed by debris material produced by debris flows (Fig. 11).

Weathering

Within this framework, weathering plays a crucial role in predisposing the rock units to slope movements and erosion phenomena. From the mineralogical, petrographical and geochemical standpoints, a number of in-depth studies (e.g. Borsi *et al.* 1976; Crisci *et al.* 1985; Del Moro *et al.* 1986; Moresi 1987; Mongelli *et al.* 1998, and references therein) have dealt with the Serre Massif (Fig. 3).

Among these studies, some discussed the chemical and mineral–petrographical features of the weathering products. Moresi (1987), working on six sampled series of tonalites, granites and granodiorites from the eastern Serre, identified four types of materials (fresh, slightly or moderately altered, highly altered and unconsolidated) based on consistency, jointing, colour, and state of conservation of micas and K-feldspar. Irrespective of the parent rock, the so-called 'unconsolidated' samples (corresponding to saprolite deposits) showed a prevailing sandy fraction. The chemical–mineralogical process responsible for the alteration of the studied lithotypes has been defined as 'monosialitic', being characterized by the prevailing presence of kaolinite as neoformation or secondary mineral, deriving from the partial destruction of the plagioclases. Further secondary minerals are Fe-hydroxides and mixed-layer biotite–vermiculite, related to the alteration of biotite. The composition of the weathering products reflects that of the parent rock: hence,

Fig. 11. Landslide- and erosion-fed alluvial fan in the Assi river, Serre Massif.

the kaolinite or Fe-hydroxides tend to increase from the tonalite to the granite weathering products. The chemical weathering trend (biotite → vermiculite; plagioclases → kaolinite) along with the behaviour of both K-feldspar and quartz (which basically tend to be fractured) were considered by Moresi to be in agreement with the general climate of the region (temperate and moderately humid). However, the absence of montmorillonite and the occasional presence of gibbsite suggest that the fairly good permeability of the sandy weathered cover favours the rapid drainage of water and, consequently, a strong Si reduction (Moresi 1987).

Engineering-geological research on the weathered granitoids of the Serre Massif was also favoured by the construction of important civil engineering works such as dams, hydraulic tunnels and roads. Such studies have documented the applicability of some consolidated weathering classifications to the Calabrian materials. For example, Calcaterra *et al.* (1993) successfully adapted the general six-fold scheme proposed by IAEG (1981) to the granitoids of the Alaco river basin, where a concrete dam and a hydraulic tunnel were under construction. The Serre granitoids, along with those cropping out in the Poro Massif, offered evidence for a tentative, indirect dating of the weathering. Ietto & Ietto (2004), moving from stratigraphical and neotectonic evidence, claimed that weathering of the southern Calabrian granitoids was already 'mature' during the Miocene, with a possible initiation between Eocene and pre-Tortonian time. The most important element described by those workers is probably the presence of Miocene transgressive sediments covering a weathered granitoid bedrock. Interesting comments were made by Ietto & Ietto (2004) on the consequences of the postulated Tertiary age of weathering of the Calabrian granitoids. Those workers considered that, assuming an average rate for the deepening of alteration such as those proposed by Migon & Thomas (2002) of 5–20 m Ma^{-1}, the actual maximum thickness of the saprolite deposits (40–50 m) could be explained by a contemporary high rate of soil erosion.

Weathering-related slope failures

A few Serre basins among those studied by Ibbeken & Schleyer were analysed in greater detail as regards the relationships between present-day slope instability and weathering processes. Calcaterra and co-workers summarized engineering-geological studies on the granitoids of the Alaco river basin, dealing with landslides there, together with those present in the conterminous Stilaro basin (Calcaterra *et al.* 1996; Calcaterra & Parise 2005). As regards mass movements, soil slips or debris flows are the dominant type in both basins (Fig. 12), with depths from 0.5 to about 15 m; these flows move along channels with variable gradients (5–45°) and involve average volumes of about 10 000 m^3, with a maximum up to >120 000 m^3.

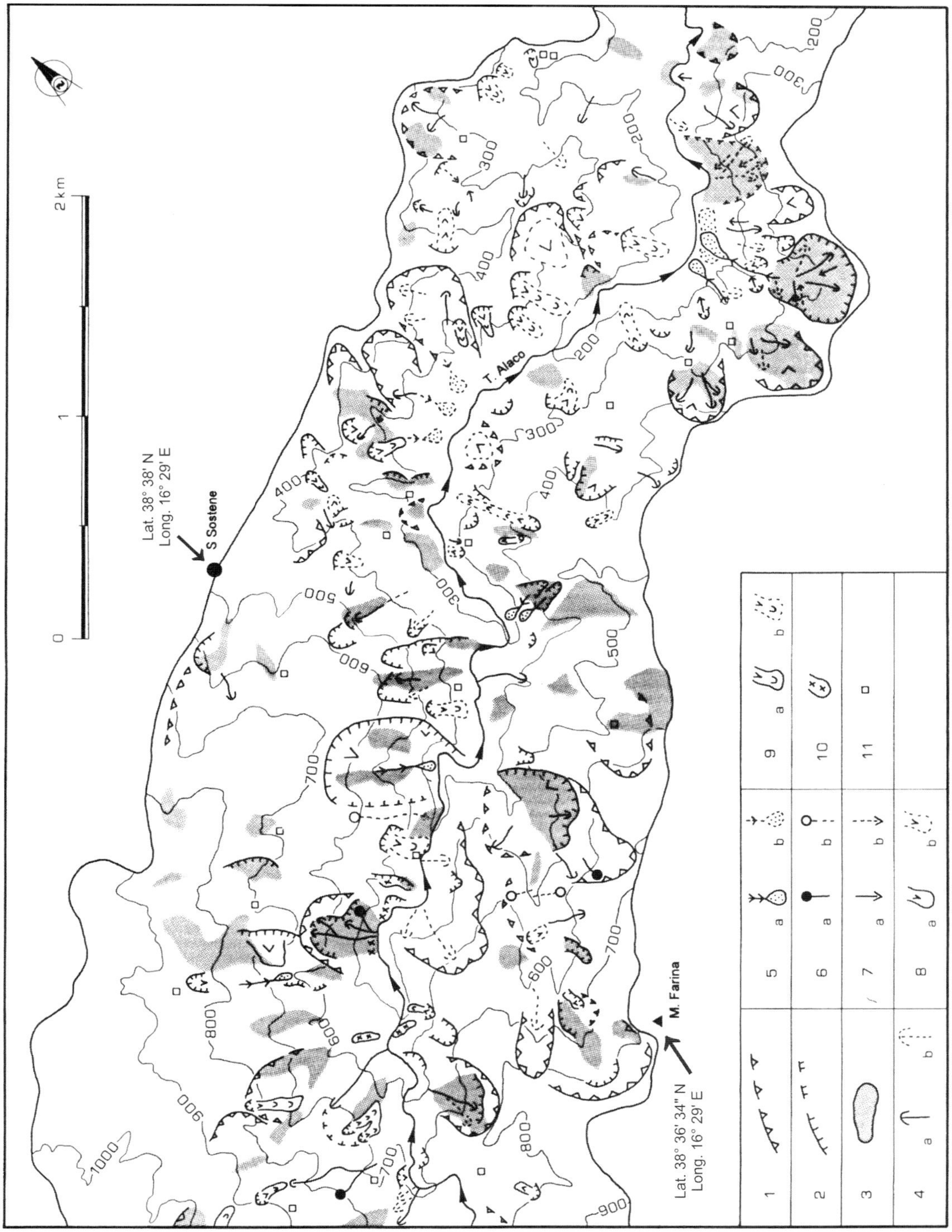

Fig. 12. Landslide inventory map of the sector from S. Sostene to S. Andrea Apostolo allo Jonio (after Calcaterra *et al.* 1996*b*). 1, Old landslide crown, partly remodelled; 2, active landslide crown; 3, complex of soil slips and debris flows; 4, active (a) and dormant (b) soil slip or debris flow; 5, active (a) and dormant (b) channelized debris flow, with related deposit; 6, active (a) and dormant (b) debris avalanche; 7, active (a) and dormant (b) surficial failure in hollow; 8, active (a) and dormant (b) slump; 9, active (a) and dormant (b) complex landslide; 10, rock failure; 11, poorly defined landslide. Contour interval 100 m.

Calcaterra and co-workers, considering a sample of 100 landslides from the Alaco and Stilaro watersheds, found that all the weathered horizons were involved in slope movements (Fig. 13), and most of the debris flow-type phenomena affect grades III to V, whereas grades II and III are involved mainly in slides, rockfalls and complex landslides. In addition, a number of cases involve grade III corestones (in complex phenomena) and rock masses (in planar and wedge failures). An important source of slope failures has been recognized in the hollows (Dietrich *et al.* 1986), downward concave morphologies where the accumulation of colluvial debris is favoured and shallow subsurface runoff can concentrate during storms.

The occurrence of a deep-seated gravitational slope deformation with an associated rock avalanche–debris flow (about 2×10^6 m^3) on the Mt. Granieri–Salincriti slope (Allaro river basin) has prompted a number of multidisciplinary studies, mostly devoted to better comprehending and characterizing deep creep-triggering mechanisms (Pellegrino & Prestininzi 2007, and references therein). The Mt. Granieri–Salincriti slope accommodates a granitic complex and biotite schists, which are placed into contact

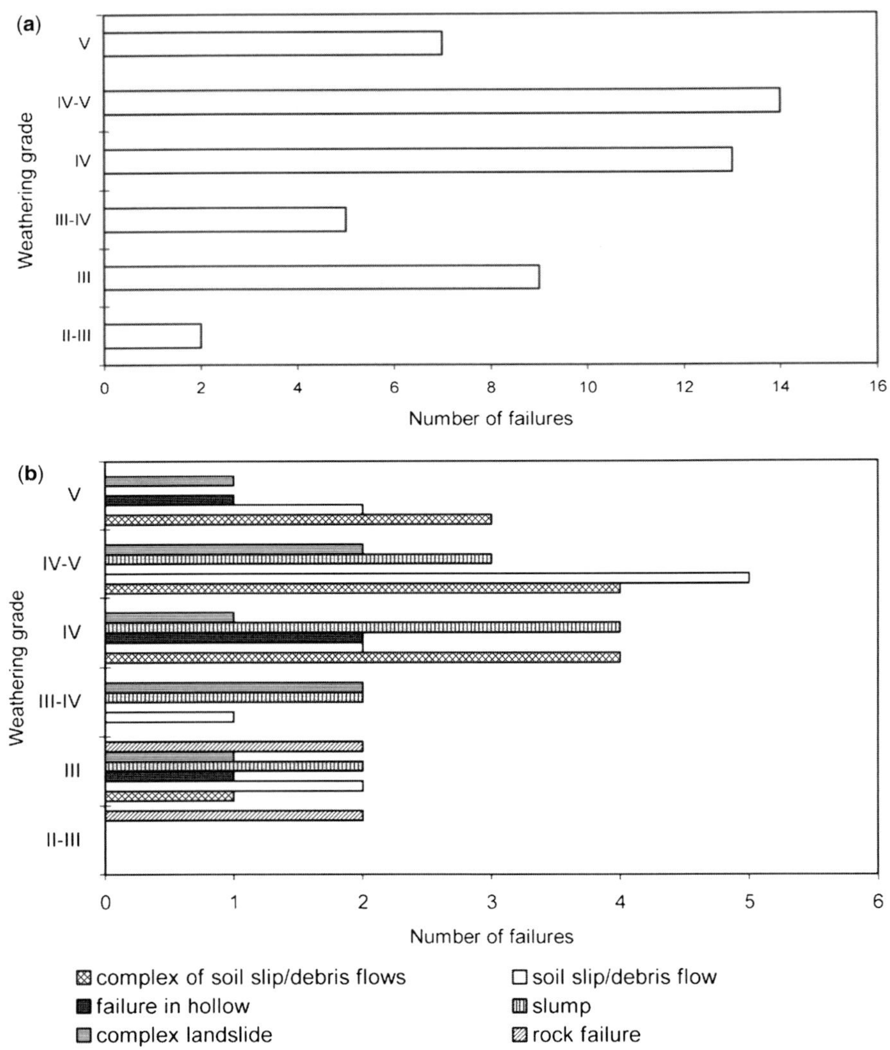

Fig. 13. Histograms showing the relationship between weathering grades and mass movements in the middle–lower reach of the Alaco basin (after Calcaterra & Parise 2005): (**a**) overall sample (50 landslides); (**b**) failures classified by type of slope movements.

by a thermal-metamorphic aureole, producing, in turn, weak belts of rocks. The distribution of the weathering horizons and geomorphological evidence imply that the occurrence of such weak belts plays a crucial role in the development of deep creep phenomena and the related slope instabilities. The proposed engineering-geological model assumes that the rock mass deformations are initially controlled by the subvertical primary anisotropies, geometry of stratigraphic boundaries, heterogeneities and relief forms. Deep geochemical processes in the metamorphic aureole and in the biotite schists near the intrusions are suggestive of a geomechanical behaviour governed by loose and argillified rocks, as shown by geomechanical data and typical weathering profiles. Conversely, the deformations and geomechanical behaviour of the granites located near the contact are controlled by soft rocks with strongly argillified facies (kaolinization). The rock mass shows a time-dependent creep fracture behaviour in agreement with the deep creep deformational mechanism, as already noted by Genevois & Prestininzi (1979). Integration of the experimental data indicates that the instabilities of the Mt. Granieri–Salincriti slope are governed by three layers with different kinematic mechanisms.

Studying a small sector of the Allaro basin, Ietto *et al.* (2007) found near the town of Fabrizia a large outcrop of fresh rock overlying thick saprolite, and interpreted it as proof of a thrust fault, although the region is one of normal faults, which occurred when the weathering mantles on the granitoids had already been formed and were mature. Those workers also dealt with landslides, which are translational slides at the boundary of fresh and weathered rock or, when they originate at the top of the slopes, rotational–translational slides. Under heavy and extensive rainfall there is a tendency for all the thickest colluvial and weathering mantles to collapse and move downward, generating mud flows and debris flows.

Aspromonte

Bedrock geology

Aspromonte can be literally translated as 'wild mountain', which provides an idea of the rugged landscape of the southernmost massif of Calabria, rising to 1956 m a.s.l. at Montalto (Fig. 3). It should be noted that such elevations are reached at a distance of only 26 km from the sea, indicating the intense uplift rates experienced by Aspromonte, of 0.9–1.5 mm a^{-1} in the last million years (Ergenzinger *et al.* 1978; Ghisetti 1981; Sorriso Valvo 1993). Given these characteristics, combined with high rainfall, which is often in the form of heavy and concentrated downpours, it is easy to appreciate the primary role played by erosion and landsliding in shaping the Aspromonte landscape. In fact, Aspromonte was described in the first decades of the last century as one of the most landslide-prone regions in the Mediterranean Basin, notwithstanding the difficulties in reaching its inner mountains at that time (because of mountain path networks in difficult conditions, combined with the fact that the mountains were perfect hiding places for local criminals).

Aspromonte lies in one of the most geodynamically active sectors in the central Mediterranean area, where complex crustal deformation is occurring as a result of the Africa–Europe collision. The core of the massif is composed of metamorphic Palaeozoic rocks, overlain by Mesozoic and Tertiary, partly allochthonous, successions. The high Aspromonte relief is dissected by a dense river network, generally tectonically controlled, formed by the so-called 'fiumare', ephemeral streams with very steep longitudinal profiles, characterized by extremely pulsatory hydraulic behaviour, and with a wide and flat braided river bed with coarse gravel material, where a small amount of water flows during the dry seasons but the whole valley bottom may become inundated during flash floods and in the rainy season.

Weathering

Since the very first studies dealing specifically with landslide phenomena in Aspromonte, the role played by weathering processes was well known; geologists and engineers working in the area were strongly impressed by the high frequency and magnitude of erosion and landslides, and this is especially true for those studies dating back to the 1950s; that is, soon after at least two of the most severe flooding events in Calabria. Montanari, in his comprehensive work dealing with landslides in the Reggio Calabria province (Montanari 1939), described carefully the presence of clay filling in the discontinuities of the rock mass, and the related swelling. Viparelli & Maione (1959), too, in their analysis of landslide occurrences and evolution in southern Aspromonte, pointed to the high degree of weathering of granite, which is often transformed into a soil with a granular structure, and of schists and phyllites, where the hydrological circulation has led to the formation of clays, which are generally concentrated along the likely surfaces of rupture in the rock mass.

Weathering-related slope failures

Notwithstanding the recognition of weathering as a crucial factor in favouring slope instability, only a few sites have been specifically analyzed so far in Aspromonte. They will be briefly described here, and consist of the Buonamico landslide and the consequent dam, the sackiüng phenomena at Platì, the impressive Vallone Colella mass-wasting processes, and the earthquake- and rainfall-induced landslides in the Bagnara Calabra area.

The Buonamico basin is located in northeastern Aspromonte. Underlain by mostly metamorphic rocks (gneisses and schists), it is among the catchments more affected by landslides in the massif, with periodic catastrophic landslides occurring in connection with extreme rainfall. On the occasion of the last major event, which occurred between December 1972 and January 1973, a large landslide (over 21×10^6 m^3 in volume) was triggered in the Costantino locality, following a 24 h precipitation of about 390 mm. The landslide involved highly weathered gneisses, and, as a result of the nature of the local geomorphological controls (narrowness of the valley at the site, high-angle impact with the opposite slope) dammed the Buonamico fiumara, forming a wide lake (Fig. 14). The lake lasted only a few hours before breaching occurred (Guerricchio & Melidoro 1973), and no significant damage was recorded at the nearby villages and towns.

At the origin of the landslide, the role played by weathering processes has to be invoked. Guerricchio & Melidoro (1973), in their study of the 1973 landslide, which also included analysis of two other nearby slope movements, noted the high degree of jointing in the rock mass, with thick discontinuity bands where the rock is completely decomposed and transformed into soils. These bands act as preferential drains for water circulation, strongly reducing the overall resistance of the rock mass. In addition, they observed evidence of hydrothermal weathering related to intrusive processes, identifying the presence of montmorillonite by petrographic analyses.

The Vallone Colella Torrent is a right tributary of the Amendolea Fiumara (the largest river in southern Aspromonte, with a drainage basin of some 152 km^2). The headwaters of Vallone Colella extend in elevation from altitudes above 1400 m to about 800 m a.s.l., with maximum local relief of 570 m. A wide area of Vallone Colella is strongly affected by landslides (Fig. 15), which produce an enormous amount of materials feeding the solid transport in the fiumara downvalley (Fig. 16). The rocks that crop out are an assemblage of metamorphic rocks (biotite schists, metarenite and phyllites, biotite–muscovite augen gneiss) in a very complex structural setting.

The first studies on the area, which were mostly dedicated to geological and morphological descriptions of the landslides, and to descriptions of the transport of debris along the main streams into the fiumara, pointed out the effects induced by weathering processes in degrading the geotechnical properties of the rock masses (Melidoro & Guerricchio 1969). Intense uplift in the area, combined with multi-phase active tectonic stress and deformations, and rock decompression, deeply degraded the rock masses, thus predisposing the whole headwater areas to active landsliding, in the form of deep-seated rotational slides evolving downvalley to debris avalanches and debris flows.

Later, the importance of weathering was particularly noted by Parise *et al.* (1997); through a combination of geomorphological studies, petrographic analyses and structural surveys, they observed the greatest intensity of weathering in the biotite–muscovite augen gneiss forming most of the walls in the main scarps of the landslide area. Weathering

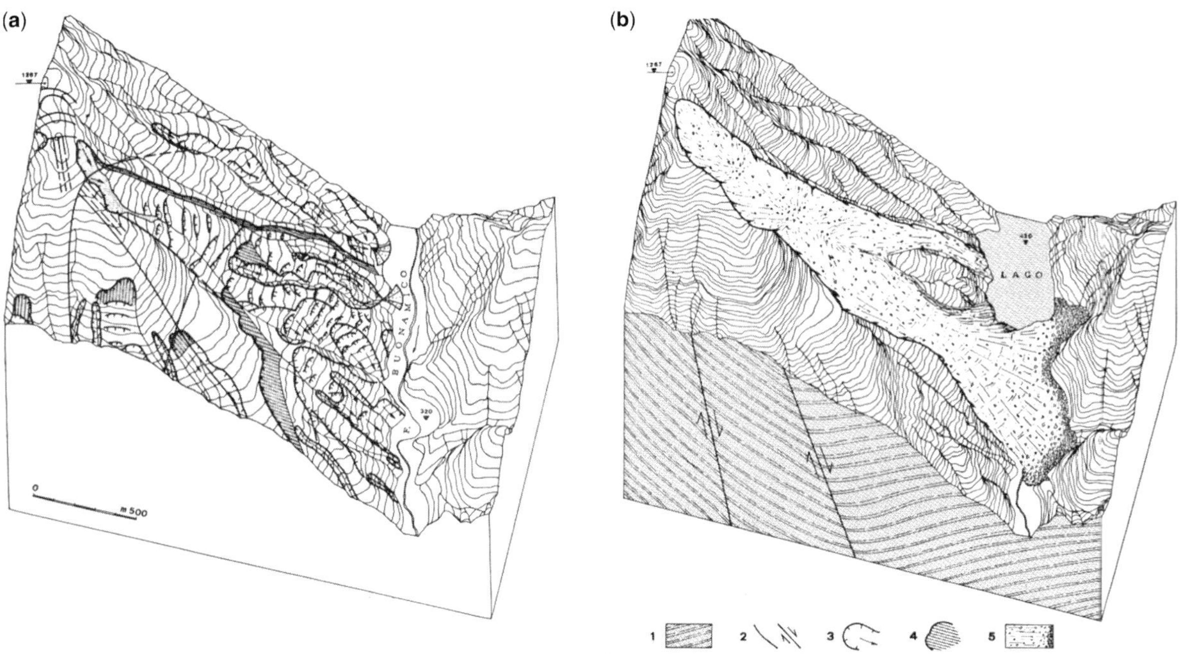

Fig. 14. Stereograms of the Buonamico landslide before (left) and after (right) the event (simplified after Guerricchio & Melidoro 1973). 1, Schists; 2, faults; 3, landslide scarps and direction of landslide movement; 4, sliding surfaces of older landslides; 5, landslide body.

Fig. 15. Overall view of the headwaters of Vallone Colella, one of the largest slope instability watersheds of the Calabrian Arc.

Fig. 16. The huge amount of solid material transported in the fiumara Amendolea, downvalley from the junction with Vallone Colella.

Fig. 17. Pseudokarst cavity in the sandy materials derived from deep weathering of gneiss at the main crown of Vallone Colella mass movements.

often reduced the whole formation (including the acidic dykes, and several cataclasite zones as well) to a coarse sand, leading to the development of pseudo-karst cavities up to 2–3 m deep produced by piping (Fig. 17). The same weathering effects also affect the granitoids (which are transformed into granitic sands) and the metarenites (which are weathered by argillization processes to grey–blue clays). Intense development of weathering combines in the Vallone Colella area to produce important structures represented by wedges along thrust ramps (Parise *et al.* 1997; Iovine & Tansi 1998). Overall, Vallone Colella hosts the main predisposing factors for landslides in Aspromonte: high relief, intense uplift, active tectonics and heavy rainfall, all exacerbated by weathering. The site may be probably considered the best example to describe the high proneness to landslides and erosion in the Calabria region.

Platì is a small town (4000 inhabitants) located at the foothills of northeastern Aspromonte. In the surroundings of the town several slopes show evidence of sackúng-type deformations (Zischinsky 1969; Radbruch Hall 1978), the main features of which are represented by ridge-top trenches at the crowns, and bulging and minor slope movements at the middle–lower slopes (Fig. 18). Metamorphic rocks, with dominant schist and gneiss and subordinate granite, characterize the slopes, whereas sedimentary rocks (conglomerates, shales and sandstones, and alluvial deposits) crop out in the bottom valleys. Although the morphological features do not look very fresh, thus testifying to a very low state of activity of the sackúngen, the phenomena have been subject of several studies in the last 20 years, particularly focused on the likely effects of a catastrophic detachment of the landslides. The local geomorphological situation is highly favourable to high-mobility rock avalanches (Nicoletti & Sorriso Valvo 1991), whose deposits might reach the downvalley infrastructure, including the town, with high destructive power. As the Platì area is both highly seismic and experiences intense and clustered precipitation, the possibility of evolution through catastrophic collapse cannot be excluded, with the failure generating a rock avalanche, and, in turn, damming a river, thus indirectly generating backwater-flooding hazards.

The above studies, as well as later papers dealing with the same area (Gullà & Nicoletti 1996; Gullà *et al.* 2004*a*), in their description of the local geology noted the presence of weathered materials in the rock mass, described following the GCO classification scheme. Schist and gneiss range from slightly weathered to highly weathered, and locally these rocks are completely weathered. Granites, which are less represented, are generally slightly to moderately weathered. The abundance of weathered rocks, together with the main features of the structural discontinuities (these were particularly highlighted by Gullà & Nicoletti 1996, and

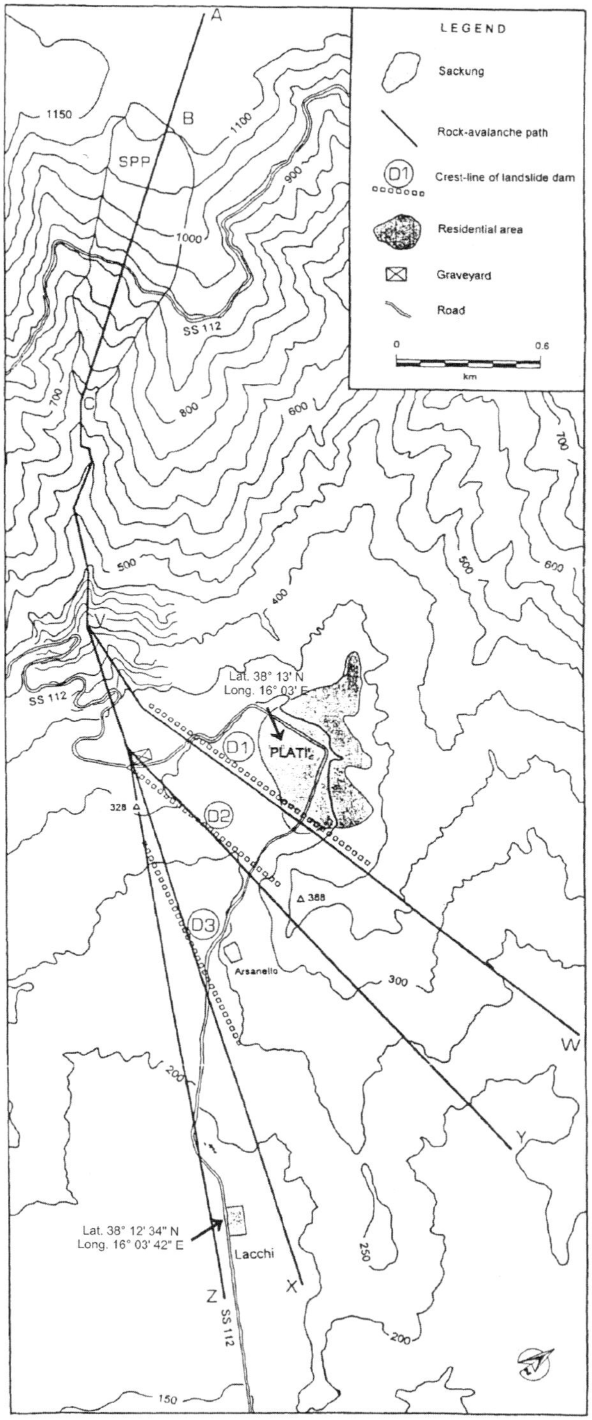

Fig. 18. Possible paths of the rock avalanche that could derive from catastrophic failure of one of the sackūng phenomena near Platì, and positions of the resulting dam (after Nicoletti & Parise 2002).

Nicoletti & Parise 2002) show the high proneness of the slopes to mass movements, the development of which appears to be strongly dependent upon the distribution and intensity of weathering in the rock masses, and the frequency and pervasiveness of schistosity and joints.

The coastal area NW of Aspromonte, in the surroundings of Bagnara Calabra, has also been described in the literature as being affected by different types of slope movements involving weathered materials as a consequence of the 1908 Messina Strait earthquake (Ms 7.1; Sauret & Bousquet 1984; Murphy 1995). Murphy (1995), in particular, reported rotational slides that occurred between Favazzina and Bagnara Calabra, and involved weathered biotite gneiss. Murphy invoked weathering as being among the predisposing conditions to slope movements in the area, and described the rocks that crop out as mostly moderately weathered rock to severely weathered, extremely weak biotite gneiss with extremely closely spaced discontinuities. An upward increase in weathering is generally observed, until, at the top of slopes, completely decomposed materials are reached.

Augen gneisses are the rocks that failed (in their more weathered sections) during the 1908 earthquake, whereas felsic gneisses and granites appear to be stable at steep slope angles. Among the possibilities to explain the development of slides in the regolith between Favazzina and Bagnara Calabra rather than in areas closer to the epicentre, Murphy listed the mineralogy of the clay fractions, with the presence of Na^+ smectite, as opposed to the interlayer clays found elsewhere (Murphy 1995).

Discussion and conclusions

Since at least the end of 19th century, Calabria has fascinated generations of geoscientists because of its intrinsic complexity and, above all, its significance as a key to interpreting the geological history of the central–western Mediterranean basin. Such a fascination is indicated also by the number of studies, especially since the 1970s, focused on the weathering-related landslides in the crystalline rocks. All the main Calabrian crystalline massifs have been analysed, even if unevenly, in terms of the characteristics of weathered materials, slope instabilities or the mutual relationships between weathering and landslides.

However, not all the crystalline massifs have been treated in the same detail. The main reasons for such differences can be attributed to the better accessibility of the study sites (especially Sila) from the main regional research centres, but also to the higher impact of weathering-related landslides upon urban centres and communications. In this respect, Aspromonte, although affected by numerous, large-scale mass movements in metamorphic rocks and granitoids, is characterized by a relatively low urban density, which was further reduced after many mountainous villages had to be abandoned, because of very difficult living conditions and, in some cases, slope instability and flooding problems. Focusing on the Sila and Serre massifs, where more than 75% of the cited studies have been conducted, several aspects were treated in detail, in many cases providing results that are of interest also to an international audience.

The mineralogical and petrographical characterization of the Calabrian crystalline rocks has been deeply investigated, highlighting well the intense transformations produced in the rock mass as a consequence of the long history of weathering. On several occasions these changes have also been related to the presence of the main discontinuity systems, and therefore to regional and/or local tectonics. The relationship between weathering-produced minerals and slope movements, on the other hand, has been less analysed so far. This might be a promising field in which to focus future research on landslides and weathering in Calabria, addressing efforts to the mineralogical characterization of those levels, within the local successions, at which failures seem to concentrate. In this sense, the role played by water also deserves more attention by researchers.

The six-fold classification introduced by GCO (1984, 1988) has been successfully applied in various settings of both the Sila and Serre Massifs, and proved to be a valid tool to produce detailed engineering-geological maps. However, when large areas were considered, the need for a simpler scheme has often been invoked (e.g. Borrelli *et al.* 2007), where the basic horizons (soil-like, mixed and rock-like) could be conveniently recognized in the field, with the aim of solving technical problems. Similar indications appeared earlier as a result of other researchers' studies in the granitoids of the Serre Massif (Calcaterra *et al.* 1993). Despite the different geological setting, further common evidence can be highlighted in the weathering of Sila and Serre. In both massifs, the depth to unweathered rocks is as great as 150–200 m from the ground surface, displaying a complex profile. For both massifs, tectonics and neotectonics have played a fundamental role: emplacement of the Calabrian Arc started during Tertiary time and its uplift is still very active. A few engineering-geological studies, such as those performed in the Alaco Torrent catchment, showed deep and repetitive 'weathering fronts' (Ollier & Pain 1996); that is, sharp contacts between fresh rocks and highly weathered saprolite.

Considering the above depths to unweathered rocks and the present-day climate in the Calabrian region, some researchers have tried to assign an age to the initiation of the physical and chemical modification of the Calabrian crystalline rocks. Following the intuition of Guzzetta (1974), Ietto & Ietto (2004) found straightforward evidence that allowed confirmation of a Tertiary age for the regional weathering. The presence of a Miocene transgressive cover, sealing an already weathered bedrock, gave pivotal proof of an 'ancient' weathering environment, acting under different, probably tropical, climatic conditions. However, the present-day thickness of the weathered cover has to be considered as the remnant of a thicker profile, dismantled by severe erosional processes. Such a complex geological and geomorphological pattern led to a variety of gravity-related processes. In Calabria, the whole sequence of slope instabilities, ranging from surficial soil slips and extremely rapid debris flows to slow-moving deep-seated gravitational slope deformations, has been recognized. The identified types have been and are being triggered by a number of factors, with moderate to strong earthquakes and intense, clustered rainstorms being the main ones. At the same time, anthropogenic actions have to be strongly invoked when analysing the activity of slope movements in Calabria, as several modifications to the environment have been induced by humans, especially in the last century.

Although the fundamental geological–geomorphological features of Calabrian slope movements have been recognized with respect to both soil-like and rock-like materials, little attention has been paid to the mechanics and kinematics of the most important instability types. In this respect, in recent times, matric suction has been usefully considered in some studies dealing with Calabrian weathered cover (e.g. Cascini *et al.* 2006), following the basic assumptions of the unsaturated soil theory (Fredlund & Morgenstern 1977). Matric suction, or negative pore pressure, increases the strength of a soil, and its role in controlling the stability of steep residual soil slopes after heavy and prolonged rainfall has already been analysed in other regional settings (e.g. Rahardjo *et al.* 2004), thus representing a promising direction for the better comprehension of the behaviour of Calabrian weathered slopes and of the related instabilities.

All of this, combined with the diffuse and deep action of the weathering processes, makes Calabria one of the most landslide-prone areas in the Mediterranean Basin, and the local river network an impressive feeder of solid material to the sea. The importance of the local physiography in the control of sediment transport and composition has been proved by several researchers, and particularly by Ibbeken & Schleyer (1991), who, in their thorough study on 19 fiumara drainage systems in the Serre and Aspromonte massifs, demonstrated how the occurrence of elevations close to 2000 m a.s.l. at short distances from the coastline (again, a consequence of intense uplift), led to remarkably immature fluvial sands, and the rapid transport of material through mass-wasting processes (Ibbeken & Schleyer 1991).

The geomorphology of Calabrian massifs controls in some ways the distribution of slope stability typologies. Because of different topography in the headwaters and middle–lower sectors of the catchments, mass movements appear to act as the main process only on those slopes where steeper gradients lead to fast transport rates of weathered materials. This was observed, for example, by Le Pera & Sorriso Valvo (2000*b*) in the Savuto River, and by Calcaterra *et al.* (2004), Garfì *et al.* (2007) and Bruno *et al.* (2008) for several catchments at the western side of the Sila Massif. The same condition applies also to the Serre and the Aspromonte, given the similarities in the first-order landforms among the Calabrian crystalline massifs (Sorriso Valvo 1993). The gentle topography of the summit plateau is thus dominated by transport-limited erosion, with development of great thickness of quartz-rich soils. Moving away from the inner highlands, a sharp increase in slope gradients marks the transition to the downstream area, where deeply incised valleys originate (the fiumare) and where the thickness of weathered materials strongly decreases, and slope movements becomes the dominant process. This occurs both as regards shallow failures, which affect the granular, unconsolidated, materials on the slopes and contained in the hollows, and deep-seated landslides, involving up to some tens of metres of differently weathered rocks, generally arranged in complex weathering profiles.

In conclusion, it can be stated that a huge effort has been made to establish the causal relationships between weathering processes and landslides in Calabria. However, because of the spatial variability and the inhomogeneous nature of weathering profiles, slope stability assessment in areas underlain by weathered rocks is a difficult task. It is therefore of paramount importance to improve and deepen the knowledge on an accurate engineering-geological model of the slopes, based upon weathering analysis and classification and aimed at an effective landslide hazard mitigation, which will also involve adequate and suitable remedial actions.

Acknowledgements. N. Golledge (British Geological Survey) and R. J. Sewell (Hong Kong Geological Survey, Civil Engineering and Development Department, Hong Kong SAR Government) are gratefully acknowledged for their constructive criticisms of the manuscript and their comments.

References

ALMAGIÀ, R. 1910. *Studi geografici sulle frane in Italia*. Memorie della Società Geografica Italiana, **14**.

AMODIO MORELLI, L., BONARDI, G. *ET AL.* 1976. L'Arco Calabro-Peloritano nell'Orogene Appenninico-Maghrebide. *Memorie della Società Geologica Italiana*, **17**, 1–60.

BAYNES, F. G. & DEARMAN, W. R. 1978. The relationship between the microfabric and the engineering properties of weathered granite. *Bulletin of the International Association of Engineering Geologists*, **18**, 191–197.

BORRELLI, L., GRECO, R. & GULLÀ, G. 2007. Weathering grade of rock masses as a predisposing factor to slope instabilities: reconnaissance and control procedures. *Geomorphology*, **87**, 158–175.

BORSI, S., HIEKE MERLIN, O., LORENZONI, S., PAGLIONICO, A. & ZANETTIN-LORENZONI, E. 1976. Stilo Unit and 'Dioritic–kinzigitic' Unit in Le Serre (Calabria, Italy). Geological, petrological, geochronological characters. *Bollettino della Società Geologica Italiana*, **95**, 1–26.

BRUNO, D. E. 2005. *Analisi della franosità e dei suoi rapporti con il weathering in aree-campione del Massiccio Silano (Calabria)*. PhD thesis, Federico II University, Naples.

BRUNO, D. E., CALCATERRA, D. & PARISE, M. 2008. Historical instability and recent slope evolution of the Ilice Torrent catchment, Calabria, Italy. *Physical Geography*, **29**, 179–194.

CALCATERRA, D. & PARISE, M. 2001. The contribution of historical information in the assessment of the landslide hazard. *In*: GLADE, T., ALBINI, P. & FRANCES, F. (eds) *The Use of*

Historical Data in Natural Hazard Assessments. Advances in Natural and Technological Hazards Research, **17**, 201–217.

Calcaterra, D. & Parise, M. 2005. Landslide types and their relationships with weathering in a Calabrian basin, southern Italy. *Bulletin of Engineering Geology and the Environment*, **64**, 193–207.

Calcaterra, D., Ietto, A. & Dattola, L. 1993. Aspetti geomeccanici ed idrogeologici di ammassi granitoidi (Serre Calabresi). *Bollettino della Società Geologica Italiana*, **112**, 395–422.

Calcaterra, D., Parise, M. & Dattola, L. 1996. Debris flows in deeply weathered granitoids (Serre Massif—Calabria, Southern Italy). *Proceedings 7th International Symposium on Landslides, Trondheim, Norway*, Vol. 1, 171–176.

Calcaterra, D., Parise, M. & Dattola, L. 1998. Weathering processes in crystalline rocks of the Sila Massif, Calabria, Southern Italy, as predisposing factor for the development of debris flows. *In*: Evangelista, A. & Picarelli, L. (eds) *The Geotechnics of Hard Soils–Soft Rocks, Naples, 12–14 October 1998*, Vol. 1, Balkema, Rotterdam, 73–84.

Calcaterra, D., Bruno, D.E., Parise, M., Silvestri, F., Critelli, S. & Capparelli, G. 2004. Effects of weathering on slope instability in gneissic rocks at Luzzi. *Proceedings 9th International Symposium on Landslides, Rio de Janeiro*, Vol. 2, 1233–1239.

Cascini, L. & Versace, P. 1988. Relationship between rainfall and landslide in a gneissic cover. *Proceedings 5th International Symposium on Landslides, Lausanne*, Vol. 1, 565–570.

Cascini, L., Critelli, S., Di Nocera, S. & Gullà, G. 1992*a*. A methodological approach to landslide hazard assessment: a case history. *Proceedings 6th International Symposium on Landslides, Christchurch*, 899–904.

Cascini, L., Critelli, S., Di Nocera, S., Gullà, G. & Matano, F. 1992*b*. Grado di alterazione e franosità negli gneiss del massiccio silano: l'area di San Pietro in Guarano (CS). *Geologia Applicata e Idrogeologia*, **27**, 49–76.

Cascini, L., Gullà, G. & Sorbino, G. 2006. Groundwater modelling of a weathered gneissic cover. *Canadian Geotechnical Journal*, **43**, 1153–1166.

Cortese, E. 1895. *Descrizione geologica della Calabria*. Regio Ufficio Geologico d'Italia, Rome.

Crisci, G. M., Paglionico, A. & Rottura, A. 1985. Tonalites from the Serre, Calabrian Arc, southern Italy. Geological, petrological and geochemical features. *Neues Jahrbuch für Mineralogie, Abhandlungen*, **153**, 59–75.

Critelli, S., Di Nocera, S. & Le Pera, E. 1991*a*. Approccio metodologico alla valutazione petrografica del grado di alterazione degli gneiss del massiccio silano (Calabria settentrionale). *Geologia Applicata e Idrogeologia*, **26**, 41–70.

Critelli, S., Di Nocera, S., Gullà, G. & Matano, F. 1991*b*. La frana di Altavilla: esempio di una tipologia di frana diffusa nelle rocce cristalline alterate della Sila Grande (Calabria settentrionale). *Geologia Applicata e Idrogeologia*, **26**, 85–110.

Del Moro, A., Paglionico, A., Piccarreta, G. & Rottura, A. 1986. Tectonic structure and post-Hercynian evolution of the Serre, Calabrian Arc, Southern Italy: geological, petrological and radiometric evidences. *Tectonophysics*, **124**, 223–238.

De Lorenzo, G. 1904. *Geologia dell'Italia meridionale*. Laterza, Bari.

Dietrich, W. E., Wilson, C. J. & Reneau, S. L. 1986. Hollows, landslides, and colluvium in soil-mantled landscapes. *In*: Abrahams, A. D. (ed.) *Hillslope Processes*. Allen & Unwin, London, 383–400.

Ergenzinger, P., Görler, K., Ibbeken, H., Obenauf, P. & Rumohr, J. 1978. Calabrian Arc and Ionian Sea: vertical movements, erosional and sedimentary balance. *In*: Closs, H., Roeder, D. & Schmidt, K. (eds) *Alps, Apennines, Hellenides*. IUCG, Scientific Report, **38**, 359–373.

Fredlund, D. G. & Morgenstern, N. R. 1977. Stress state variables for unsaturated soils. *Journal of the Geotechnical Engineering Division, ASCE*, **103**, 447–466.

Garfì, G., Bruno, D. E., Calcaterra, D. & Parise, M. 2007. Fan morphodynamics and slope instability in the Mucone River basin (Sila Massif, southern Italy): significance of weathering and role of land use change. *Catena*, **69**, 181–196.

GCO 1984. *Geotechnical manual for slopes*. Geotechnical Control Office, Hong Kong.

GCO 1988. *Guide to rock and soil descriptions*. Geotechnical Control Office, Hong Kong, Geoguide, **3**.

Genevois, R. & Prestininzi, A. 1979. Time dependent behaviour of granitic rocks related to their alteration degree. *Proceedings International Congress on Rock Mechanics, Montreaux*, 153–159.

Ghisetti, F. 1980. Caratterizzazione dei blocchi della Calabria meridionale in base alle velocità di sollevamento nel Plio-Pleistocene: una proposta di zonazione neotettonica. *CNR, Progetto Finalizzato Geodinamica, Naples, Contributi alla realizzazione della Carta Neotettonica d'Italia*, **356**, 775–809.

Ghisetti, F. 1981. Upper Pliocene–Pleistocene uplift rates as indicators of neotectonic pattern: an example from southern Calabria (Italy). *Zeitschrift für Geomorphologie Neue Folge*, **40**, 93–118.

Guerricchio, A. & Melidoro, G. 1973. Segni premonitori e collassi delle grandi frane della valle della fiumara Buonamico (Aspromonte, Calabria). *Geologia Applicata e Idrogeologia*, **8**, 315–346.

Gullà, G. & Matano, F. 1997. Surveys of weathering profile on gneiss cutslopes in Northern Calabria, Italy. *Proceedings International Symposium on Engineering Geology and the Environment, Athens*, Vol. 1, 133–138.

Gullà, G. & Nicoletti, P. G. 1996. Rock weathering and discontinuity network in a sackung in the Aspromonte Massif (Calabria, Italy). *Proceedings 7th International Symposium on Landslides, Trondheim*, 971–976.

Gullà, G., Borrelli, L. & Greco, R. 2004*a*. Weathering of rock mass as possible characterizing factor of predisposition to slope instabilities. *Proceedings 9th International Symposium on Landslides, Rio de Janeiro*, Vol. 1, 103–108.

Gullà, G., Aceto, L., Antronico, L., Cilento, M., Niceforo, D., Perna, E. & Terranova, O. 2004*b*. Failure and post failure conditions of a landslide involving weathered and degraded rocks. *Proceedings 9th International Symposium on Landslides, Rio de Janeiro*, Vol. 2, 1241–1246.

Guzzetta, G. 1974. Ancient tropical weathering in Calabria. *Nature*, **251**, 302–303.

IAEG 1981. Rock and soil description and classification for engineering geological mapping. Report by the IAEG Commission on engineering geological mapping. *Bulletin of International Association of Engineering Geology*, **24**, 235–274.

Ibbeken, H. & Schleyer, R. 1991. *Source and Sediment*. Springer, Berlin.

Ietto, A. 1969. I problemi geologico-tecnici delle gallerie nel cristallino-metamorfico della Calabria. *Bollettino della Società dei Naturalisti in Napoli*, **78**, 309–354.

Ietto, A. 1976. Geologia e pianificazione territoriale e urbana in Calabria. *Memorie della Società Geologica Italiana*, **14**, 421–490.

IETTO, A. & CALCATERRA, D. 1988. Deformazioni gravitative profonde e tettonica presso Tropea (M.te Poro, Calabria). *Memorie della Società Geologica Italiana*, **41**, 911–915.

IETTO, A. & IETTO, F. 2004. Age and history of the weathering of granitoids in southern Calabria. *Geografia Fisica e Dinamica Quaternaria*, **27**, 37–45.

IETTO, F., DONATO, F. F. & IETTO, A. 2007. Recent reverse faults and landslides in granitoid weathered profiles, Serre Mountains (southern Calabria, Italy). *Geomorphology*, **87**, 196–206.

IOVINE, G. & TANSI, C. 1998. Gravity-accommodated 'structural wedges' along thrust ramps: a kinematic scheme of gravitational evolution. *Natural Hazards*, **17**, 195–224.

IRFAN, T. Y. & DEARMAN, W. R. 1978. The engineering petrography of a weathered granite in Cornwall, England. *Quarterly Journal of Engineering Geology*, **11**, 223–244.

JACKSON, L. E. JR, KOSTASCHUK, R. A. & MACDONALD, G. M. 1987. Identification of debris flow hazards on alluvial fans in the Canadian Rocky Mountains. *In*: COSTA, J. E. & WIECZOREK, G. F. (eds) *Debris Flows/Avalanches: Process, Recognition, and Mitigation*. Geological Society of America, Reviews in Engineering Geology, **7**, 115–124.

LE PERA, E. & SORRISO VALVO, M. 2000*a*. Weathering and morphogenesis in a Mediterranean climate, Calabria, Italy. *Geomorphology*, **34**, 251–270.

LE PERA, E. & SORRISO VALVO, M. 2000*b*. Weathering, erosion and sediment composition in a high-gradient river, Calabria, Italy. *Earth Surface Processes and Landforms*, **25**, 277–292.

LE PERA, E., CRITELLI, S. & SORRISO VALVO, M. 2001. Weathering of gneiss in Calabria, southern Italy. *Catena*, **42**, 1–15.

LUMB, P. 1962. The properties of decomposed granite. *Géotechnique*, **12**, 226–243.

MARCHI, L., PASUTO, S. & TECCA, P. R. 1993. Flow processes on alluvial fans in the Eastern Italian Alps. *Zeitschrift für Geomorphologie Neue Folge*, **37**, 447–458.

MELIDORO, G. & GUERRICCHIO, A. 1969. Geologia, fenomeni franosi e problemi di difesa del suolo nel Vallone Colella (Aspromonte). *Geologia Applicata e Idrogeologia*, **4**, 63–83.

MIGON, P. & THOMAS, M. F. 2002. Grus weathering mantles, problems of interpretation. *Catena*, **49**, 5–24.

MONGELLI, G., CULLERS, R.L., DINELLI, E. & ROTTURA, A. 1998. Elemental mobility during weathering of exposed lower crust: the kinzigitic paragneiss from the Serre, Calabria, southern Italy. *Terra Nova*, **10**, 190–195.

MONTANARI, G. 1939. Studio generale sui movimenti franosi in provincia di Reggio Calabria. *Annali dei Lavori Pubblici*, **2–3**, 3–42.

MONTANARI, G. 1940. Studio generale sui movimenti franosi in provincia di Catanzaro. *Annali dei Lavori Pubblici*, **2–3**, 3–48.

MONTANARI, G. 1941. Note alla carta delle frane della provincia di Cosenza. *Annali dei Lavori Pubblici*, **1**, 3–29.

MORESI, M. 1987. L'alterazione dei graniti delle Serre orientali (Calabria). *Rendiconti SIMP*, **42**, 237–248.

MOYE, D. G. 1955. Engineering geology for the Snowy Mountain scheme. *Journal of the Institution of Engineers, Australia*, **27**, 287–289.

MURPHY, W. 1995. The geomorphological controls on seismically triggered landslides during the 1908 Straits of Messina earthquake, Southern Italy. *Quarterly Journal of Engineering Geology*, **28**, 61–74.

NICOLETTI, P. G. & PARISE, M. 2002. Scenarios of risk by rock avalanche and backwater flooding in the hypothesis of the failure of a sacking: the care of plati (Calabria, Southern Italy). *Proceeding International Symposium 'Landslide Risk Mitigation and Protection of Cultural and National Heritage', Kyoto (Japan)*, 635–660.

NICOLETTI, P. G. & SORRISO VALVO, M. 1991. Geomorphic controls of the shape and mobility of rock avalanches. *Geological Society of America Bulletin*, **103**, 1365–1373.

NICOTERA, P. 1959*a*. Lo stato di dissesto idrogeologico della Penisola calabrese. *Geotecnica*, **6**, 3–40.

NICOTERA, P. 1959*b*. *Rilevamento geologico del versante settentrionale del Monte Poro (Calabria)*. Memorie e Note Istituto di Geologia Applicata, Napoli, **7**.

NICOTERA, P., DE RISO, R. & TONNETTI, G. 1968. *Indagini geologiche e tecniche per lo studio di uno sbarramento sul T. Jassa (Calabria)*. Memorie e Note Istituto Geologia Applicata, Napoli, **10**.

OLLIER, C. D. 1971. Causes of spheroidal weathering. *Earth-Science Reviews*, **7**, 127–141.

OLLIER, C. D. & PAIN, C. 1996. *Regolith, Soils and Landforms*. Wiley, Chichester.

PARISE, M. & CALCATERRA, D. 2000. Debris flow-related fans in weathered crystalline rocks, and the potential hazard in Calabria, Italy. *In*: WIECZOREK, G. F. & NAESER, N. D. (eds) *Proceedings 2nd International Conference on 'Debris-Flow Hazards Mitigation: Mechanics, Prediction, and Assessment', Taiwan, 16–18, August 2000*, 203–211.

PARISE, M., SORRISO VALVO, M. & TANSI, C. 1997. Mass movement and tectonics in the Aspromonte massif (Southern Italy). *Engineering Geology*, **47**, 89–106.

PELLEGRINO, A. & PRESTININZI, A. 2007. Impact of weathering on the geomechanical properties of rocks along thermal-metamorphic contact belts and morpho-evolutionary processes: the deep-seated gravitational slope deformations of Mt. Granieri–Salincriti (Calabria—Italy). *Geomorphology*, **87**, 176–195.

RADBRUCH HALL, D. H. 1978. Gravitational creep of rock masses on slopes. *In*: VOIGHT, B. (ed.) *Rockslides and Avalanches, 1. Natural Phenomena*. Developments in Geotechnical Engineering, **14A**, 607–657.

RAHARDJO, H., AUNG, K. K., LEONG, E. C. & REZAUR, R. B. 2004. Characteristics of residual soils in Singapore as formed by weathering. *Engineering Geology*, **73**, 157–169.

RUXTON, B. P. & BERRY, L. 1957. Weathering of granite and associated features in Hong Kong. *Geological Society of America Bulletin*, **68**, 1263–1292.

SAURET, B. & BOUSQUET, J. C. 1984. Manifestations d'instabilité du sol dans la zone epicentrale du seisme de Messine de 1908. Le role de la liquefaction. *In*: FLAGGOLLET, J.-C. (ed.) *Proceedings 'Mouvements de terrains', Colloque de Caen, 22–24 March 1984*. BRGM Documents, **83**, 63–74.

SORRISO VALVO, M. 1993. The geomorphology of Calabria: a sketch. *Geografia Fisica e Dinamica Quaternaria*, **16**, 75–80.

SORRISO VALVO, M. 1997. Landsliding during the Holocene in Calabria, Italy. *In*: MATTHEWS, J. A., BRUNSDEN, D., FRENZEL, B., GLASER, B. & WEISS, M. (eds) *Rapid Mass Movement as a Source of Climatic Evidence for the Holocene*. European Science Foundation, Special Issue of ESF Project, European Palaeoclimate and Man, **12**, 97–108.

SORRISO VALVO, M. & SYLVESTER, A. G. 1993. The relationship between geology and landforms along a coastal mountain front, northern Calabria, Italy. *Earth Surface Processes and Landforms*, **18**, 257–273.

SORRISO VALVO, M., GULLÀ, G., ANTRONICO, L., TANSI, C. & AMELIO, M. 1999. Mass movements, geological structure and morphologic evolution of the Pizzetto–Greci slope (Calabria—Italy). *Geomorphology*, **30**, 147–163.

TANSI, C., MUTO, F., CRITELLI, S. & IOVINE, G. 2007. Post-collisional strike-slip tectonics in the central Calabrian Arc (Southern Italy). *Journal of Geodynamics*, **43**, 393–414.

TORTORICI, L., MONACO, C., TANSI, C. & COCINA, O. 1995. Recent and active tectonics in the Calabrian arc (southern Italy). *Tectonophysics*, **243**, 37–55.

TWIDALE, C. R. 1990. The origin and implications of some erosional landforms. *Journal of Geology*, **98**, 343–364.

VIPARELLI, M. & MAIONE, U. 1959. Natura ed evoluzione di alcune frane nell'Aspromonte meridionale. *Geotecnica*, **4**, 3–12.

ZISCHINSKY, U. 1969. Über Sackungen. *Rock Mechanics*, **1**, 30–52.

Weathering processes, structural degradation and slope–structure stability of rock-hewn churches of Lalibela (Ethiopia)

G. Delmonaco*, C. Margottini & D. Spizzichino

ISPRA—Geological Survey of Italy, Via Curtatone 3, 00185 Rome (Italy)

**Corresponding author (e-mail: giuseppe.delmonaco@apat.it)*

Abstract: Lalibela is located in the northern–central part of Ethiopia, *c.* 600 km north of Addis Ababa in Northern Wollo (Ahmara Region). The town, which has about 12 000 inhabitants, is situated at an altitude of 2500 m. In its centre a unique complex of 11 rock-hewn Christian Orthodox churches is located, cut out of the rock some 800 years ago. Their construction is attributed to King Lalibela (1167–1207). All the churches exhibit widespread evidence of structural damage and weathering processes, which are posing a serious problem for the preservation of the monuments. Preliminary geological field surveys, *in situ* geotechnical analyses and laboratory tests have been carried out in the area to understand the cause of weathering and the likely consequences on the structural stability of the churches. Laboratory tests on volcanic materials detected a deep alteration of rocks as a result of the widespread presence of montmorillonite. This is the primary cause of the progressive deterioration of the physical and mechanical characteristics of the slope-forming rocks in Lalibela. Major consequences are the alveolar weathering of churches' façade, the degradation of the roofs and a reduction of rock strength, which are causing a progressive structural instability in some rock-hewn monuments. Also, sliding of façades along bedding joints was detected. Understanding the weathering processes affecting the Lalibela churches allows the detection of problems and the timely implementation of medium- to long-term protection strategies for conservation of the monuments.

Lalibela is located in the northern–central part of Ethiopia, *c.* 600 km north of Addis Ababa in Northern Wollo, Amhara Region (Fig. 1). The town, which has about 12 000 inhabitants, is located at an altitude of 2500 m above sea level (a.s.l.). A unique complex of 11 rock-hewn churches lies in the centre of the town, cut out of the bedrock some 800 years ago during the reign of King Lalibela (1167–1207) of the Zagwe dynasty. The churches are still used daily for religious practices and ceremonies, and during major religious occasions large crowds of believers and pilgrims travel to the site. Since 1978, the churches and their surrounding areas have been included in UNESCO's World Heritage List.

Some of the churches are sculpted out of solid highly weathered basalts (Fig. 2). Some are quarried enlargements of caves located in the softer portions of volcanic deposits. They are almost all connected by long underground tunnels and passages. The maze of churches offers a wide variety of architectural styles: Grecian pillars, Arabesque windows, ancient swastika and Star of David carvings, arches, and Egyptian-like buildings. The main cluster of 11 churches is located in the village centre: Biet Golgotha–Biet Mia'el, Biet Maryam, Biet Meskel, Biet Danaghel, Biet Medhane-Alem, Biet Amanuel, Biet Merkorios, Biet Abba Libanos, Biet Gabriel-Rufa'el and Biet Gyorgis.

Over the years, the churches have been exposed to wind, rainfall, thermal changes and human activities. This exposure has resulted in the severe degradation of the monuments, most of which are now considered to be in a critical condition. The major cause of degradation is the large concentration of montmorillonite (an expansive clay belonging to the smectite family), and related weathering, in the silica-poor volcanic rock.

However, because of the outstanding value and the nature of the monuments, the conservation of the Lalibela rock-hewn churches attracted long ago the interest of the international community. A first restoration attempt was carried out in 1920. Currently, almost all of the churches are protected by shelters, to prevent the erosive effects of rain and direct water circulation over the top layers, and temporary scaffolding to prevent collapse of the vulnerable structures (Fig. 3).

Geological setting of the area

The town of Lalibela is located on the western flank of a high mountainous area whose principal summits are Mount Abune Yosef, Mount Degosach, and Mount Asheten Mariam. The region is characterized by outcrops of Tertiary volcanic rocks (Fig. 4), comprising basalt, trachyte and tuff (Kazmin 1973; Mohr & Zanettin 1988). The geological series of the area, from the oldest and stratigraphically lower unit, consist of the following formations (ACEL-SAVA 1997).

From: Calcaterra, D. & Parise, M. (eds) *Weathering as a Predisposing Factor to Slope Movements.* Geological Society, London, Engineering Geology Special Publications, **23**, 131–147.
DOI: 10.1144/EGSP23.8 0267-9914/10/$15.00

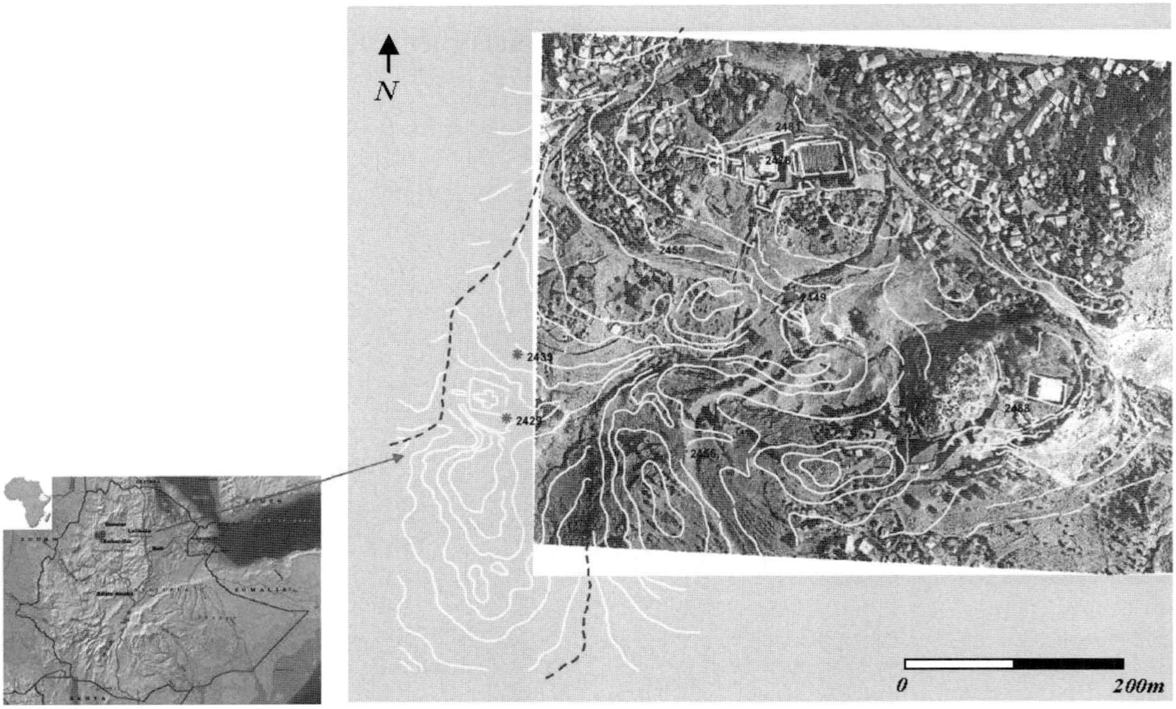

Fig. 1. Location of Lalibela and contour map superimposed on aerial photograph of the area of the churches.

Fig. 2. View of Biet Gyiorgis.

Fig. 3. View of Biet Abba Libanos. Shelter, temporary scaffolding and channel pipe are visible.

(1) The Ashangi Formation (Oligocene–Miocene) which mainly comprises olivine basalt alternating with agglomerates and tuff (Merla *et al.* 1979).

(2) The Amba Aiba basalts (Oligocene–Miocene) which mainly consist of flood basalt with rare tuffs. The flows are characterized by columnar jointing and have thickness up to tens of metres (Merla *et al.* 1979).

(3) The Amba Alaji rhyolites (Miocene) generally overlie the Aiba basalts. This formation consists of a succession of alkaline to perialkaline rhyolites, alkaline and trachytic alkaline flood basalt. The acidic rocks are thick ignimbrites and tuffs, which crop out widely in the area and form typical landscape elements such as steep walls and cones. Furthermore, the flat surface of the Ethiopian Plateau is commonly formed by the upper part of thick ignimbrites of the Amba Alaji rhyolites (Merla *et al.* 1979).

(4) The Termaber Formation basalts (Miocene to Pliocene) are at the top of the local stratigraphical series. The

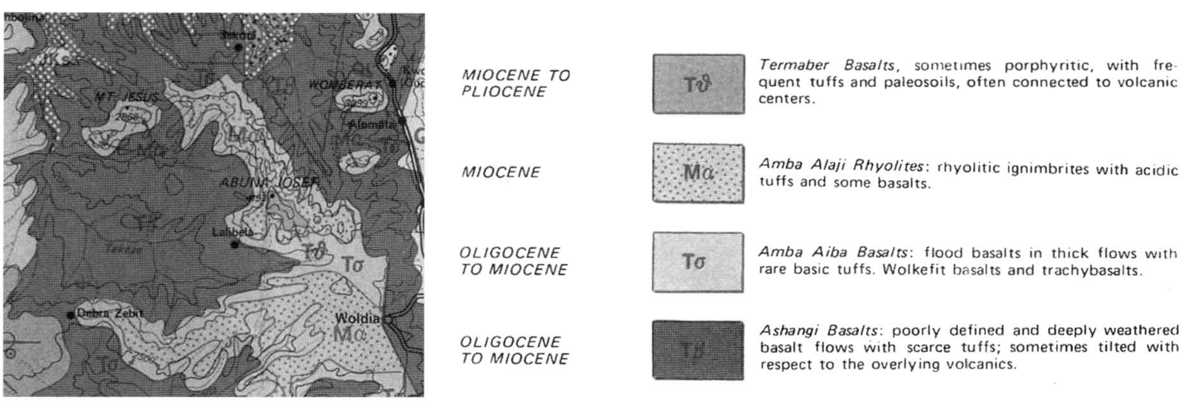

Fig. 4. Extract from the geological map of the Lalibela region (after Merla *et al.* 1979).

Termaber Formation comprises lenticular basalts with a large component of tuffs, scoriaceous lava flows and typical red palaeosols (Merla *et al.* 1979).

According to ACEL-SAVA (1997), the churches are carved within the Amba Aiba Formation, in two distinct units, the lower basalt and the basic tuff. On the top, the fill material consisting of silt-sized particles to angular tuff fragments, which was deposited during the excavation of the churches and tunnels, can be clearly seen. Also, a regolith, the product of weathering of the basaltic rock, was identified by ACEL-SAVA (1997) along the northern side of Biet Medane-Alem Church.

ICCROM (1978) detailed the local geological components of the Amba Aiba basalts (from the oldest and stratigraphically lower), as follows.

(1) Dark grey basalt: a fine-grained, very tough rock, containing minerals such as olivine, pyroxene and plagioclase, which in a fault-brecciated zone disintegrate rapidly. Chloritization is prominent in the fault breccia.

(2) Scoria: an angular reddish (iron-enriched) volcanic agglomerate. The particle size ranges from a fine ash to large lava bombs up to 2 m in diameter. Secondary mineralization is prevalent, with calcite found along discontinuities and a large amount of salt precipitation within the vesicles. The scoria is heavily and differentially weathered, with basalt pyroclasts standing out from the more easily weathered components. Secondary silica produced by metasomatic fluid has precipitated quartz in scoria vesicles.

In synthesis, the Amba Aiba Formation (basic tuff and basalt) described by Merla *et al.* (1979) and ACEL-SAVA (1997) was defined by ICCROM (1978) as basalt and scoria and later subdivided by TEPRIN (2002), based on geomechanical properties (from bottom to top) as fresh basalt, fractured basalt, moderately weathered tuff, highly weathered tuff and fill sediments, as also mentioned by ACEL-SAVA (1997). A summary of the stratigraphical and lithotechnical units in the Lalibela churches, as provided by the above references, is shown in Table 1. The present investigation shows that the so called 'weathered tuff' is simply the hydrothermally-altered with incipient lateritization nearly aphyric vesicular unit of the massive to slightly fractured basaltic lava.

Engineering geology investigation

The general geology of the area (ACEL-SAVA 1997; TEPRIN 2002) is characterized by the presence of a basalt bedrock, which is fresh in the lower part and fractured in the upper part of the formation. The basalt is overlain by a red vesicular lava, which is moderately weathered in the lower part and highly weathered in the upper part. On the top a weathered regolith is described from the northern side of Biet Medane-Alem Church and, finally, a covering of fill sediment was produced during the excavation of churches and trenches (Fig. 5).

The basalt can be considered as a good engineering material, with bulk density ranging between 22.06 and 23.83 kN m^{-3}. The uniaxial compressive strength (UCS or σ_c) obtained in the laboratory (TEPRIN 2002) has values of 7.8 MPa in the fractured unit and 21.4 MPa in the fresh material, sampled in the area of the churches. The UCS obtained in the field with a Schmidt hammer (ISRM 1978) had higher values between 33 MPa (Biet Gyorgis) and 55 MPa (Biet Medane-Alem), both in the fractured unit.

The vesicular lava outcropping in the Biet Abba Libanos site appears to belong to the highly weathered unit of the basalts. The density was established at about 16.5–17.1 kN m^{-3}, whereas the bulk density was estimated to range from 20.00 to 21.18 kN m^{-3} (TEPRIN 2002).

Geotechnical and geophysical analyses were carried out on Lalibela vesicular lava samples from the Biet Abba Libanos area. A UCS (σ_c) laboratory test on a weathered lava sample gave values ranging from 5.53 to 7.93 MPa. P-wave and S-wave velocities were also measured in laboratory, providing respectively values of 1.846–2.026 km s^{-1} and 1.100–1.144 km s^{-1}. The laboratory data are reported in Table 2.

Figure 6 shows the stress–strain plot for UCS from sample 1. σ_c was also measured in the field with a Schmidt hammer (ISRM 1981), all over the site. The results reveal higher values than obtained in the laboratory, ranging from 27 to 42 MPa. The vesicular lava exhibits an average uniaxial compressive strength of about 36 MPa in the dry

Table 1. *Reconstruction of stratigraphical units in the Lalibela area*

Age	*Merla* et al. *(1973)*	*ICCROM (1978)*	*ACEL-SAVA (1997)*		*TEPRIN (2002)*
Holocene–recent			Fill Weathering regolith	Fill Weathering regolith	Fill
Oligocene to Miocene	Amba Aiba basalt	Scoria	Basic tuff	Highly weathered reddish tuff Moderately weathered tuff	Highly weathered tuff Moderately weathered tuff
		Basalt	Basalt	Basalt	Fractured basalt Fresh basalt

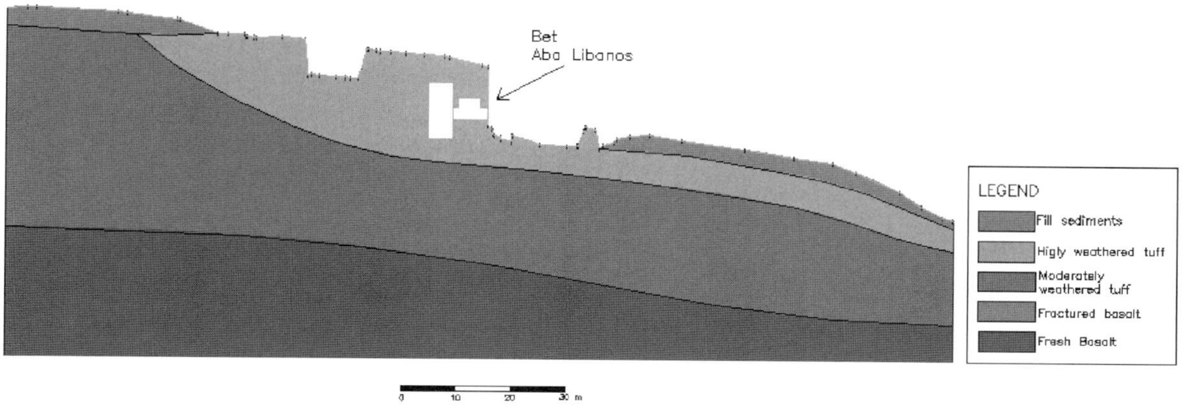

Fig. 5. Geological section of the church of Biet Abba Libanos.

condition ($\gamma = 16\ \mathrm{kN\ m^{-3}}$), decreasing to 28 MPa when saturated. The basalt has a ucs of about 44 MPa ($\gamma = 20\ \mathrm{kN\ m^{-3}}$). All the collected data, according to material type, are reported in Figure 7. The main results of the field survey can be summarized as follows.

(1) Even when the tests were performed apparently on the same unit, there is a scatter of 70% in the results, confirming the impression of a large internal inhomogeneity as a consequence of the mode of deposition of the original pyroclastic flow.

(2) The weathered (alveolized) lava generally has the same values as nearby fresh material, or slightly lower ones; nevertheless, alveoli are more evident where the average values of σ_c are lower (Fig. 7). This supports the finding that the alveoli are mainly connected with joint chemical–physical processes in specific strata.

Table 2. *Geotechnical and geophysical analysis on Lalibela vesicular lava samples*

Sample	*Dimensional characteristics and density*								
	Height (H) (mm)	*Diameter (D)* (mm)	*H/D*	*Area* ($\mathrm{mm^2}$)	*Volume* ($\mathrm{mm^3}$)	m_{dry} (G)	ρ_d ($\mathrm{Mg\ m^{-3}}$)	γ_d ($\mathrm{kN\ m^{-3}}$)	
1	55.4	41.7	1.33	1365.7	75660.9	131.91	1.74	17.1	
2	48.2	41.0	1.18	1320.3	63636.3	107.06	1.68	16.5	
3	46.0	41.2	1.12	1333.2	61325.6	103.55	1.69	16.6	
	Uniaxial compressive strength								
	Value (kg)	*Load* (kN)	σ_1 (MPa)	E_{sec} (GPa)	E_{tan} (GPa)	E_{tan0} (GPa)	ν_{sec}	ν_{tan}	
1	1101	10.80	7.906	1.717	1.538	1.89	0.163	0.238	
2	744	7.29	5.526	–	–	–	–	–	
3	771	7.56	5.671	–	–	–	–	–	
	Analysis of P and S waves								
	*t*_P (μs)	*t*_S (μs)	V_P ($\mathrm{km\ s^{-1}}$)	V_S ($\mathrm{km\ s^{-1}}$)	V_P/V_S	Pois_dyn	*G*_dyn (GPa)	*K*_dyn (GPa)	*E*_dyn (GPa)
1	29.8	49.4	1.859	1.121	1.657	0.214	2.192	3.102	5.323
2	26.1	43.8	1.846	1.100	1.678	0.224	2.037	3.021	4.990
3	22.7	40.2	2.026	1.144	1.771	0.266	2.211	3.986	5.597

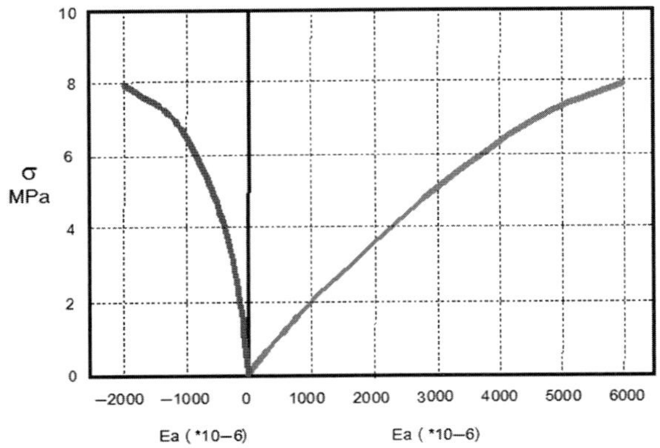

Fig. 6. Stress–strain plot for sample 1.

(3) Wetting of the basalt probably greatly reduces the strength of the material (e.g. Biet Marqorewos).

From a structural point of view, the basalt formation exhibits three main discontinuity orientations: (1) subhorizontal discontinuities of syngenetic origin, probably as a consequence of lithostatic load or different pyroclastic flows; these discontinuities are spaced about 5–10 m apart, with a low dip that probably follows the original morphology at the time of the volcanic eruption; (2) subvertical discontinuities, mainly parallel to the slope, generated by cooling effects and/or lateral unloading; (3) medium-angle dipping joints as a result of shear stress caused by overload acting along vertical structures (e.g. Biet Abba Libanos).

Because water was found to play an important role in the degradation of vesicular lava, a slaking test was performed on a sample from Biet Abba-Libanos; this showed that such material was not susceptible to slaking. Shear strength parameters were determined to investigate the potential resistance to sliding of rock-walls, such as at Biet Abba-Libanos. A tilt test was performed on site, and gave a friction angle of about 45° (Fig. 8). The roughness coefficient (JRC) is between 12 and 14.

Petrographical and mineralogical analyses

Laboratory tests were carried out on superficial samples of volcanic rocks, which were mainly weathered. Five thin sections were obtained, one from basalt and four from hydrothermally altered scoriaceous unit samples, and analysed using polarized light microscopy (Fig. 9).

The fresh basalt exhibits a typical porphyric holocrystalline structure, a slightly fluidal texture with phenocrysts of clinopyroxene, olivine partially altered to iddingsite and aggregates of phyllosilicates. The volcanic glass is composed of plagioclase, olivine, Fe-oxides (hematite, limonite), magnetite and ilmenite.

The vesicular lava, macroscopically a black to reddish brown welded scoria, has a pumiceous structure, a heterogeneous and partly oriented texture, and is moderately porphyric. The rock includes lava clasts with a high-vacuole

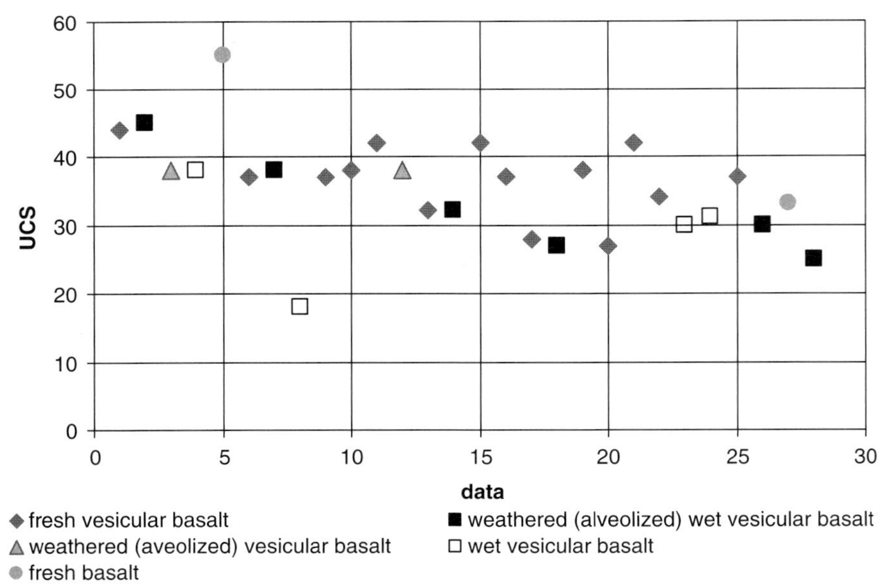

Fig. 7. Distribution of UCS for different material types.

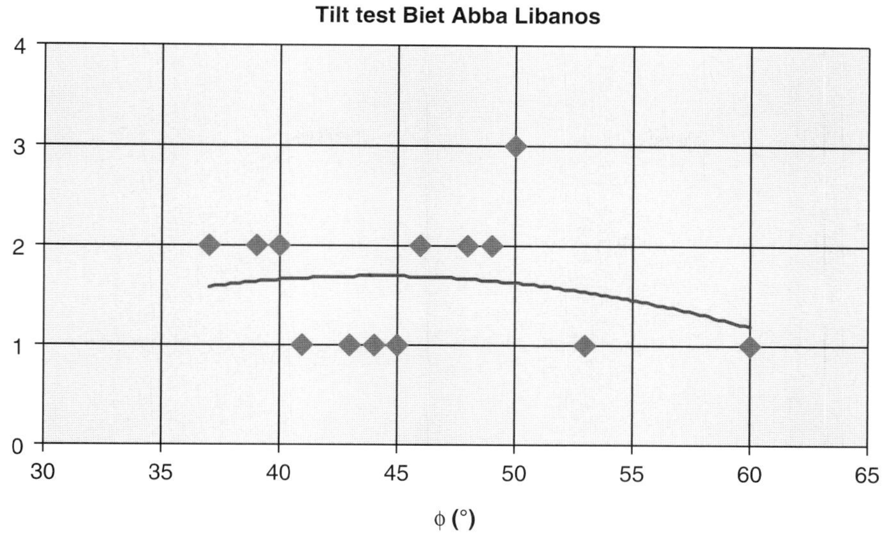

Fig. 8. Friction angle (basic + first-order asperities) resulting from tilt test.

structure and fluidal texture as a result of the orientation of small isotropic elongated crystals. Primary minerals are opaque minerals (*c.* 10%), plagioclase (*c.* 5%), pyroxene (*c.* 2%) and olivine (*c.* 2%). The percentage of voids is *c.* 30%, represented by vesicles that are partially or totally filled by secondary minerals. The black glass, representing around 50% of the total, is glassy in the pumiceous component and hypohyaline in the lava clasts. Secondary minerals are calcite, quartz, orthoclase and zeolites. Most of the rock composed of black glass shows alteration to clay, recognizable in the yellow–brownish fine-grained aggregate (montmorillonite).

To better detect the presence and clarify the composition of clay minerals produced by alteration of the minerals forming the volcanic rocks in Lalibela, the same samples were analysed by X-ray diffractometry (XRD) and the results were interpreted by standard procedures. The XRD instrument allowed the analysis of all of the sample without any special preparation to produce the <2 μm 'clay fraction' material. The mineralogy of the total sample, as derived from the XRD scan of the bulk material, is reported in Figure 10. The diffractogram was interpreted using software that allows a rapid comparison between a peak position in each spectrum and that of reference

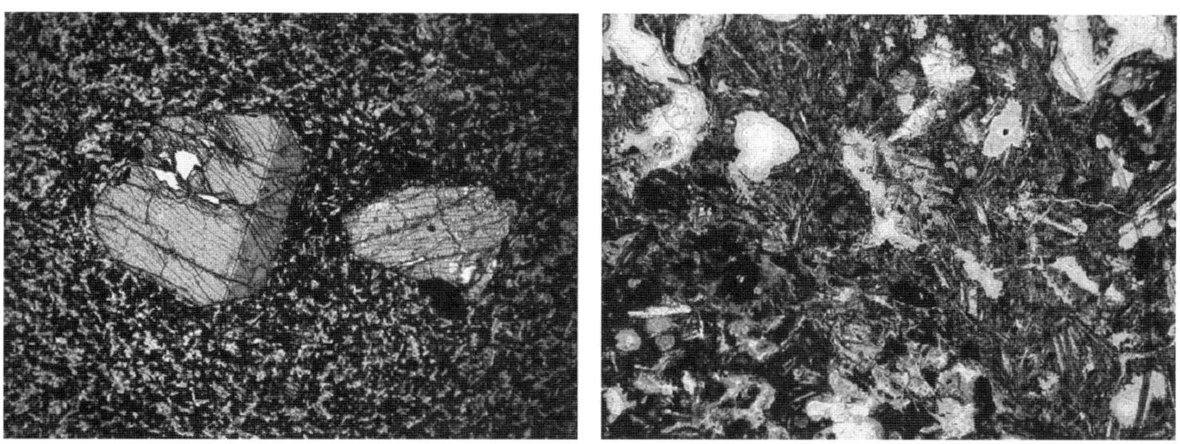

Fig. 9. Thin section of fresh basalt (left) and weathered basalt (right) samples.

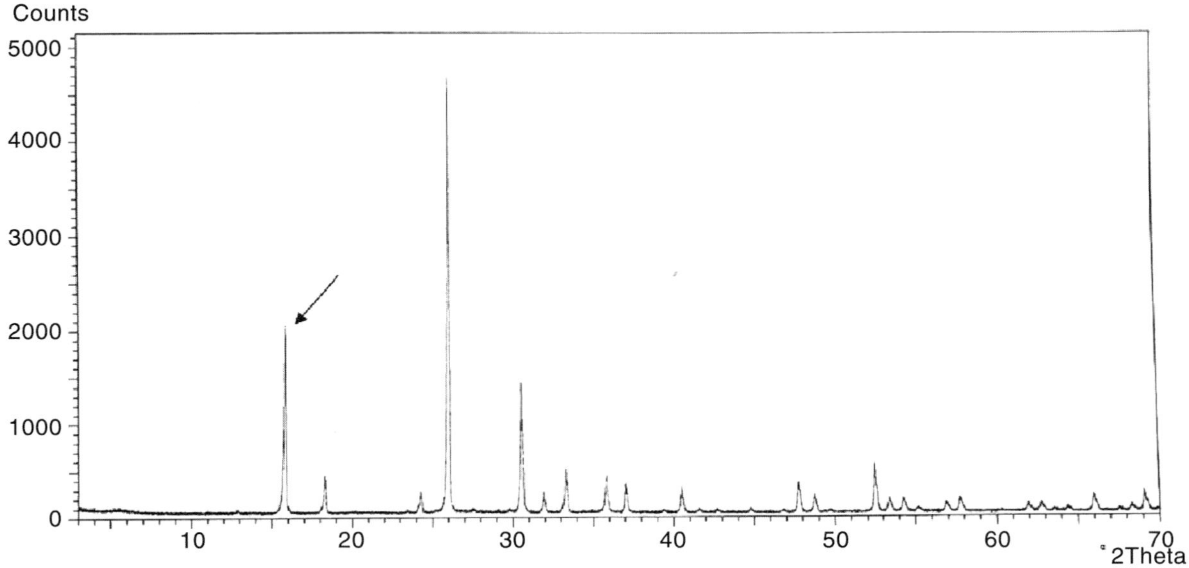

Fig. 10. Diffractogram of vesicular lava sampled in Lalibela. The arrow marks the peak interpreted as montmorillonite (smectite).

crystalline elements reported in a specific database. An important finding is the detection of montmorillonite, which exhibits a spacing of about 15.5 nm; other recognized minerals are analcime, heulandite and traces of calcite. Montmorillonite is a clay mineral of the smectite family. It typically forms microscopic or at least very small platy micaceous crystals. The water content is variable, and when water is absorbed by the crystals they tend to expand to several times their original volume. In general terms, expansion–contraction processes may induce a variation of volume up to 10–20 times the original size but depending on the chemical composition, the expansion can be very high.

Weathering processes and damage analysis

Weathering processes have been analysed in various studies that have recognized distinct main types of rock degradation. According to ICCROM (1978), cracking of basalt is mainly due to water absorption; alternate wetting and drying of the scoriae results in hydration and dehydration as well as crystallization of salts at the evaporation surfaces. In the case of salts such as nitrates and sulphates, their presence results in a significant change of volume, accompanied by generation of pressures capable of disaggregating rock surfaces. In addition, salt solubility produces an expansive reaction under wetting and drying cycles, causing disintegration of volcanic agglomerates and scoriae.

The presence of salts has been studied in a highly weathered vesicular lava sample taken in Biet Abba Libanos. The chemical analysis, undertaken by CNR-Pisa, provided the results reported in Table 3.

The results show the presence of calcium nitrate and traces of Ca–Mg sulphates and NaCl.

Another study, by UNESCO (1995), reported that it was not possible to determine the relative physical impact on

Table 3. *Chemical analysis of salts concentration in Biet Abba Libanos weathered vesicular lava*

	mg	*%*	*mequiv*	*Total mequiv*
Anion				
Cl^-	0.9	0.3	0.025	
SO_4^-	1.3	0.4	0.027	
NO_3^-	13.9	4.6	0.224	
				0.28
Cation				
Ca^+	4.8	1.6	0.240	
Mg^{2+}	0.58	0.19	0.048	
K^+	0.07	0.02	0.002	
Na^+	0.4	0.13	0.017	0.31
Total	21.95	7.3		

Fig. 11. Examples of weathering and induced degradation processes affecting Lalibela churches: (**a**) rainfall mechanical erosion (Biet Medane-Alem); (**b**) surface alteration: pedogenesis (Biet Marqorewos); (**c**) alveolization (Biet Gabriel Rufael); (**d**) alveolization and corrosion (Biet Abba Libanos), the dotted-line arrow shows surface water path; (**e**) crumbling of rock (Biet Golgota); (**f**) efflorescence (Biet Medane-Alem); (**g**) biological colonization (Biet Gyiorgis); the main affected areas are circled.

monuments of each potential deterioration factor because of the heterogeneity of the volcanic rock from which the churches have been carved. That study also affirmed that other potential factors of rock degradation were still unknown.

Weathering caused by hydration and cyclic hot and cold (e.g. thawing and freezing) air temperature leading to easy decomposition of feldspar mineral constituents of the volcanic rocks in Lalibela was recognized as the main potential factor by the ACEL-SAVA study (1997).

The inspection conducted in field surveys in 2004 and 2006 on the 11 churches revealed, apparently, a variety of weathering phenomena acting in the area (Fig. 11). The recognized weathering processes affecting the volcanic rocks are the following: (1) chemical weathering of silica-poor basalt and vesicular lava from hot water circulation, with transformation of some original minerals (mainly plagioclase and Mg-minerals) to montmorillonite; (2) mechanical erosion produced by raindrops; (3) soil evolution (pedogenesis) in the superficial lava layers; (4) degradation (alveolization) of some particular sensitive, and/or critically exposed, basalt strata (UNESCO 1995; Panizza & Piacente 2003); (5) corrosion of the basalt formation as a result of water percolation; (6) biological colonization by lichens and vegetation causing damage to the affected rocks; (7) minor chemical weathering by salts (efflorescence).

Looking in detail at the state of conservation of the churches the following processes can be observed. According to the mineralogical and petrographical analysis described above, the volcanic tuff contains a high percentage of montmorillonite. The excavation of the churches has exposed the deeper tuff layers to atmospheric effects. Weathering and instability of the churches, mainly on the exposed facades and cliff, are caused by diffuse superficial runoff that promotes alveolization and corrosion as a consequence of expansion and contraction of montmorillonite caused by hydration, which is causing crumbling of rock in areas where the concentration of this mineral is higher.

On top of the cliff the alteration of rock and generation of soil (pedogenesis, the rate of which is increased by the hydro-expansive clay mineral) is posing serious problems to the static conservation of the rock-hewn churches located just below the surface topographic level. In some cases the collapse of the roof has already occurred, as in the church of Biet Marqorews.

In such a situation, the combination with other weather-related factors becomes very important because, for instance, the alteration and consequent weakening of rock by pedogenesis is amplifying the erosion caused by the mechanical action of raindrops, which mainly affects roofs. Erosion caused by raindrop impact plays a very important role in the degradation of many roofs, especially where the concentration of montmorillonite is very high (e.g. Biet Medane-Alem).

Only a few cases of efflorescence have been surveyed; it can be probably related either to chemical spreading of guano or to the diffusion of cement alteration products used in past restoration.

In many cases the presence of biological colonization by lichen and other major plants is evident. Water infiltration the from surface is also promoted by the presence of plant roots extending into pre-existing joints. This causes the enlargement of joints and pore pressure increase, reducing the shear strength of the rocks.

Besides weathering processes, the exposed rocks are affected by discontinuities caused either by the erosion of the Amba Aigi Formation at isothermal conditions, with generation of subhorizontal or low-angle joints, or unloading after the excavation of churches that promoted tensile stress along vulnerable structures such as walls, facades and pillars, with formation of medium- to high-angle jointing surfaces.

In some cases, when there is the congruence between the dip of the joint and the dip of the slope, with the dip direction of both being $\pm 15^\circ$ and the friction angle of rock being greater than the slope angle, the occurrence of planar sliding is possible. This has already occurred in Biet Abba Libanos in the 19th century (Delmonaco *et al.* 2005) and it might occur in the future in the same place or elsewhere.

The presence of some crack patterns is also an important factor for water infiltration and, probably because of the hydration of montmorillonite, for damaging the interior of churches, including statues and reliefs. Again, the contemporary action of many weathering factors is evident. Similarly, the weathering of montmorillonite is weakening the mechanical strength of the original rock and at present the original stability of walls has not yet been ensured. The structural vulnerability of the monuments is aggravated by the original poor design of some structures (e.g. overheavy roof), which amplifies some of the degradation effects. A typical example is in Biet Emanuel: the initial engineering design was probably not appropriate, as the thickness of the wall was underestimated and is incapable of supporting the heavy weight of the roof. Over time, the weathering of montmorillonite is producing the decay of the mechanical properties of rock leading to the present critical stage.

Analysis of present condition of the churches

The churches were carved in the basalt unit of the Amba Aiba Formation, some 800 years ago. Since then, they have been exposed to wind, rain and thermal changes as well as to human activities. This has resulted in a severe degradation of the monuments, most of which are considered to be in a critical condition. Several causes of physical and chemical decay in the hewn rock, and their interrelations, have been identified, such as the characteristics of the basalt and vesicular lava, evaporation of salts, expansion of montmorillonite, decomposition of feldspars, the heavy weight of hewn roofs compared with the lighter bearing capacity of the walls, dilatation of the large monolithic structures, mechanical erosion produced by raindrops, basalt alteration as a result of weather and climatic factors, and discontinuities in the rock formation with direction and dip capable of generating sliding.

The detailed investigation of the monuments revealed that some churches are apparently affected by one dominant mechanism whereas most of them are affected by a combination of factors leading to the damage of structures. A brief description of the major problems for static stability is given below. Also, for all the churches a discontinuity pattern has been carefully reconstructed, updating the work of UNESCO (1995). A structural damage analysis has been undertaken, classifying all the churches according to the presence, type and number of cracks (Burland *et al.* 1977), from grade 1 (very slight) to grade 5 (very severe), to identify the highest priority of intervention with respect to their static conditions. A scheme of the damage conditions of Lalibela churches is given in Figure 12.

Biet Medane-Alem (grade 4)

This is the largest church in the area (Fig. 13). The roof is completely altered, as a consequence of mechanical erosion by rainfall, degradation and poor restoration with epoxide resins, which were removed in 1965–1968. The two lateral arcades inside the church exhibit a continuous long crack throughout the structure. In the NE corner water percolation and discontinuities occur. The external facades show a wide set of discontinuities, mainly concentrated at the windows and at the corners. Specifically, the NW and SW corners are at the point of collapse. The major reason for the static instability can be identified in the high overload caused by the great thickness of the roof (about 6 m).

The general conditions were aggravated in the past, before the installation of the shelter, by water percolation from the roof. In particular, in the NW and NE corners, the situation was aggravated by water running over the roof in these lower parts, without any system of discharge. The existing shelter has clearly avoided major and further damage. As far as alteration is concerned, the northern face is the most deteriorated and the southern face seems to be least affected.

Biet Maryam (grade 2)

This has been largely restored to protect it from weathering processes, which are apparent on the natural slope close to the church. At present, the small entrance on the west and the two entrances on the north and south facades are affected by degradation. In addition, the existing shelter provides only partial protection from rainfall. The general static condition is relatively good. There is no evidence of major instability; the NE and SE corners probably collapsed in the past and have been reconstructed. The south façade

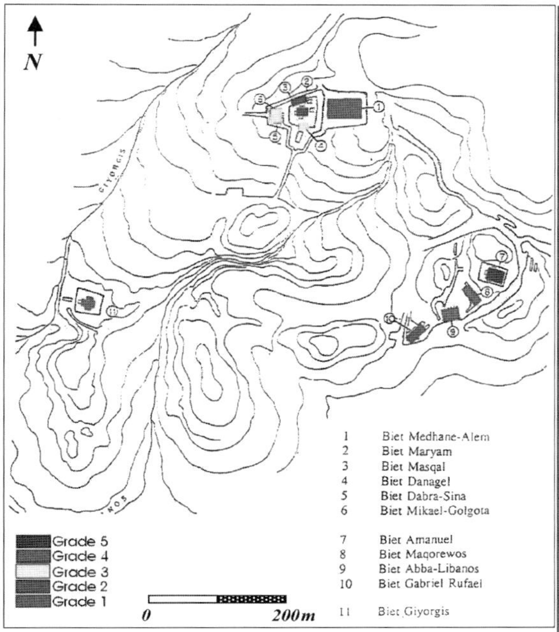

Degree of damage	Description of typical damage #	Approximate crack width (mm)
1. Very slight	Fine cracks which can easily be treated during normal decoration. Perhaps isolated slight fracture in building. Cracks in external brickwork visible on close inspection.	< 1 *
2. Slight	Cracks easily filled. Re-decoration probably required. Several slight fractures showing inside of building. Cracks are visible externally and some re-pointing may be required externally to ensure watertightness. Doors and windows may stick slightly.	< 5 *
3. Moderate	The cracks require some opening up and can be patched by a mason. Recurrent cracks can be masked by suitable linings. Restoring of external brickwork and possibly a small amount of brickwork to be replaced. Doors and windows sticking. Service pipes may fracture. Watertightness often impaired.	5 to 15* or a number of cracks > 3
4. Severe	Extensive repair work involving breaking-out and replacing sections of walls, especially over doors and windows. Windows and door frames distorted, floors sloping noticeably. Walls leaning or bulging noticeably, some loss of bearing in beams. Service pipes disrupted.	15 to 25* but also depends on number of cracks
5. Very severe	This requires a major repair job involving partial or complete rebuilding. Beams lose bearing, walls lean badly and require shoring. Windows broken with distortion. Danger of instability.	Usually > 25 * but depends on number of cracks.

#In assessing the degree of damage, account must be taken of its location in the building or structure.

* Crack width is only one aspect of damage and should not be used on its own as a direct measure of it.

Fig. 12. State of conservation of Lalibela churches and degree of damage with particular reference to ease of repair of plaster and brickwork or masonry (Burland *et al.* 1977).

Fig. 13. Geological profile of the church and cliff (above) of Biet Medane-Alem, and discontinuity pattern on the church façade (below) (Bianchi Barrivera 1963.)

seems to have deteriorated less whereas the north façade has been significantly affected by alteration. Various parts of the monument reveal losses of material through crumbling and granular disintegration (UNESCO 1995).

Biet Mascal (grade 2)

The church is hewn in very good material. Only one important discontinuity (about 2 mm wide) has been detected, extending from the top of the entrance to the middle of the structure. Minor discontinuities are also present. At the middle level of the church a horizontal discontinuity, which originated from the cooling of the original lava flow, is evident. No major intervention is required. Stabilization of the crack can be easily achieved with a simple emergency plan. Alteration phenomena were detected and repaired in the past.

Biet Danagel (grade 3)

The church exhibits a major discontinuity crossing the entire structure and dipping toward the facade. Kinematic analysis confirmed that the facade is almost at the limit condition for sliding (Fig. 14), according to the value of the friction angle determined from a tilt test on site material. However, the crack dips slightly into the ground, and this is contributing to stability. The joint roughness coefficient (JRC) has been estimated with Burton equipment as about 12. Two other large syngenetic discontinuities cross the church horizontally at 1 m from the floor and at the top. The entire structure is probably in a condition to survive for some time but a simple intervention with stainless steel anchors and nails can be foreseen, especially to stabilize the discontinuity crossing the facade.

Biet Dabra Sina and Biet Mikael-Golgota (grade 3)

A major crack, affecting the statue of St. Peter, and a similar one located a few metres below, cutting a back statue, require attention. The same discontinuity is evident on the cliff in front of the main entrance (dip direction 290°; dip 30°). The St. Peter fracture was monitored by Angelini in 1966 (IFM 1967) and at present seems to have enlarged by about 1 mm. In addition, the presence of an important crack,

Fig. 14. Kinematic analysis (left) and external view of the major discontinuity (right).

5 mm wide and clearly visible for more than 1 m in the middle of Biet Dabra Sina church, in the vault, requires attention. The north wall of Biet Mikael-Golgota is completely weathered and statues are irreparably damaged. Water flowing from the upper level should be avoided to prevent surface basalt corrosion.

Biet Amanuel (grade 5)

This is probably the monument in the worst condition. The weight of the roof, the bearing capacity of walls and weathering processes are acting together negatively for future conservation. The pattern of discontinuities (Fig. 15) clearly shows the number of cracks affecting the facades and the weathering processes. A long fracture, about 10 cm wide, crosses the entire church. The vaults between the ground floor and the mezzanine, on the right arcade, exhibit a network of cracks, and the first vault from the entrance is badly disconnected. Almost all the arches have cracks in the centre, in both the longitudinal and transverse direction.

A comparison of this pattern with that at Biet Medhan Alem clearly shows the dramatic situation of the church. The main reason for the structural situation is the static load of the roof (which is 3 m thick). The axonometry designed by Bianchi Barriviera (1963) clearly depicts the thickness of the roof and the walls below.

All the facades are affected by weathering, with the exception of the western one, which was restored by Angelini in 1966 (IFM 1967). Efflorescence was also observed, probably connected with weathering of concrete after modern restoration (UNESCO 1995). A 3D mathematical model reconstructing the static conditions of the church was developed using SAP2000®. The model has been implemented with elasticity modulus of 17.2×10^6 kN m^{-2}, density of 16.7 kN m^{-3} and Poisson modulus of 0.2. All the horizontal, vertical and shear components of stress have been calculated (Fig. 15). The model reveals that the highest concentration of stress is located in the central pillar of the church, where the static load reaches a value of about 540 kN m^{-2} (Fig. 16). A simple comparison with the UCS of intact rock shows that such a static load is about 10% of the ultimate rock strength (5500–7900 kN m^{-2}). According to the model, the pattern of discontinuities can be referred to the general stress affecting the church. For instance, the south facade shows a concentration of tensile stress along the horizontal direction; this is in good agreement with the pattern of discontinuities surveyed in the field, specifically for joints dipping towards the SW and SE corners (Fig. 16). The above consideration presents some limitation: in general, discontinuities are propagating downward where they become compressive but, as the adopted model is in the linear elastic domain, it provides a simple indication of areas where tensional stress is triggered. This means that the evolution of tensile stress (e.g. crack development) cannot be directly extrapolated by graphic outputs but needs to be compared with direct field investigation. Also, in this case, the tensile stress is not significant (*c.* 70 kN m^{-2}) but coupled with weak and weathered portions of the rock could have caused the observed damage.

Coupling the static mathematical model with discontinuity patterns directly surveyed in the field, the critical condition of the church is clear.

Biet Marqorewos (Mercurios) (grade 4)

The edifice is composed of chambers supported by pillars opening onto a longitudinal courtyard of irregular shape. The roof of the church had partially collapsed and recently a further small portion fell down. The reason lies in the alteration of the basalt (pedogenesis) because of the limited thickness of the roof (Fig. 17); it was found that the UCS measured in the field reaches, at this site, the minimum values typical of wet material. The remaining roof is completely weathered and water is flowing inside, especially in the western wall. Some of the external columns show weathering phenomena in the lower part, and an internal pillar is affected by a wide subhorizontal crack.

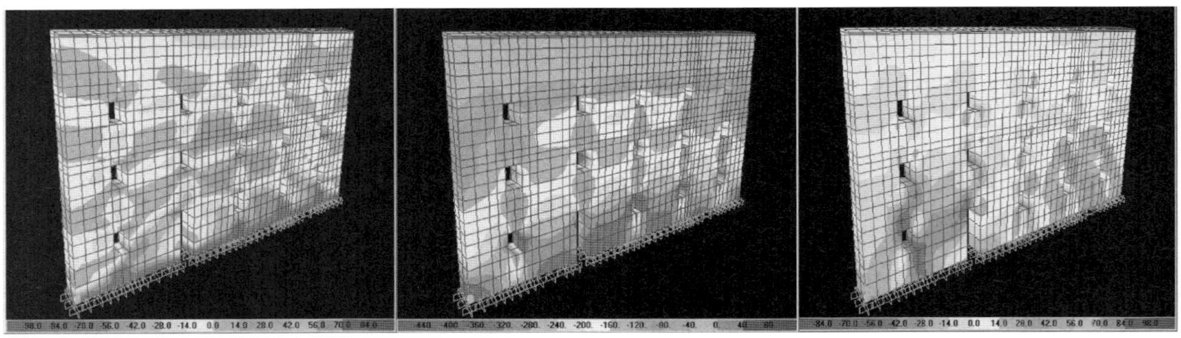

Fig. 15. Biet Amanuel southern façade: comparison between horizontal, vertical and shear components of stress. The values (kN m^{-2}) are shown in the coloured legend.

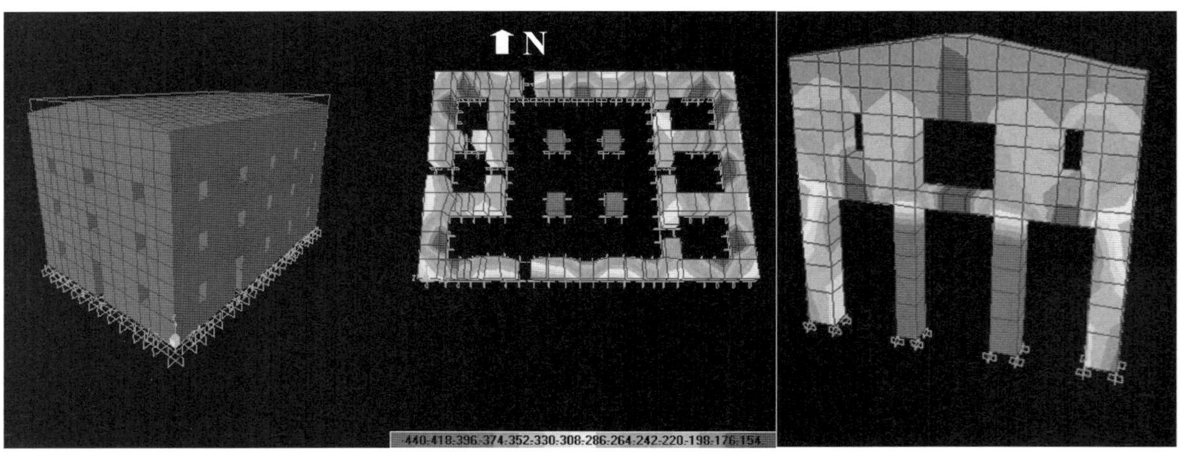

Fig. 16. Evaluation of static conditions of Biet Amanuel. The most critical conditions (in red) are in the central pillars, under a static load of 420–350 kPa, about 10% of the uniaxial compressive strength. (Courtesy of P. Clemente and G. Buffarini.).

Fig. 17. Section of Biet Marqorewos (Bianchi Barriviera 1963), with the central part collapsed, and photograph showing the present-day situation.

Biet Abba-Libanos (grade 4)

This is a monolithic church anchored to the rock in which it was carved. Because of its excavation within the rock face (the height of the circular tunnel is 7 m) and its small dimensions, it imitates a cave church. Two major phenomena affect the monument: alveolization in the lower part of the edifice and the presence of a subhorizontal joint, dipping outward, that affects the walls, potentially promoting wall sliding (Fig. 18). Alveolization is producing a fine sand, a sample of which was taken for comparison (by X-ray analysis and scanning electron microscope) with fresh material. According to UNESCO (1995) the sand produced by crumbling of the vesicular lava has been considerably enriched with sodium and calcium nitrates [$NaNO_3$ and $Ca(NO_3)_2$]. It can therefore be assumed initially that the crystallization of soluble salts was the cause of the crumbling and alveolization of the vesicular lava. Water is probably the driving agent. Also interesting is the weathering of the indoor columns, which is probably connected with the flow of fresh air from the entrance.

A serious concern derives from the relationship with local discontinuities. The frontal facade partially collapsed in the past and has been reconstructed. The reason for the collapse was sliding along the detected discontinuity. Also, now, according to the shear strength of the discontinuity based on tilt test, the kinematic analysis (plot in Fig. 18) clearly shows that all the conditions leading to a possible slide are satisfied. To aggravate the situation, two lateral walls prone to sliding are connected by an iron bar (tie rod). Thus the most unstable sides were tied together, when a correct intervention would have been to fix the unstable walls to stable ground (the direction of the bar should be perpendicular to the present one). The JRC of the main discontinuity has been estimated with Burton equipment as 14. Crack gauges installed by Angelini in 1966 (IFM 1967) do not show any displacement.

A preliminary estimation of the strength conditions of the rock-forming material, using the techniques of rock slope stability, was conducted (Delmonaco *et al.* 2005). Planar sliding was evaluated according to the analytical solution of Hoek & Bray (1981) for the plane failure mode in rock

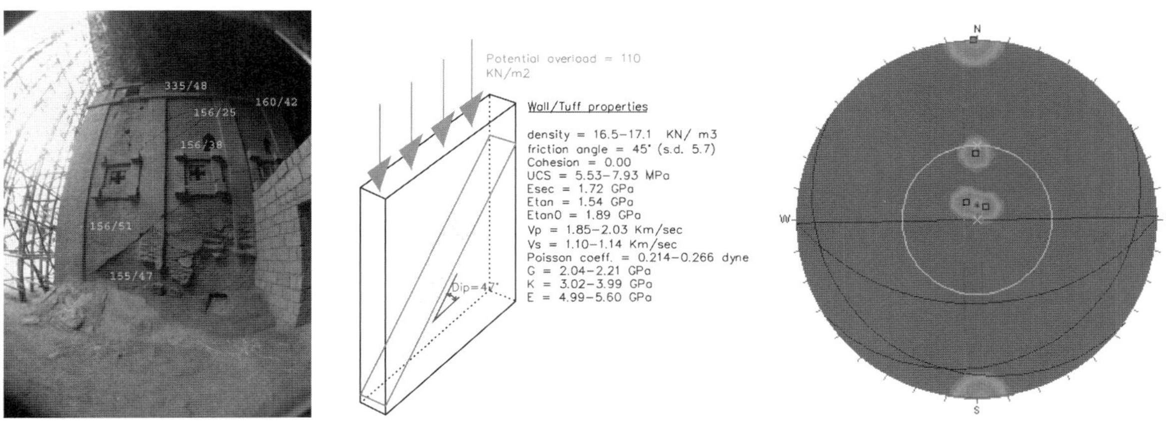

Fig. 18. Section (left) and lateral prospect of Biet Abba Libanos (middle); the plot on the right shows the dipping discontinuities that kinematically lead to possible sliding. A value of $F_s = 1.03$ has been estimated without overload (roof structure).

slopes. The model assumes the presence of a crack, as surveyed in the field, with null cohesion and a friction angle of 45°. Input shear strength parameters were obtained by *in situ* and laboratory tests (Table 2 and Figs 7 & 8). The upper part of the cliff, above the church, was not considered in the first iteration, as the church is detached from the ceiling. In following simulations the rock-roof was considered in the model as an overload. The results of the calculation without the roof overload show that the safety factor is 1.03; that is, around the limit of equilibrium. When considering the overload, the safety factor drops abruptly, confirming that the load of the cliff is not affecting the static load of Biet Abba Libanos. The reason is probably related to the lateral 'arch effect', caused by the excavation of the tunnel surrounding the church. Because of this geometry, the upper part of the cliff is statically sustained by the lateral part of the cliff and not by the church itself. This analysis undertaken in Biet Abba Libanos Church can be considered as a first attempt to apply the slope stability concept to a man-made structure carved in a natural rock-slope.

Biet Gabriel-Rufael (grade 3)

This church is a very impressive structure, because of its dimensions and distinct architectural style. In this area weathering processes are particularly developed, especially in the two courtyards that surround the edifice to the south and north. Especially in the north courtyard the presence of water is indicated by a deep underground cistern (Fig. 19). This cistern provided water even during the Sahel drought of 1980. One large horizontal crack crosses the church. Other fissures are visible inside; the largest has a displacement of 10 mm. Also, the roof exhibits a vertical crack, as does one of the internal pillars. Water is infiltrating from the roof and dampness is widespread. Also, the mechanical property of basalt inside the church seems to reflect the general wetting of materials (UCS = 30 MPa inside; UCS = 34 MPa outside).

Biet Giyorgis (grade 1)

The structure is in generally good condition. Small minor cracks affect the external walls. The major problem identified is the weathering on the south wall as a consequence of water falling on the roof and being channelled into a short gutter and thus penetrating the church.

Conclusions

The monumental complex of rock-hewn Orthodox churches of Lalibela shows damage of varying degrees of severity as a result of various types of weathering processes acting in the area. The main weathering forms have been detected and described after visual analysis undertaken during field surveys in 2004 and 2006.

The main outcomes of this research suggest that water represents the main causative factor of damage for the highly vulnerable rock materials and structures, in terms of direct rainfall, soil infiltration, capillarity and widespread humidity. An important role in cliff or rock wall decay is also played by the presence of syngenetic geological fractures and superimposed unloading and structural fractures connected with the excavation of the churches.

Laboratory tests have shown the widespread presence of montmorillonite (smectite) clay, especially in vesicular lava layers, in which most of the churches have been excavated and which is extremely sensitive to water. The process of expansion and contraction of rock materials promoted by weather-related factors, such as heavy rainfall, high humidity and seasonal temperature variation, has resulted in a progressive degradation of the mechanical

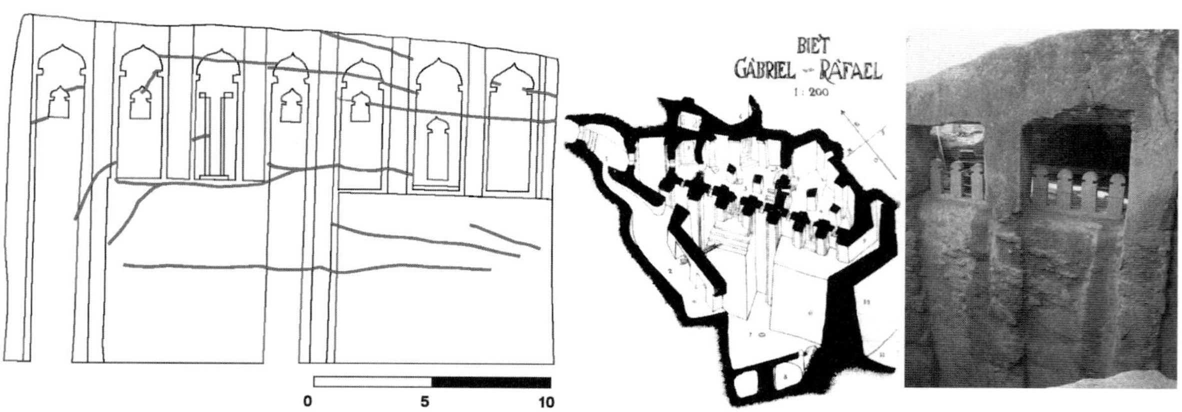

Fig. 19. Biet Gabriel-Rufael: discontinuity pattern (left) from UNESCO (1995), 3D sketch of the north courtyard (Bianchi Barrivera 1963) with the deep cistern (centre), and view of the weathering processes on the cliff and wall (right).

properties of the volcanic rocks. An engineering-geological analysis, *in situ* and from laboratory tests, has verified the poor geotechnical properties of some areas where volcanic materials are affected by wetting, and intense alteration of rocks that results in high concentrations of montmorillonite. These phenomena are affecting even the mechanical properties of rock, probably causing a degradation of geomechanical properties and the formation of a secondary group of discontinuities since the excavation of the Churches. This process is at present affecting the further stability of some churches, such as Biet Ammanuel. The presence of a syngenetic pattern of discontinuities is causing a rare phenomenon of slope instability of a man-made structure, as in the case of Biet Abba-Libanos.

Besides geological and structural aspects, the intense use of the site for religious reasons and the lack of long-term structural measures to mitigate the effects of uncontrolled water circulation at the site are the main factors of vulnerability that affect the churches, as found by the field analysis of damage. The recovery of the original drainage system, for instance, is fundamental to mitigate the negative impacts of water on the monuments.

Temporary structures have been constructed to preserve the churches from degradation: shelters for the protection of roofs from direct erosion by rainfall, small channels to remove superficial water from outer walls, and wood scaffolds to prevent unstable structures from collapse. The design of sustainable long-term rehabilitation strategies requires deeper studies on weathering processes acting in the area to address better solutions to preserve the churches from structural and slope-structural problems and to implement restoration of damaged artefacts.

Acknowledgements. The work has been carried out under the auspices of UNESCO, World Heritage Centre, to whom authors the express gratitude for involvement in the research activities for the conservation of the World Heritage site of Lalibela. Special thanks go to N. Al-Hassan of UNESCO for proving helpful support. J. Haile Mariam, General Director of the Ethiopian Ministry of Culture and Tourism, is gratefully acknowledged for local support and help in Lalibela. The authors would like to express their gratitude to P. Manetti and S. Battaglia (CNR), F. G. Guidi (ENEA) and G. Trojsi (University of Naples) for their contribution to X-ray and petrographical analyses and interpretation. A special acknowledgment goes to P. Clemente (ENEA) and G. Buffarini (ENEA) for their contribution to the evaluation of static conditions of Biet Amanuel. P. Laureano is gratefully acknowledged by the authors for providing important inputs for this research and its further development.

References

ACEL-SAVA 1997. *Geology and engineering geology of Lalibela Churches* (unpublished).

BIANCHI BARRIVIERA, L. 1963. *Le chiese in roccia di Lalibela e di altri luoghi del Lasta*. Rassegna di Studi Etiopici.

BURLAND, J. B., BROMS, B. B. & DE MELLO, V. F. B. 1977. Behaviour of foundations and structures. *In*: *Proceedings of 9th International Conference on Soil Mechanics and Foundation Engineering, Tokyo*. Japanese Society of Soil Mechanics and Foundation Engineering, Tokyo.

DELMONACO, G., MARGOTTINI, C. & SPIZZICHINO, D. 2005. Slope–structure stability modeling for the rock hewn church of Biet Abba Libanos in Lalibela (Ethiopia): preliminary results. *In*: SASSA, K., FUKUOKA, H., WANG, F. & WANG, G. (eds) *Landslides: Risk Analysis and Sustainable Disaster Management*. Springer, Berlin, 213–218.

ICCROM 1978. *Proceedings of the Symposia of International Experts on Stone Conservation. Lalibela* (unpublished).

IFM 1967. *Lalibela—Phase 1. Adventure in restoration*. International Fund for Monuments, New York.

ISRM 1978. International Society of Rock Mechanics, Commission on Standardisation of Laboratory and Field Tests. Suggested methods for the quantitative description of discontinuities in rock masses. *International Journal of Rock Mechanics and Mining Science, and Geomechanics Abstracts*, **15**, 319–368.

ISRM 1981. International Society of Rock Mechanics, Commission on the Classification of Rocks and Rock Masses. Basic geotechnical description of rock masses. *International Journal of Rock Mechanics and Mining Science, and Geomechanics Abstracts*, **18**, 85–110.

KAZMIN, V. 1973. *Geological map of Ethiopia*. Geological Survey of Ethiopia, Addis Ababa.

MERLA, G., ABBATE, E. *ET AL*. 1979. *A geological map of Ethiopia and Somalia (1973) 1:2 000 000 and comment with a map of major landforms*. Consiglio Nazionale delle Ricerche, Rome.

MOHR, P. & ZANETTIN, B. 1988. The Ethiopian flood basalt province. *In*: MACDOUGALL, J. D. (ed.) *Continental Flood Basalt*. Kluwer, Dordrecht, 63–100.

PANIZZA, M. & PIACENTE, S. 2003. *Geomorfologia culturale*. Pitagora, Bologna.

TEPRIN 2002. *Shelters for five churches in Lalibela project*. Tender dossier (unpublished).

UNESCO 1995. *Preservation of the Rock-hewn Churches in Lalibela. Vols I, II and III*. UNESCO Paris (unpublished).

The Hunters Crossing weathered rock slide, Haywood County, North Carolina, USA

Rebecca S. Latham[1], Richard M. Wooten[1]*, Edward D. Billington[2], Kenneth A. Gillon[1], Anne C. Witt[1], Jennifer B. Bauer[1], Stephen J. Fuemmeler[1], Thomas J. Douglas[1], David Kinner[3] & Cheryl Waters-Tormey[3]

[1] North Carolina Geological Survey, 2090 US Highway 70, Swannanoa, NC 28778, USA
[2] Schnabel Engineering, 11 A Oak Branch Drive, Greensboro, NC 27407, USA
[3] Department of Geosciences and Natural Resources Management, Western Carolina University, Cullowhee, NC 28723, USA

**Corresponding author (e-mail: Rick.Wooten@ncdenr.gov)*

Abstract: The Hunters Crossing landslide is a slow-moving, weathered rock slide affecting a small community of condominiums in the town of Waynesville in the Blue Ridge Mountains of Haywood County, North Carolina. In November 2005, studies were begun to assess the characteristics of this landslide and the potential for further movement and damage to structures. Work included drilling several boreholes, performing seismic velocity surveys, and surveying benchmarks among other investigations. Data indicate that the potential failure surface is located no more than 11 m below the ground surface, possibly at the contact between saprolite and partially weathered rock. However, inclinometers installed at two locations on the slope have not detected enough movement to corroborate that assessment. Studies continue at this site to determine the location of the failure surface, to identify the mechanisms that accelerate movement, and to relate these findings to a broader understanding of weathered rock slides elsewhere in the southeastern USA.

In 2005, the Hurricane Recovery Act authorized the North Carolina Geological Survey (NCGS) to produce landslide hazard maps of 19 counties in the western part of North Carolina. Part of this work included developing a landslide inventory that now contains over 3100 recorded landslides that have occurred within the past 130 years. Of these, 64% (2003 entries) are classified as debris flows (as defined by Cruden & Varnes 1996) whereas only 2.8% (87 entries) are weathered rock slides. It is important to understand the mechanisms, causes, and triggers that lead to weathered rock slides because they typically affect larger tracts of land and move at slower rates over longer time frames than debris flows, allowing them to cause considerable damage to infrastructure throughout western North Carolina and particularly in Haywood County. Damage to homes is a particular hardship because homeowner's insurance policies typically do not cover losses from landslides as is the case at Hunters Crossing.

Landslides involving rock of varying degrees of weathering and decomposition present unusual challenges for classification and prediction. Because material contained in these types of landslides has properties of both soil and rock with relict mineralogy and bedrock structure, neither the term 'debris' nor 'rock', as used in accordance with Cruden & Varnes (1996), unambiguously describes the displaced material prior to movement. We use the term 'weathered rock' in this paper as originally described by the Unified Rock Classification System (URCS; Williamson 1984) and now standardized in American Society for Testing and Materials (ASTM) method D5878 (ASTM 2008), and apply it to the landslide material described in this paper. Using the URCS, weathered rock is defined as rock in the partially decomposed state or the completely decomposed state. Completely decomposed rock can be remoulded with hand pressure to finer than gravel-sized material (passes a No. 4 sieve), but still preserves some relict mineralogy and bedrock structures. Partially decomposed rock can be remoulded with hand pressure but only into particles that are gravel sized or greater, and also preserves some relict mineralogy and structures. Williamson suggested classifying both partially and completely decomposed state material using the URCS along with the Unified Soil Classification System.

Similarly, according to Sowers (1954), weathered rock can be classified into two categories: the intermediate zone, analogous to rock in the completely decomposed state; and the partially weathered zone, analogous to rock in the partially decomposed state. He found that weathered rock in the intermediate zone is typified by soil composed of micaceous sandy silt or silty sand. Sowers (1954) designated the partially weathered zone as a transition zone characterized by soil similar to the intermediate zone, but with more gravel and layers of hard bedrock.

From: Calcaterra, D. & Parise, M. (eds) *Weathering as a Predisposing Factor to Slope Movements.* Geological Society, London, Engineering Geology Special Publications, **23**, 149–165.
DOI: 10.1144/EGSP23.9 0267-9914/10/$15.00

Lambe (1996) provided another scheme for classifying material in the weathering profile based on Deere & Patton (1971). The C Horizon of the residual soil, often referred to as saprolite, is typically micaceous and composed of silty to sandy soil with few corestones. According to Lambe, this type of material retains many of the relict structures but lacks the bond strength of the parent rock. In the weathered rock zone of the weathering profile, he described transition material and partly weathered rock. Transition material can resemble rock or soil, is composed of 10–90% corestones, and typically has higher permeability than the overlying saprolite. Partly weathered rock is soft to hard rock with some stained joints and alteration of feldspars and mica minerals. For the purpose of this paper, Lambe's description of saprolite is synonymous with Williamson's completely decomposed rock and Sowers' intermediate zone; Lambe's partly weathered rock is considered the same as Williamson's partially decomposed rock and Sowers' partially weathered zone. The terms completely decomposed rock and saprolite are used interchangeably here.

Background

There are few published reports on weathered rock slides in the Blue Ridge of western North Carolina. Sams & Gardner (1974) discussed landslides in weathered rock that developed in mica-bearing, foliated metamorphic rock during the construction of Interstate 26 in nearby Polk County, North Carolina. The slope failures in Polk County occurred on the south slope of Miller Mountain, which is characterized by gneisses of varying composition overlain by the partially weathered and intermediate zones, and mantled with colluvium (material transported downslope by creep or other mass-wasting processes). Curved and leaning trees, along with hummocky topography, indicate that past slope movements had taken place here. The degree and depth of weathering of the rock was largely controlled by the composition of the bedrock and the intensity and abundance of discontinuities, creating a highly irregular contact between the partially and/or completely decomposed bedrock (intermediate and partially weathered zones) and fresh bedrock. The axial planes of gentle folds and steeply dipping, irregular joints in the mica-bearing units probably allowed water to infiltrate and decompose the bedrock at faster rates than the surrounding rock units of varying composition. Sams & Gardner also noted Sowers' (1954) reference to the importance of the mica content of the rock in the intermediate zone, something that will be discussed here in reference to the material found at the Hunters Crossing slide.

Lambe & Riad (1990) also discussed landslides in partially weathered rock and saprolite in an attempt to quantify the shear strength properties of these types of geological materials. One of the three slides that they discussed is a weathered rock slide 7.4 km WSW of Hunters Crossing in Haywood County. Movement of this failure, named the Balsam Gap II landslide, started in autumn 1980. The slide had an estimated volume of about 15 300 m^3 of material, and a basal rupture surface near the colluvium–saprolite contact at an approximate depth of 4.7 m. The parent bedrock here was classified as coarse granitic gneiss interlayered with mica schist (Lambe & Riad 1990). Several similarities exist between the Balsam Gap II and Hunters Crossing landslides.

Site location and geological setting

The Hunters Crossing development is located in Haywood County, North Carolina in the Blue Ridge Mountains about 36 km from the Tennessee border (Fig. 1). This development is a small community of three, two-unit condominiums constructed between 1991 and 1998 within the extraterritorial jurisdiction of the town of Waynesville. The homes are positioned on a 20–40° slope at an average elevation of 925 m. The development is located upslope of several houses roughly 50–60 years in age, and constructed in, or near, the narrow floodplain of Browning Branch along Lickstone Road (Fig. 2).

The Hunters Crossing slide is located within the mountainous southern Blue Ridge physiographic province, along the SE edge of the NE–SW-trending Waynesville–Canton intermontane basin (Hack 1982). The mountainous ridges bounding this basin have altitudes that commonly exceed 1500 m, and a relief in excess of 600 m. The Hunters Crossing landslide occurs at the toe of a west-facing, convex mountain slope (crest elevation of 1233 m). Browning Branch, a third-order stream, delineates the toe of the active slide. Aerial photography from 1953 reveals that unusually steep toe slopes existed prior to the construction of all but one of the older houses along the toe of the slide on Lickstone Road. It is possible that this oversteepened toe could be the result of undercutting of the base of the slope by Browning Branch during an associated episode of flooding; however, there is no evidence at this time to suggest that this was the case, indicating that the slope may have been oversteepened by human activity. Previously active landslide scarps exist within the bounding lateral scarps of the current slide (Fig. 2).

The site is located in the northernmost portions of the Late Proterozoic Cartoogechaye terrane, a Central Blue Ridge tectonic unit characterized by Hatcher *et al.* (2004) as consisting of metasandstone, pelitic schist, mafic and ultramafic rocks, and minor Grenville age basement gneisses. Rocks of the Cartoogechaye terrane were metamorphosed during the Ordovician to assemblages as high as granulite facies (Hatcher *et al.* 2005). This terrane lies in migmatitic, biotite gneiss interlayered with biotite–garnet gneiss and amphibolite of uncertain stratigraphic position (North Carolina Geological Survey; NCGS 1985). Lithological units underlying the Hunters Crossing landslide are generally the same as those described by Lambe & Riad (1990) at the Balsam Gap II slide. Road cuts at Hunters Crossing

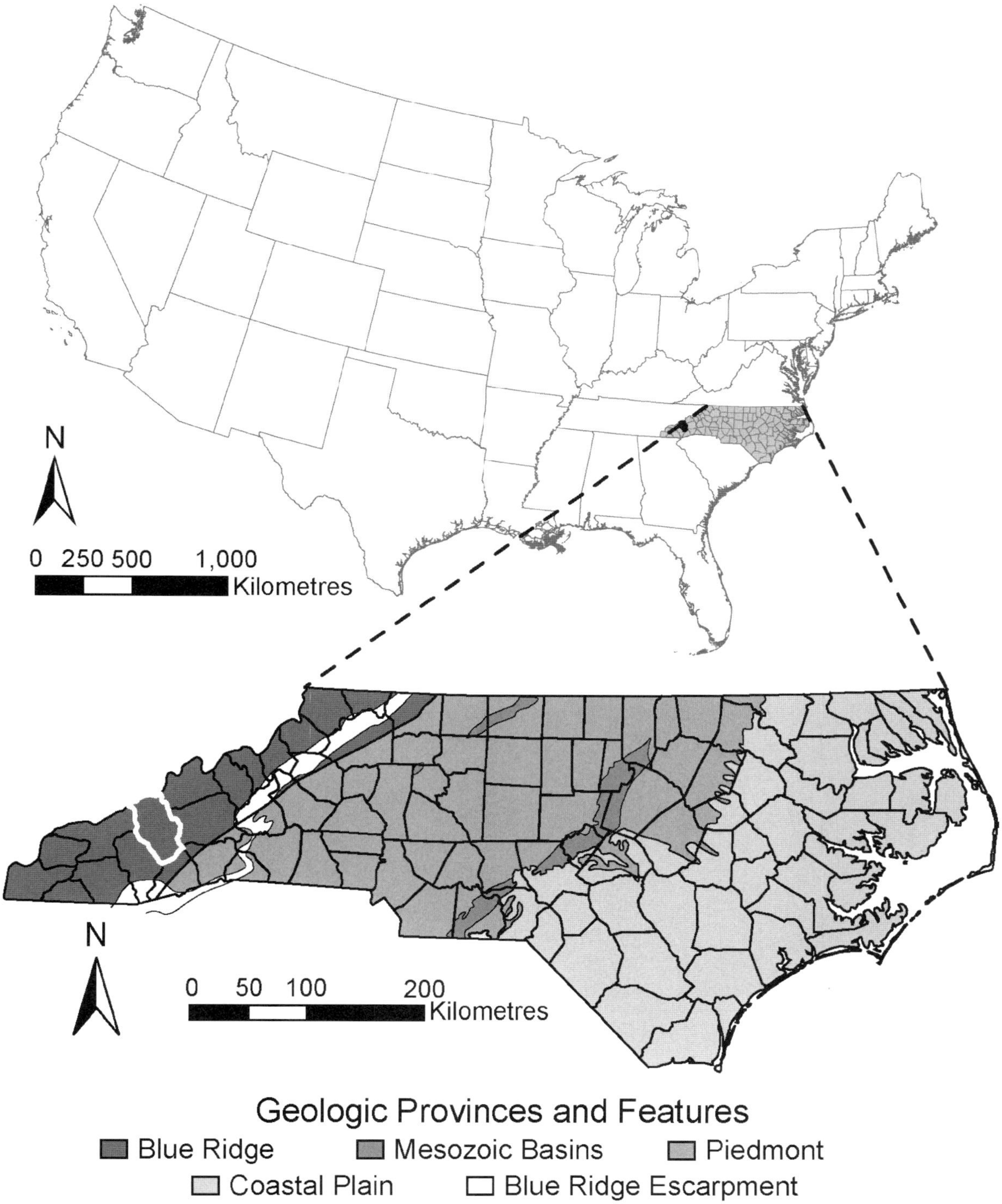

Fig. 1. Location and geological provinces of North Carolina, USA, and the location of Haywood County (outlined in white in lower figure).

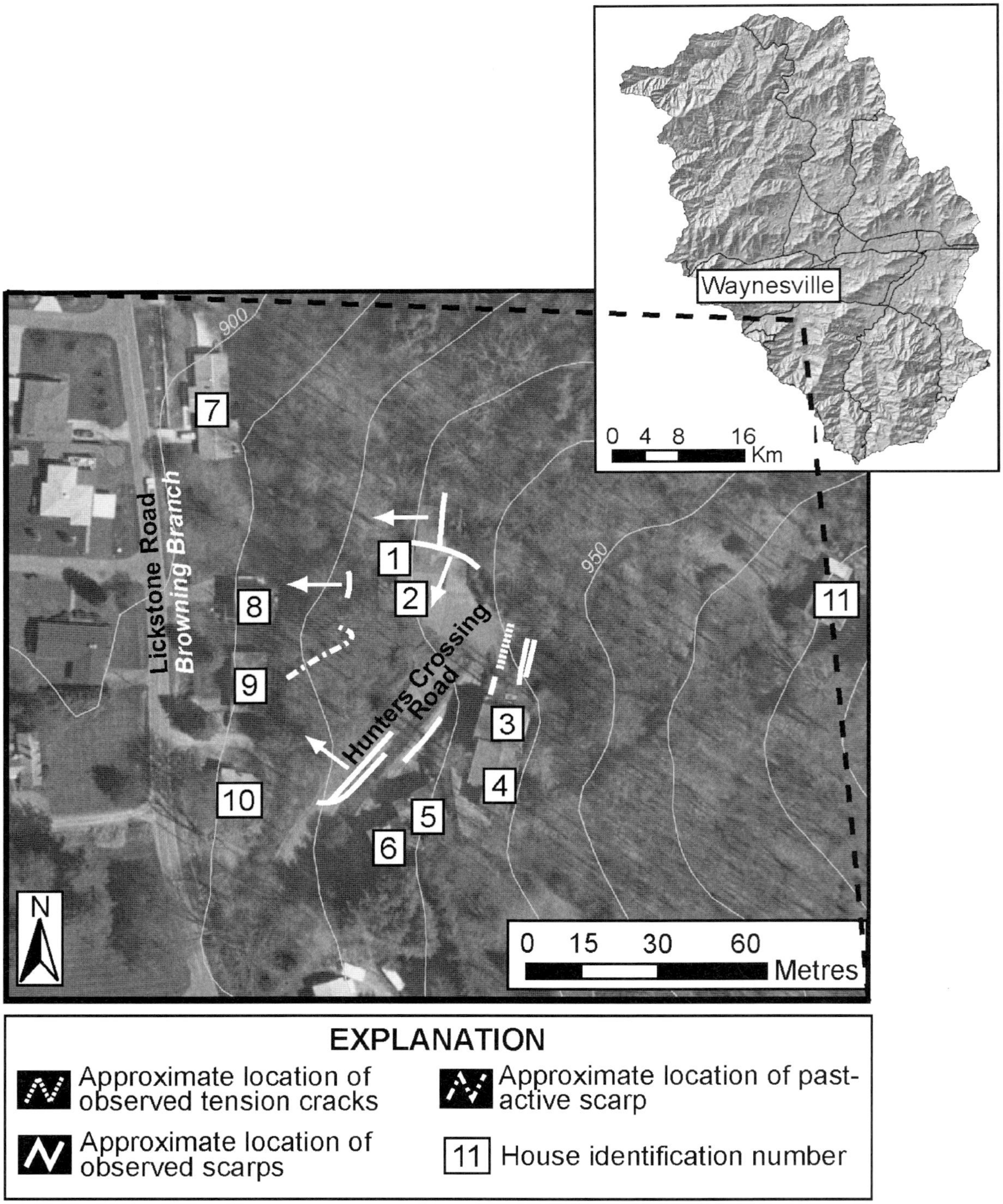

Fig. 2. Aerial photograph (2004) of the Hunters Crossing development. House numbers 1–6 are condominiums, and numbers 7–11 are single-family houses. Arrows indicate direction of relative ground movement. Photograph courtesy of Haywood County. Inset is 6 m digital elevation model shaded relief map of Haywood County.

expose completely decomposed rock composed of migmatitic metagreywacke and schist containing minor pods of amphibolite. Outcrops of fresh bedrock near house number 11 (arbitrary numbers have been assigned to all residences discussed in this paper; see Fig. 2) above the slide area are medium- to coarse-grained, well-foliated, garnet–biotite–quartzofeldspathic gneiss.

Several regional joint sets and lineaments persist through the Southern Blue Ridge and are prominent in the area around Hunters Crossing. Hatcher *et al.* (2005) noted both the Late Palaeozoic NW–SE- and NE–SW-trending joint sets and Mesozoic north–south fracture set. They also mentioned the Fontana–Rhodiss lineament, which trends ENE–WSW from the western portion of the Blue Ridge through Haywood County well into the Piedmont.

Bedrock foliations measured in the vicinity of the slide area typically strike NE–SW and dip greater than 75° NW or SE reflecting steep, tight to isoclinal folding typical of the area. Compositional layering measured in weathered bedrock outcrops typically strikes NW–SE, with steep (greater than 75°) dips to the SW. Fractures measured at several locations showed variable strikes, with mostly shallow dips of 20–50°. Figure 3 shows representative structural data measured at Hunters Crossing. Variations in the orientations of foliations, compositional layering, and fractures across the site record multiple ductile and brittle deformation events. The nearly vertical, planar structural discontinuities are likely to have provided avenues for infiltrating surface water, and thus, locally increase the weathering rates and the resulting thickness of the decomposed bedrock layer that Sams & Gardner (1974) described elsewhere as the intermediate zone (saprolite).

Initial investigations

Several Hunters Crossing homeowners first observed signs of distress in structures at the site in September 2005. Cracks began to develop in the foundations of condominium units 1 and 2, as well as in a retaining wall adjacent to these units (Fig. 2). The residents also reported hearing sounds in their homes that probably resulted from distress of the structures. At the request of the owner of unit 2, NCGS staff visited the property on 10 November 2005 to assess the seriousness of the situation and to determine whether the structural distress was related to landslide activity. The NCGS confirmed the presence of foundation cracks on the west sides of units 1 and 2, which extended along and across mortar lines for nearly the full height of the basement wall (Fig. 4). Both units also showed leaning exterior support posts. Cracks and bulging of the ground surface were visible beneath the carport on the east side of unit 2, and cracks were present in the vegetated slope immediately uphill from this carport. Vertical displacement and separation (20–50 mm) were also observed along the wall, which joins the two condominiums.

NCGS staff members visited houses 8 and 9, located along Lickstone Road, on 18 November 2005 and observed cracked exterior masonry walls, particularly in areas near the chimneys. The owner of house number 9 also reported that a large bulge had developed in the driveway located on the west side of his house, preventing vehicular access to the garage (Fig. 5). This bulge was probably caused by the landslide pushing the concrete slab foundation of the garage against the asphalt driveway.

Based on the November 2005 field investigations, the NCGS postulated that the failure was a deep-seated (greater than 15 m), rotational slide that began upslope of units 1 and 2. The slide was considered to extend beyond the base of the current slope into the floodplain of Browning Branch.

Subsurface analysis

Data acquisition and analysis

In response to the November 2005 investigations by the NCGS and other local authorities, the Hunters Crossing homeowners association hired an engineering firm to conduct a geotechnical study of the property. As part of their investigation, they drilled four boreholes at the site in early December 2005 using a hollow stem auger. The borehole locations are shown in Figure 3. Samples were taken from the boreholes at *c.* 1.5 m intervals using a split-spoon sampler following guidelines in the American Society for Testing and Materials (ASTM) method D6151 (ASTM 1997) and ASTM D1586 (ASTM 1999).

Hole B-1 was drilled in the driveway of units 1 and 2 near where slide movement was initially observed and was terminated at 15.25 m without encountering competent bedrock or groundwater. Material in this hole was described as consisting principally of micaceous sandy silt and silty sand with rock fragments (ML and SM, in accordance with ASTM D2488; ASTM 2000). Above 6 m, blow counts were less than 20 blows per foot (304.8 mm, bpf), but dramatically increased to 40 bpf or higher below this depth. Sample recovery in this borehole averaged 66% of the sampled interval, possibly indicating a high degree of weathering and fracturing of the material.

Hole B-2 was drilled in the driveway adjacent to unit 5. This borehole contained material similar to that found in B-1: micaceous–silty sand saprolite with some rock fragments. Blow counts were less than 26 bpf above a depth of 10.7 m and were around 50 bpf in the last sampling interval collected at a depth of *c.* 12 m. This boring was terminated at 14 m without encountering competent bedrock or groundwater. Sample recovery averaged 66%.

Holes B-3 and B-4 were drilled at the toe of the slope adjacent to house number 9. These two borings encountered silty sand described as saprolite with rock fragments. However, the descriptions made no mention of mica present in either of these boreholes (Connors & Moore 2006). Both borings

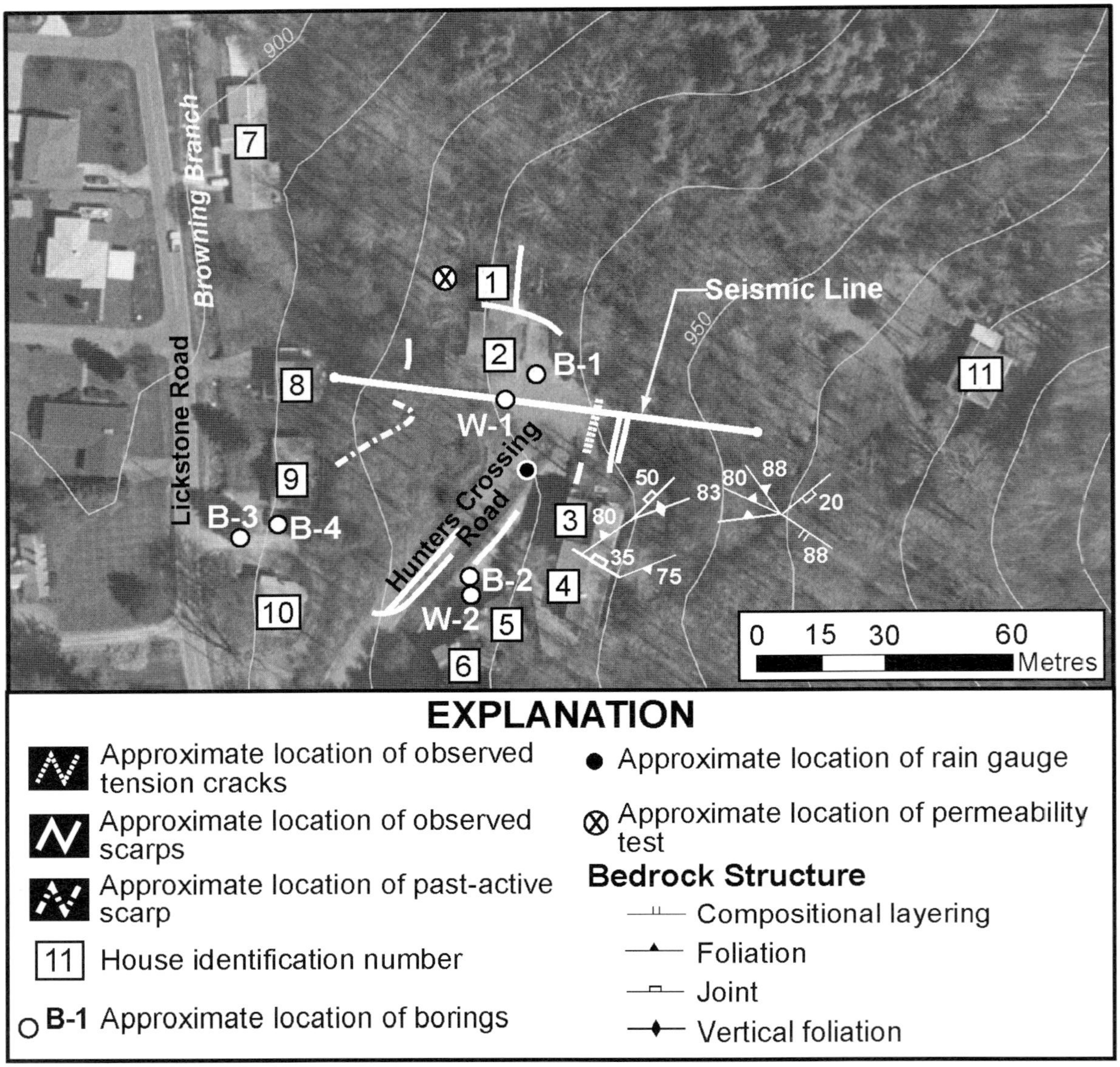

Fig. 3. Aerial photograph (2004) of the Hunters Crossing development showing the locations of selected monitoring and data acquisition sites, and representative bedrock structural data. Photograph courtesy of Haywood County. North is towards the top of the photograph.

were terminated between 2.5 and 3 m at auger refusal. Blow counts were relatively low (less than 12 bpf) for the two samples taken from B-3 and the one sample taken from B-4. The relatively thin soil profile encountered in these two borings suggests either that the flat area below the over-steepened toe of the slope is the result of erosion during flooding events of Browning Branch, or that a thicker, former soil profile was removed by grading of the ground surface during house construction.

Additional boreholes were drilled using a hollow stem auger in January 2008 for inclinometer installation. Hole W-1 lies on the line of the seismic survey line (described below) near B-1, and hole W-2 is adjacent to B-2 (see Fig. 3). In hole W-1, samples were taken at *c*. 1.5 m intervals according to the previously listed ASTM standards. Material was primarily micaceous–silty–sandy saprolite (below 4 m) or completely decomposed–micaceous–biotite granitic gneiss. Blow counts were less than 15 bpf above a depth

Fig. 4. (**a**) Crack in west-facing foundation wall of unit 1 (arrow points to crack); (**b**) cracked and offset retaining wall adjacent to units 1 and 2 (10 November 2005 NCGS staff photographs).

of 9.1 m and increased to between 25 and 50 bpf to a depth of 25.5 m at auger refusal. Coring continued to a depth of 28 m in granitic gneiss with thin layers of schist interlayered with migmatite where the boring was terminated. Rock quality designation (RQD) (ASTM D6032; ASTM 2002) was 24% in the first core run and 68% in the second run. Foliation

Fig. 5. Bulge and cracked pavement (indicated by arrows) that developed in the driveway of house number 9. The home-owner reported that this feature developed around the time of observed landslide movement upslope (18 November 2005 NCGS staff photograph).

in the cored rock dips 30° or slightly flatter. The 24 h depth to water measurement was 20.4 m, whereas the 48 h depth to water measurement was 8.9 m.

Material in W-2 was similar to that in W-1, B-1, and B-2: primarily micaceous, silty, sandy saprolite derived from schist, biotite granitic gneiss, and pegmatite. Split-spoon samples were only taken every 3 m in W-2 as opposed to the 1.5 m sampling interval in the other borings. Blow counts were less than 15 bpf above a depth of 11.9 m and increased to between 28 and 75 bpf to a depth of 20.9 m where auger refusal was met and rock coring began. Coring continued through granitic gneiss interlayered with migmatite to a depth of 22.6 m where the borehole was terminated. One run of rock core was recovered with an RQD of 32% indicating a highly fractured interval. Similar to W-1, foliation in the cored rock dips *c.* 30°. The 48 h depth to water was 17.8 m.

Discussion

It is uncertain whether the initial four borings actually encountered the failure surface, but because the latter two were drilled into competent bedrock, they probably crossed the failure surface. If the failure surface is relatively thin, it may occur in an unsampled interval and also be difficult to identify based on the material encountered in the borehole. It is also possible that the change in blow counts described above in boreholes B-1, B-2, W-1 and W-2 represents the failure surface. Each of these borings showed an increase in blow counts at depths of between 7 and 11 m, which probably represents the contact zone between partially decomposed bedrock and the underlying completely decomposed bedrock that may correspond to the failure surface. It is also possible that the failure surface coincides with the contact between competent bedrock and the overlying partially to completely decomposed rock, a scenario that appears to fit similar failures described elsewhere in the region by Latham *et al.* (2007).

Geophysical investigations

Seismic velocity surveys may be used to help evaluate the spatial extent of this type of landslide if the movement of the soil and weathered rock mass disrupts its internal structure and lowers the seismic velocity of the displaced material, relative to the deeper, undisturbed material. Seismic surveys conducted over debris slides have shown distinct velocity contrasts between the materials above and below failure surfaces. For example, a seismic refraction investigation by Larsen (1995) of a translational slide in Puerto Rico showed compressional-wave velocities in the colluvium and saprolite above the failure surface to be 350 m s^{-1}, compared with 1000–1500 m s^{-1} in the saprolite below the failure surface. The same study evaluated a rotational slide and found compressional-wave velocities of less than 400 m s^{-1} above the failure surface and 870 m s^{-1} just below the failure surface. The results of velocity surveys can then be used to help select the locations and depths of boreholes, inclinometers and drains, and to assist remedial design.

To explore the velocity structure of the Hunters Crossing weathered rock slide, we conducted seismic refraction and surface-wave seismic surveys in July 2007. Seismic data were collected along a line about 105 m long, extending down the approximate centre of the slide from above the uppermost, observed scarp to the toe of the slope (Fig. 3).

Data acquisition

Seismic refraction data were collected using a 24-channel seismograph and 4.5 Hz geophones spaced 3.05 m apart for an array, or 'spread' length, of 70.1 m. Two spreads were used to span the survey line. The energy source was one or more 400 grain, black powder, 8 gauge shotgun shells fired using a Betsy Seisgun into a shallow auger hole (Fig. 6). Five source locations ('shots') were used per spread. Surface-wave data were collected using the multichannel analysis of surface waves (MASW) method, utilizing the same equipment as described above, with a geophone spacing of 0.61 m and a shot spacing of 1.22 m. The first 24 channels were used for the first array location with a length of 14.0 m. Subsequent array locations were moved two geophones down the spread, consistent with the shot spacing. Two energy sources were used to generate the surface wave energy for the survey: a 40 kg accelerated weight drop source and a 7.3 kg sledgehammer. Some preliminary processing of the seismic data was performed in

Fig. 6. Firing a 400 grain black powder shell using a Betsy Seisgun for the seismic refraction survey. The seismic cable and geophones are in the foreground. View is looking downslope (west). Photograph taken in July 2007.

the field for quality control. A level survey provided relative elevations for data analysis.

Data analysis

The seismic refraction data were processed by picking first arrivals of energy for each trace on each shot record ('first breaks'). The first break times, distance between geophones, and geophone elevations were used to generate a compressional-wave velocity model of the subsurface using the program SeisImager (copyright Oyo Geosciences 2004). Surface-wave seismic data were processed using the MASW method, with the program SurfSeis, version 2.05 (copyright Kansas Geological Survey 1998–2007). Processing steps included conversion of time series data to frequency domain, selection of a dispersion curve (variation of velocity with frequency), and inversion modelling. The inversion step produces a shear-wave velocity model of the subsurface to match the observed dispersion curve. A one dimensional (1D) shear-wave model was generated for each of the 14 m long array locations. The 1D models were combined to generate a 2D cross-section of the approximate subsurface shear-wave velocity. The SurfSeis program does not incorporate elevation changes, so the model was rotated to match the general slope of the survey line and the model elevations were adjusted to match the approximate elevations along the seismic survey line.

Discussion of seismic velocity modelling results

The seismic refraction compressional-wave velocity model shows velocities ranging from a low of 100 m s^{-1} in the near surface to a maximum of 4500 m s^{-1} at depth (Fig. 7). The depth to hard bedrock, assumed to be 2000 m s^{-1}, is 14 m at the upslope end of the model, reaches a maximum depth of 24 m at station 20, and then decreases to 8 m depth at the toe of the slope. The change in blow counts at 9.1 m depth in boring W-1, located on the line at approximately station 54, corresponds to a compressional-wave velocity of about 500 m s^{-1}.

The modelled compressional-wave velocities in the upper 7 m (near surface) show some general correlation with the site features (Fig. 7). Near-surface velocities decrease relatively suddenly between horizontal stations 10 and 20 m, upslope of the observed active scarp. This suggests that slide movement may extend further upslope than the mapped active scarp. Close to the toe of the slope, at station 95 m, the near-surface velocities increase in the

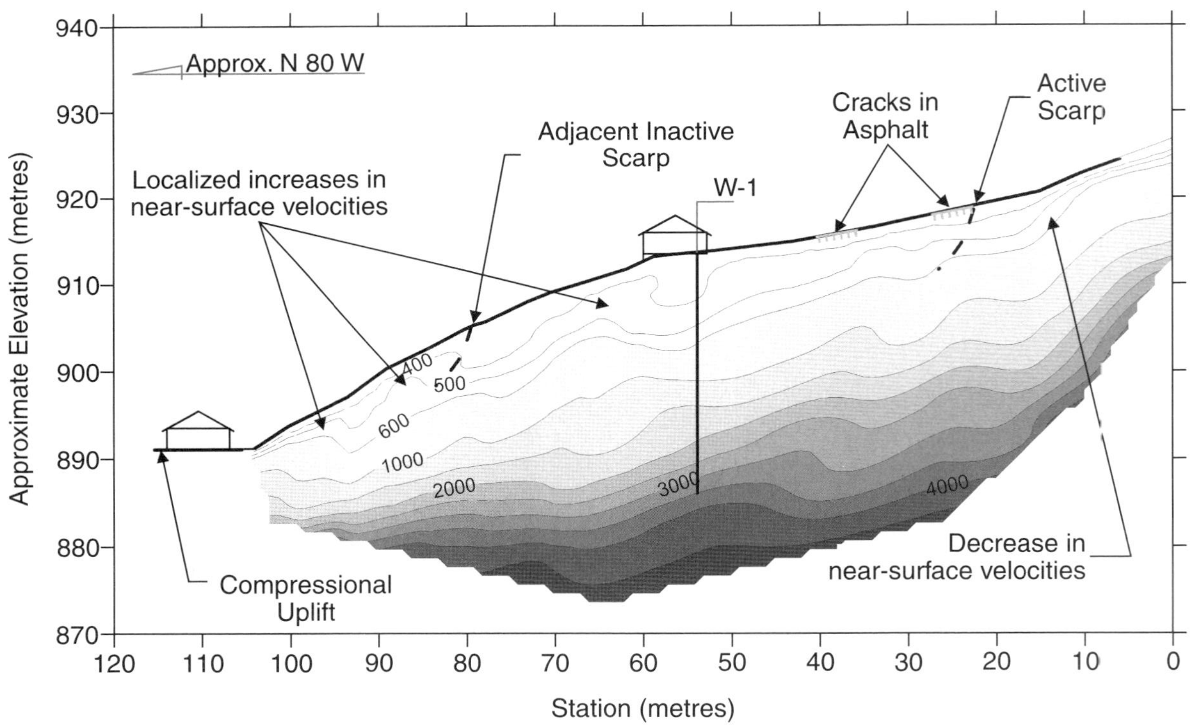

Fig. 7. Seismic refraction compressional-wave velocity model with site observations superimposed (contoured velocities are in m s^{-1}).

downslope direction, suggesting that this area may be in compression. Near-surface velocities also increase between stations 60 and 70 m and between stations 80 and 90 m. These increases may be due to localized variations in weathering or some other factors.

The results of the surface-wave analysis are shown as a shear-wave velocity model (Fig. 8). Because shear-wave velocities are a direct indication of the stiffness of soil and rock, the model should indicate the extent of less dense materials. The modelled velocities range from 100 m s^{-1} to a maximum of 700 m s^{-1} at depth. Some consistency is seen for the zone with a velocity of less than 200 m s^{-1}, generally considered to indicate soft soil. This low-velocity zone ranges in depth from 3 to 11 m, reaches a maximum depth from ground surface at about station 51.5 m, and appears to extend upslope of the active scarp, suggesting potential, future slope movement above the observed active scarp, as was also suggested by the compressional-wave velocity model.

The change in blow counts at 9.1 m depth in boring W-1 corresponds to a shear-wave velocity of about 175 m s^{-1} at station 54. The shear-wave velocities at this location show a significant increase just below this depth, where they quickly increase from 200 m s^{-1} to 400 m s^{-1}. The increase in blow counts and shear-wave velocity at about 9–10 m below ground surface at this location may represent the base of the slide movement.

The shear-wave model shows some localized increases in near-surface velocities within the low-velocity zone (Fig. 8). The largest increase is seen in the area between stations 55 and 70 m, consistent with that observed in the compressional-wave velocity model, indicating that this area is stiffer, perhaps as a result of a lithology change and/or less weathering.

A possible failure surface or maximum depth of movement is proposed based on the extent of the low-velocity zone in the shear-wave velocity model. This failure surface is drawn as a dashed line on the model in Figure 8 and reaches a maximum depth of about 10.5 m below ground surface near station 54 m. Comparison with the compressional-wave velocity model shows that the possible maximum depth of movement corresponds to the compressional-wave velocity contour of 600 m s^{-1}.

A more comprehensive understanding of the seismic velocity models would require more study, including inclinometer data showing the measured depth(s) of landslide movement at selected locations.

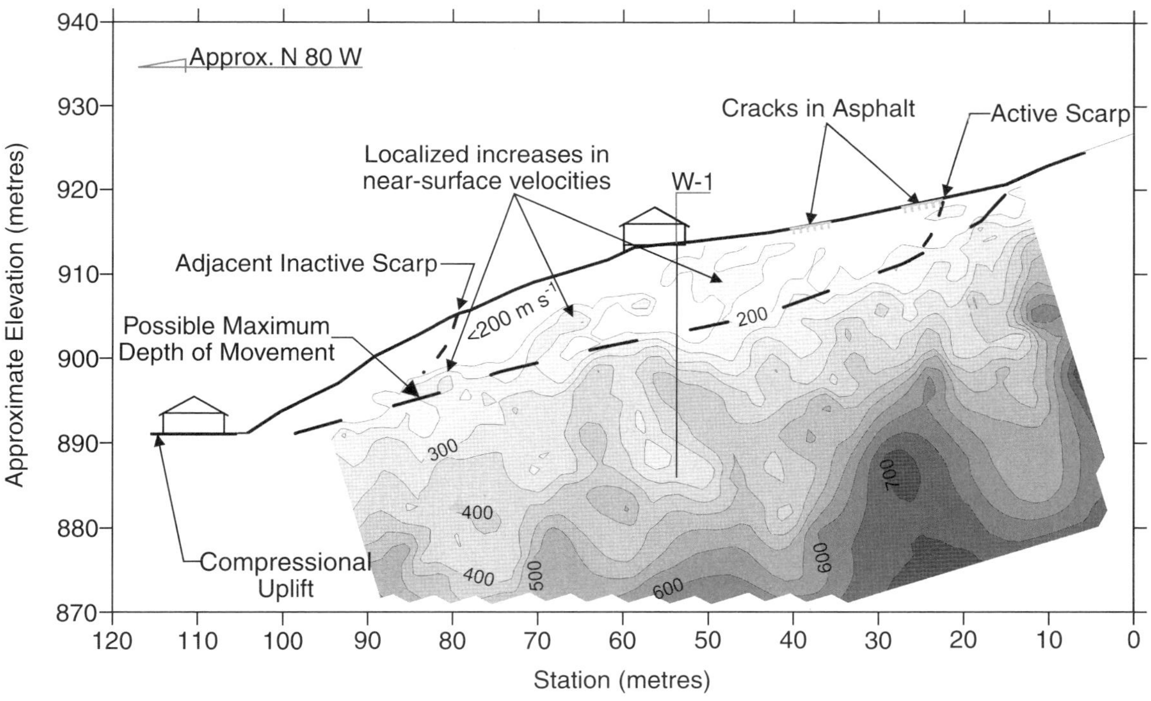

Fig. 8. Shear-wave velocity model developed from surface-wave seismic data using MASW method with site observations superimposed (contoured velocities are in m s^{-1}). Area shaded white between the ground surface and the 200 m s^{-1} contour has a velocity between 95 and 200 m s^{-1}. Long-dashed line indicates possible maximum depth of movement.

Other data collected

Survey data

As part of the initial contracted engineering study, a local surveying firm established benchmarks along the slope to measure horizontal and vertical displacement at locations on and near the landslide. The initial survey on 10 January 2006, and successive measurements on 16 January 2006 and 3 March 2006 were made using a total station survey instrument. When compared with the baseline survey made on 10 January, the 16 January readings appear to better represent the actual, observed movement than the 3 March readings, which showed some unrealistic movement. Figure 9, therefore, shows the vector displacement observed between the 10 January and 16 January surveys. These data indicate that the slide was active during the measurement period, when stations at the base of the slope increased in elevation whereas points near the head scarp decreased in elevation. The highest rates of horizontal movement (*c.* 50 mm or greater) were concentrated in the northern portion of the slide mass, suggesting that the main head scarp may be migrating to the north, which, in turn, could eventually affect more property.

Western Carolina University (WCU) began additional site monitoring as part of a project for an annual, undergraduate course. To renew the site monitoring, students attempted to

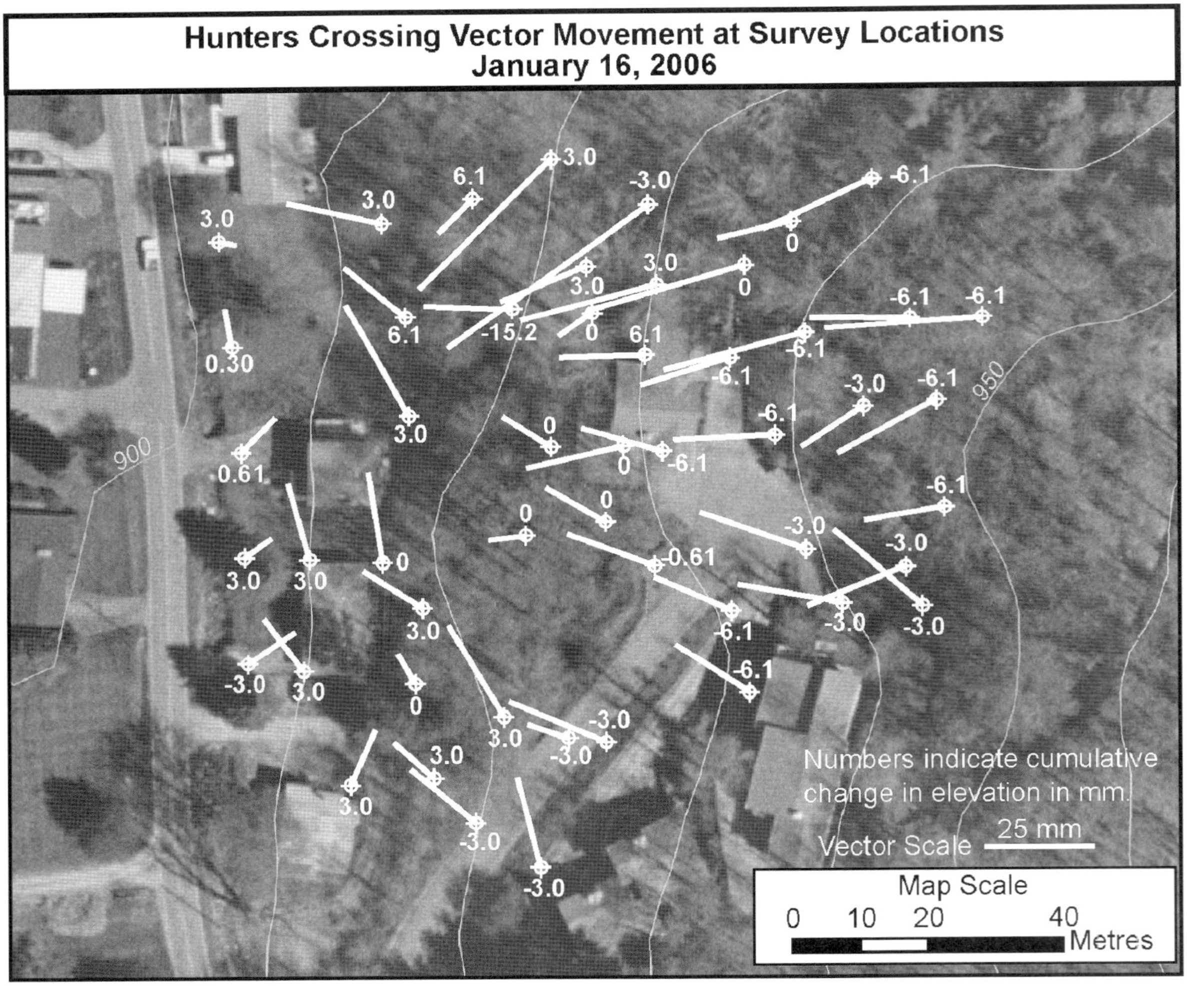

Fig. 9. Aerial photograph (2004) of the Hunters Crossing development showing the location of survey stations and vector displacement measured between the 10 January 2006 and 16 January 2006 surveys. Photograph courtesy of Haywood County. North is towards the top of the photo.

reoccupy the pre-existing survey points in February 2007 to assess movement since the 3 March 2006 survey. Initial data analysis indicated that horizontal movement continues particularly in the northern part of the slide, similar to the results of the 16 January survey.

Crack monitors

During their site investigation, the contracted engineering firm installed 10 crack monitors on the foundations of units 2, 3 and 4, and on houses 8 and 9 on 21 December 2005 (Fig. 2) to measure additional movement along cracks in these structures. They revisited the monitors in January and February 2006 and recorded roughly 0.5 mm of horizontal movement at unit 2, and roughly 0.5 mm of vertical movement at unit 3. WCU began monitoring eight of the 10 crack monitors in early 2007. Significant horizontal displacement in the north–south direction was observed at unit 2, with smaller amounts of offset noted at house 8 (north–south movement) and house 9 (east–west displacement). Six of the eight crack monitors showed roughly 0.5 mm, negative (downslope), vertical displacement. The two exceptions were unit 3, which showed 1.0 mm of negative, vertical displacement and unit 4, which showed a positive vertical displacement of 0.5 mm.

The crack monitor data indicate that the structures constructed on the slide continue to move. It is difficult to determine whether this displacement indicates continued movement of the slide or only reflects stress release after a stress build-up in the relatively rigid buildings caused by past slide movement. Although the crack monitor data indicate that movement of the slide probably continues, they do not necessarily provide an accurate representation of the continued rate of movement of the landslide.

Constant head permeameter

WCU faculty and students utilized an Amoozemeter™, a compact, constant head permeameter, in spring 2007 to measure *in situ* saturated hydraulic conductivity in the vadose zone at Hunters Crossing. One test was completed in the slide material downslope of unit 1 (Fig. 3) where the permeability (at an approximate depth of 210–370 mm) was calculated as 1.0×10^{-3} cm s^{-1}. The calculation, using the Glover solution (Amoozegar 1989), may be in error because of the below-optimum hydraulic head (H) to radius (r) of the borehole ratio during the test ($H\ r^{-1} = 4.4$; optimum $5 < H\ r^{-1} < 10$). However, the derived value does correspond to the permeability of silty sand to well-sorted sand (Fetter 1994), which is similar to what was observed in the upper metre of the closest borehole (B-1).

The permeability rate at this location indicates that at least the upper metre of material at this one location in the middle to upper portion of the slide is fairly permeable, and water should flow through this zone relatively quickly. No groundwater was encountered in the boreholes B-1 and B-2 in the upper reaches of the slide. However, as mentioned above, borehole W-1 had a 24 h groundwater measurement of 20.4 m below the ground surface and a 48 h groundwater measurement of 8.9 m, whereas borehole W-2 had a 48 h depth to groundwater measurement of 17.8 m. The 48 h reading in W-1 corresponds to a zone of moist soil observed during drilling between depths of 8.5 and 10.5 m. Samples were reported as 'moist' again below a depth of 19.2 m and wet below a depth of 22.3 m. This perched 'moist' zone between 8.5 and 10.5 m could indicate either a zone of lower permeability material (e.g. a zone of partially decomposed rock within the saprolite) or a zone of higher fracture intensity.

However, with limited hydrological information below 1–2 m, it is difficult to interpret the groundwater conditions at greater depths and at other locations across the failure. Thus, the role that subsurface infiltration and groundwater plays in slope movement at this site is unclear.

Inclinometers

Slope inclinometer casing was installed in boreholes W-1 and W-2 in January 2008, shortly after drilling was completed. Hole W-1 was cased to a depth of 26.2 m, and hole W-2 was cased to a depth of 21.3 m. Measurements were made on five occasions between January 2008 and December 2008, and indicate settlement of the casing.

There are zones of potential deflection along both axes in both boreholes. However, it is too early to determine if this is definitive slide movement because all of the current observed casing movement is still within the diameter of the borehole. Continued monitoring is planned at roughly 2-month intervals with variations in monitoring intervals as deemed necessary by observed movement or lack of movement.

Discussion

Landslide characteristics

Based on the data collected, the boundaries of the head and toe of slide have been determined, whereas the lateral extent of the movement is variable and continues to increase, as indicated by the propagation of tension cracks to the north. The approximate areal extent of the displaced mass is 6100 m^2. The observed damage to structures indicates that the failure contains some rotational component to the movement. This movement is expressed by extension at the head and compressional uplift at the toe. Assuming that the depth of the failure surface is no greater than 11 m below the ground surface at its deepest location as suggested, an approximation of the volume of the displaced mass is roughly 33 550 m^3.

The material encompassed in the slide involves saprolite and/or weathered rock of varying degrees of decomposition.

The change in blow counts between 8 and 11 m could indicate a possible transition from saprolite (completely decomposed rock) to partially decomposed rock. The configuration and composition of the failure surface is poorly understood at this time. It is possible that movement is occurring along remnant structures that dip parallel to the slope. Foliation and joint orientations that form dip slopes could form portions of the failure surface. The boring logs record manganese staining along some relict bedrock structures, indicating that water travelled along these features, which would decrease the strength of the rock mass by accelerated weathering. Some of the material in the boreholes is partially decomposed to completely decomposed pegmatite, a material associated with other slope failures in Haywood County. However, it is not currently understood why decomposed pegmatites tend to be less stable than the surrounding parent rock.

According to Cruden & Varnes (1996), this slide falls into velocity class 2, very slow with rates of movement between 16 and 1.6 m a^{-1}. Ground observations indicate that movement continues but at a slow to intermittent rate. It should be noted that western North Carolina has been in a period of extreme or exceptional drought since mid-2007, and during this time, movement appears to have continued at the site. Therefore, it is conceivable that the movement rate could increase given a return to average or higher rainfall conditions.

Potential causes and triggers of the Hunters Crossing weathered rock slide

Causes of a landslide are the long-term processes that make the slope marginally stable and create the conditions for future instability. Examples of causes include the geology, geomorphology and hydrology of a site. Triggers are those actual, short-term events or conditions that cause a slope to move from stable to unstable (Wieczoreck 1996). The most common example of a trigger is water from a source added to the slope. At Hunters Crossing, possible causes are the geology and hydrology of the site. The thick accumulation of highly micaceous, weathered rock, in combination with the steeply dipping bedrock discontinuities, provides avenues for water to infiltrate into the soil profile. Another possible cause of the recent movement relates to the history of instability at this site. The past active scarp located in the middle of the current sliding mass (Fig. 2) indicates past instability on this slope. Although it is uncertain when this failure occurred, it is possible that this past activity decreased the shear strength of the weathered rock on the slope leaving it more susceptible to future movement. This past activity, compounded by any slope cuts made at the toe of the slope along Lickstone Road, would probably have left the site marginally stable.

One Hunters Crossing resident reported three water supply line leaks and subsequent repairs between August and November 2005. Water use records indicate that consumption increased from 57 kilolitres (kl) to 725 kl from 26 August 2003 to 29 September 2003 for house 11 at the top of the hill (Fig. 2). An increase from 57 to 246 kl was recorded in a 26 September 2005 metre reading for the same water supply line. An average home requires 28–45 kl of water each month (Hyatt 2005). This line is separate from the water supply line that feeds the Hunters Crossing condominiums and averages roughly 40 kl of water usage per month. Records also indicated above average water usage of 951 kl during October 2005 by the residents of the condominiums (Hyatt 2005). Because these data indicate that, at a minimum, the slide movement was affecting the water line, the town of Waynesville discontinued water service to the development on or around 20 November 2005 in an attempt to prevent increased slide movement and reduce the potential for catastrophic failure. It is unclear, however, whether the slide movement caused the water line to leak or whether the leaking water line triggered movement of the slope. Regardless, the addition of water to a marginally stable slope would have decreased slope stability. It is possible that water from the supply line infiltrated the saprolite and created a perched aquifer at the potential boundary between saprolite and partially weathered rock around a depth of roughly 8–11 m. At a minimum, the leaking water line would have added water to the subsurface during the time of leaking. This increase in weight and pore-water pressure could have caused movement to start or accelerate.

Although high-intensity rainfall is a common triggering mechanism for landslides, the relationship between rainfall and slope movement is unclear at this location, particularly because of the lack of movement associated with a major storm event. In September 2004, heavy rainfall from the remnants of Hurricanes Frances and Ivan triggered flooding and landslides across western North Carolina. The US Geological Survey (USGS) rain gauge at the Lake Logan dam, *c.* 7.4 km ESE of Hunters Crossing in Haywood County, recorded 250 mm of rainfall from 6 to 8 September 2004 and 170 mm of rainfall between 16 and 17 September 2004 (USGS 2007). This rainfall event could have been the trigger that reactivated this potentially landslide-prone area. However, movement was not reported until a year after the rainfall event, by which time, the effects of a short-term rainfall event had probably dissipated. Other high-rainfall events might also be responsible for triggering slope movement at Hunters Crossing. In August 2005, a localized, intense storm cell caused several landslides in neighbouring Jackson County, *c.* 1 month before movement was observed at Hunters Crossing. The North Carolina Division of Emergency Management reported flooding along Allen Creek, which inundated Browning Branch Road, an area just north of the Hunters Crossing community, during the same storm event. The USGS rain gauge at Lake Logan recorded 40 mm of rainfall for this storm event (USGS 2007). In each case, there was a relatively short period between the rainfall event and ground movement.

Eschner & Patric (1982) established that in the forested, eastern USA, 125 mm of rainfall in a 24 h period is enough to trigger debris flows. We are not aware of any clear evidence in western North Carolina, however, to show how slow-moving, weathered rock slides respond to high-intensity rainfall events. In the 3 months preceding the first indications of slope movement at Hunters Crossing, rainfall measurements at the Lake Logan dam (USGS 2007) were above average (National Weather Service 2007). These low-precipitation (less than 60 mm), but frequent storm events may have provided the long-term rainfall necessary to trigger movement.

Efforts to monitor precipitation at Hunters Crossing continue. WCU installed a tipping bucket rain gauge in early 2007 at Hunters Crossing near unit 3 in an area clear of trees (see Fig. 3 for location). Because collection primarily occurred in the winter months, little rainfall was recorded. As previously stated, movement continues despite the fact that this region has been in a state of extreme to exceptional drought for nearly 2 years. This movement is probably due to a continuing decrease in the shear strength of the weathered rock perpetuated by strain softening along the rupture surface coincident with additional movement.

Results of investigations

The investigations described above, in combination with continued NCGS visual inspections conducted to date, indicate that the Hunters Crossing weathered rock slide continues to move a few centimetres per month or per year. Displacement has been observed in each of the site monitoring events, and damage continues to man-made structures at the site. However, some questions remain. First, what is the configuration and location of the failure surface? Second, what triggered the 2005 slide movement? Third, what causes the slide to continue to move? The results of the geophysical analysis along with the borehole data suggest that the failure surface may be between 8 and 11 m below the ground surface, although a distinct failure surface was not recognized in the borehole samples. However, this interval does roughly correspond to an increase in blow counts observed during drilling. The change in blow counts in turn could indicate a change in material (e.g. the boundary between saprolite and partially decomposed bedrock). This proposed, more shallow, failure surface would indicate a more rotational–translational or purely translational sliding surface as opposed to the previously postulated deep-seated, rotational failure surface. It is possible that the failure surface is not as discrete in weathered rock slides as it is in other types of landslides. Lambe & Riad (1990) located the failure surface of the Balsam Gap II landslide using closely spaced inclinometer data rather than relying on the techniques employed to date at this location. Similarly, Cornforth (2007) warned against interpreting the location of the failure surface based on borehole data rather than inclinometer readings.

Questions also remain about the development of thick, in-place accumulations of partially to completely decomposed bedrock. As of early 2009, 12 other weathered rock slides have been identified in Haywood County. Many of these lie on the prominent NE–SW-trending lineaments that pervade the topography of Haywood County and can be seen locally at Hunters Crossing. The increased fracture frequency could provide additional avenues for water infiltration, creating an increased rate of bedrock weathering. Also, many of the 12 weathered rock slides originate in the Cartoogechaye terrane. These more mafic rocks may contain a higher percentage of biotite and hornblende, which, in turn, may weather faster than the minerals in neighbouring rock units. These weathered rock slides also occur in a portion of the county with lower relief, more rounded mountains indicative of less resistant, more weathered rocks. They also are typically located near the toes of slopes in areas more likely to have upward groundwater gradients and groundwater discharge.

There is evidence of regional weathering within Haywood County. Hadley & Goldsmith (1963) and Mills & Allison (1994) mapped multiple, composite, large-scale debris deposits in the Hazelwood and Dellwood 7.5 min quadrangles, just west of Hunters Crossing. It is possible that the same conditions that led to the deposition of these large-scale deposits could have been responsible for accelerating the weathering of the bedrock in this region. As part of the continuing landslide hazard mapping by the NCGS, other weathered rock slides in the region will probably be identified. With significantly more data, statistical evidence may suggest correlations between the failures and factors that contribute to landslides in weathered bedrock. Detailed geological mapping may also provide more information on the structural domains present as well as rock types that might be prone to the development of thick weathering profiles.

Remedial options

In November 2005, the Code Enforcement Officer for the town of Waynesville declared condominium units 1 and 2 unsafe for occupancy, and the residents have since relocated. The geotechnical investigation report (Connors & Moore 2006) presented several options to remediate the slope so that residents might return to their homes. One proposed solution was to construct a soil berm along the toe of the slope where houses 8, 9 and 10 are located (Fig. 2) to act as a cantilever to the observed motion. This proposed berm would be roughly 21 m wide and between 6 and 9 m high. However, it was noted that settlement would probably continue for some unknown period of time after construction of the berm while void space in the slide material would decrease. This settlement would prevent homeowners from returning to their homes for some undetermined time, and the berm would not guarantee protection of the structures from further damage (Connors & Moore 2006). Another suggested slope repair technique would involve using anchored tie-backs drilled into the soil or rock at depths beyond the failure surface. Such a system would require more studies to determine the location of the slip surface, would cost an estimated minimum of one million US dollars, and would have no guarantees of success. The final recommendation involved relocating the homes to stable ground, an option that a local contractor estimated would cost, at a minimum, $58 000 to $63 000 per house. The general consensus of the homeowners association in early 2006 was not to attempt to stabilize the slide because of the high costs and uncertain outcomes.

Conclusions

Based on the data collected to date, it appears that the failure surface is potentially located at a maximum of 11 m below the ground surface, possibly at the contact between saprolite or completely decomposed rock and partially decomposed rock. This proposed failure surface is indicated by the seismic velocity survey and corroborated by blow counts and groundwater observed in the boreholes. However, this conclusion is based on limited subsurface information, particularly given the areal extent of this slide (i.e. very few boreholes used to represent a large area). More work needs to be done at the site to properly assess the configuration of the slide, to provide a better understanding of the earth materials present, and to determine the effect of short- and long-term changes in groundwater conditions on slide movement.

Acknowledgements. The authors would like to gratefully acknowledge the North Carolina General Assembly, the North Carolina Division of Emergency Management, Alpha Environmental Sciences, Inc. in Waynesville, NC, the North Carolina Department of Transportation, and the homeowners affected by the Hunters Crossing landslide, who have graciously allowed us to study the area during a difficult time in their lives.

References

AMOOZEGAR, A. 1989. Comparison of the Glover solution with the simultaneous equations approach for measuring hydraulic conductivity. *Soil Science Society of America Journal*, **53**, 1362–1367.

ASTM 1997. Standard D6151—Standard practice for using hollow stem augers for geotechnical exploration and soil sampling. *In*: *Annual Book of ASTM Standards, v. 04.09, Soil and Rock (II)*. American Society for Testing and Materials, West Conshohocken, PA.

ASTM 1999. Standard D1586—Standard test method for penetration test and split barrel sampling of soils. *In*: *Annual Book of ASTM Standards, v. 04.08, Soil and Rock, Building Stones, Geotextiles*. American Society for Testing and Materials, West Conshohocken, PA.

ASTM 2000. Standard D2488—Standard practice for description and identification of soils (visual manual procedure). *In*: *Annual Book of ASTM Standards, v. 04.08, Soil and Rock (I)*. American Society for Testing and Materials, West Conshohocken, PA.

ASTM 2002. Standard D6032—Standard test method for determining rock quality designation (RQD) of rock core. *In*: *Annual Book of ASTM Standards, v. 04.09, Soil and Rock (II)*. American Society for Testing and Materials, West Conshohocken, PA.

ASTM 2008. Standard D5878—Standard guides for using rock-mass classification systems for engineering purposes. *In*: *Annual Book of ASTM Standards, v. 04.09, Soil and Rock (II)*. American Society for Testing and Materials, West Conshohocken, PA.

CONNORS, J. K. & MOORE, R. D. 2006. *Slope Evaluation for Hunters Crossing in Waynesville, North Carolina*. Alpha Environmental Services, Inc., Waynesville, NC, Report **5664.01**.

CORNFORTH, D. H. 2007. Seven deadly sins of landslide investigation, analysis, and design. *In*: TURNER, A. K. & SCHUSTER, R. L. (eds) *Landslide and Society: Keynote and Invited Presentations at the 1st North American Landslide Conference, Vail, Colorado*. AEG Special Publication, **22**, 149–177.

CRUDEN, D. M. & VARNES, D. J. 1996. Landslide types and processes. *In*: TURNER, A. K. & SCHUSTER, R. L. (eds) *Landslides: Investigation and Mitigation*. Transportation Research Board Special Report **247**, 36–75.

DEERE, D. U. & PATTON, F. D. 1971. Slope stability in residual soils. *In*: *Proceedings of the Fourth Panamerican Conference on Soil Mechanisms and Foundation Engineering*, American Society of Civil Engineering **1**, 87–170.

ESCHNER, A. R. & PATRIC, J. H. 1982. Debris avalanches in eastern upland forests. *Journal of Forestry*, **80**, 343–347.

FETTER, C. W. 1994. *Applied Hydrogeology*, 3rd edn. Prentiss Hall, Upper Saddle River, NJ.

HACK, J. T. 1982. *Physiographic Divisions and Differential Uplift in the Piedmont and Blue Ridge*. US Geological Survey Professional Paper, **1265**.

HADLEY, J. B. & GOLDSMITH, R. 1963. *Geology of the Eastern Great Smoky Mountains, North Carolina and Tennessee*. US Geological Survey Professional Paper, **349-B**.

HATCHER, R. D. JR, BREAM, B. R. & MILLER, C. F. 2004. Paleozoic structure of internal basement massifs, southern Appalachian Blue Ridge, incorporating new geochronologic, Nd and Sr isotopic, and geochemical data. *In*: ROLLO, R. P., CORRIVEAU, L., MCLELLAND, J. & BARTHOLOMEW, M. J. (eds) *Proterozoic Tectonic Evolution of the Grenville Orogen in North America*. Geological Society of America, Memoirs, **197**, 525–547.

HATCHER, R. D. JR, MERSCHAT, A. J. & THIGPEN, J. R. 2005. Blue Ridge primer. *In*: HATCHER, R. D. JR & MERSCHAT, A. J. (eds) *Blue Ridge Geology Geotraverse East of the Great Smoky Mountains National Park, western North Carolina*. Carolina Geological Society Annual Field Trip Guidebook, 5–6 November 2005, 1–24.

HYATT, V. 2005. Homeowners seek solution. *The Mountaineer* (Waynesville, NC), 30 November.

LAMBE, P. C. 1996. Landslide types and processes. *In*: TURNER, A. K. & SCHUSTER, R. L. (eds) *Landslides: Investigation and Mitigation*. Transportation Research Board Special Report **247**, 507–524.

LAMBE, P. C. & RIAD, A. H. 1990. *Determination of Shear Strength for Design of Cut Slopes in Partly Weathered Rock and

Saprolite. Federal Highways Administration Report Project **23241 88–3**.

LARSEN, M. C. 1995. Use of seismic refraction techniques for investigating recent landslides in a tropical rain forest in Puerto Rico. *In*: MILLER, R. L., ESCALANTE, J. A., REINEMUND, J. A. & BERGIN, M. J. (eds) *Energy and Mineral Potential of the Central American–Caribbean Regions*. Springer, Berlin, 411–414.

LATHAM, R. S., WOOTEN, R. M., WITT, A. C., GILLON, K. A., DOUGLAS, T. J., FUEMMELER, S. J. & BAUER, J. B. 2007. Big slow movers: a look at weathered rock slides in western North Carolina. *In*: SHAEFER, V. R., SCHUSTER, R. L. & TURNER, A. K. (eds) *Conference Presentations, 1st North American Landslide Conference, Vail, Colorado*. AEG Special Publication, **23**, 545–557.

MILLS, H. H. & ALLISON, J. B. 1994. Controls on the variation of fan-surface age in the Blue Ridge Mountains of Haywood County, North Carolina. *Physical Geography*, **15**, 465–480.

NATIONAL WEATHER SERVICE 2007. Greenville Spartanburg Forecast office, normal climate data for Asheville, NC, 2007. http://www.weather.gov/climate/index.php?wfo = gsp

NCGS 1985. *Geologic Map of North Carolina*. North Carolina Geological Survey, Raleigh, NC.

SAMS, C. E. & GARDNER, C. H. 1974. Engineering Geology of I-26 Landslides, Polk County, North Carolina. *In*: *Proceedings of the 25th Annual Highway Geology Symposium, 22–24 May 1974*. Raleigh, NC, 200–253. http://www.highwaygeologysymposium.org/upload/postmeetings/hgs%2025.pdf

SOWERS, G. F. 1954. Soil problems in the southern Piedmont Region. *Proceedings, American Society of Civil Engineers, Soil Mechanics and Foundations Division*, **80**, Separate 416.

USGS 2007. USGS real time water data for USGS 03455773 Lake Logan at Dam near Hazelwood, NC. http://waterdata.usgs.gov/nc/nwis/dv/?site_no = 03455773&agency-cd = usgs &referred_module = sw.

WIECZOREK, G. F. 1996. Landslide triggering mechanisms. *In*: TURNER, A. K. & SCHUSTER, R. L. (eds) *Landslides: Investigation and Mitigation*. Transportation Research Board Special Report **247**, 76–90.

WILLIAMSON, D. A. 1984. Unified rock classification system. *Bulletin of the Association of Engineering Geologists*, **XXI**, 253–254.

Shear strength of soils derived from the weathering of granite and gneiss in Brazil

Willy A. Lacerda

Civil Engineering Programme, Federal University of Rio de Janeiro, Rio de Janeiro, Brazil
(e-mail: willyl@globo.com)

Abstract: Research on the properties of residual soils has been continuing in Brazil since the pioneering work by Vargas in 1953. A great number of earth dams built since then and the initiation of graduate courses since 1960 have increased the rate of in-depth studies on this theme. This contribution considers the more recent research on the properties of these soils, mainly shear strength parameters.

The classification and main properties of tropical residual soils, such as their compressibility, permeability and shear strength, have been treated extensively in the near past (Vargas 1953, 1974, 1985; Sandroni 1977, 1981, 1991; Bell 1981; Carvalho *et al.* 1985; Lacerda *et al.* 1985; Nogami *et al.* 1985; Lacerda & Almeida 1995; Fookes 1997; Futai *et al.* 2004, 2006). The 1985 International Conference on the theme remains the key reference of the subject (Anonymous 1985). The microstructures of these soils have also been studied to a certain degree (e.g. Sandroni 1977; Collins 1985). Saturated colluvial slopes in tropical regions have been reported by Lacerda (2004), who showed that these slopes have a natural angle about half the residual friction angle of these materials. The initiation of slides in saprolites and colluvium has been reported by Lacerda (2007). The present contribution will focus on the more recent advances in the research of microstructure and shear strength of these soils.

Weathering profiles

A large part of Brazil is composed of rocks dating from the Precambrian, mainly granites and gneisses (Fig. 1). This 'shield' encompasses the southern and southeastern regions, which include mountain ranges up to 2000 m in height and in which there are many major cities. Basalt lava flows occupy a large part of the mid-western and southern regions. The weathering of these rocks is thus of great concern regarding the problem of stability of slopes, as these rocks are subjected to a tropical climate, with rainfall reaching a total of 2000 mm or more each year.

The weathering profile of residual soils in Brazil proposed by Vargas (1974) shows the sequence of residual soil layers derived from granites, gneisses and basalts. He called 'mature residual soil' what is now known as lateritic soil and 'young residual soil' the saprolitic soil. The terms lateritic soil and saprolitic soil were adopted during the International Conference on Tropical Soils (Anonymous 1985). Pastore (1995) proposed the weathering profile shown in Table 1.

The term saprolite 'refers to that part of the weathering profile where the soil largely preserves the microfabric and volume of the parent rock' (Aydin 2006). Saprolitic soil is saprolite with more advanced weathering, but that has not undergone the process of laterization. The uppermost section of the saprolitic soil is considered as 'mature', or lateritic soil, as a result of the process of laterization. The weathering profile in a tropical region very often shows a very narrow or indistinct V horizon.

Figure 2 shows a typical residual soil profile of a gneissic rock. The superficial, reddish layer is lateritic, and the light yellow layer is saprolitic. In the saprolitic soil and in the saprolite horizon relict structures of the original rock are visible. Figure 3 shows a saprolite horizon of banded gneiss, showing mica-rich planes involved in a slide. The slickenside of this failure plane is shown in Figure 4. The role of these field-scale heterogeneities in the stability of slopes was studied by Aydin (2006).

In Figure 5 a well-drained residual soil profile of gneiss is shown with some geotechnical parameters. The B horizon (with a thickness of about 3 m) comprises the lateritic soil, which has a porous structure and therefore a high void ratio ($e = 1.5$), and a high clay content (about 50%), contrasting with the lower clay content of the saprolitic soil below, with a void ratio less than unity. The predominant clay mineral in the saprolitic soil is kaolin. The phreatic level is 8 m below the ground surface, and the natural water content lies between the plastic and liquid limits.

From: Calcaterra, D. & Parise, M. (eds) *Weathering as a Predisposing Factor to Slope Movements*.
Geological Society, London, Engineering Geology Special Publications, **23**, 167–182.
DOI: 10.1144/EGSP23.10 0267-9914/10/$15.00

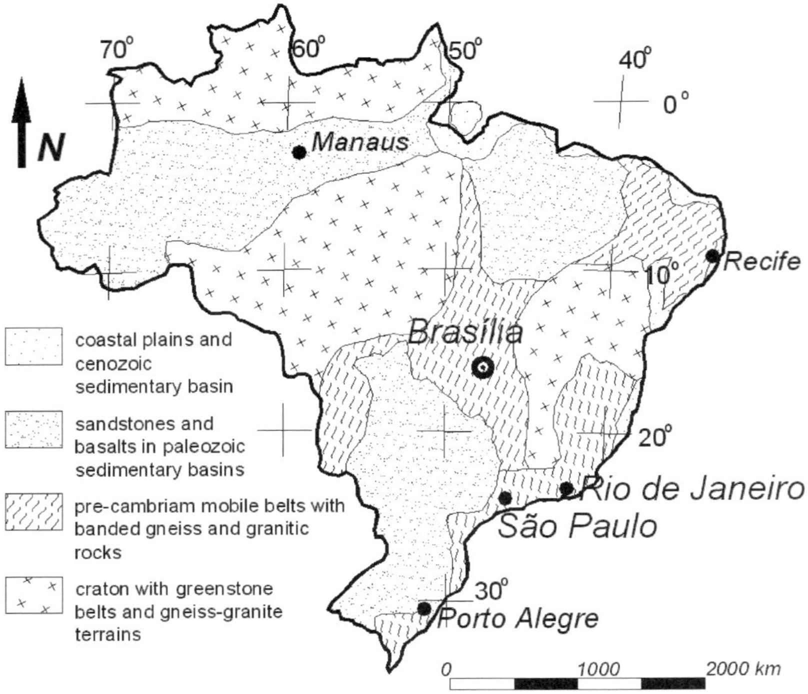

Fig. 1. Simplified geological map of Brazil (after Pires 1998).

Laterites, lateritic and saprolitic soils

Laterites and lateritic soils form a group comprising a wide variety of red, brown, and saprolitic soils that are the result of deep weathering of the parent rock that has not yet undergone laterization. The soil name 'laterite' was coined by Buchanan (1807) in India, from the Latin word 'later' meaning brick. Such soils are characterized by forming hard, impenetrable and often irreversible pans when dried. They are found at shallow depths, in low-grade slopes, and are so tough that they are used in Brazil as concrete aggregate, construction stone for pavements and rip-rap for the protection of dam slopes from the wave action of reservoirs. Figure 6 shows a typical laterite. The rounded forms of the oxides enveloping the grain structure should be noted. These 'stones' are widely used in the northern regions of Brazil, in which rock outcrops are rare, as construction material, and even as concrete aggregate.

Laterites may vary from a loose material to a massive rock. Because of this confusion, most workers now prefer to use definitions based on hardening, such as 'ferric' for iron-rich cemented crusts, 'alcrete' or bauxite for aluminum-rich cemented crusts, 'calcrete' for calcium carbonate-rich crusts, and 'silcrete' for silica-rich cemented crusts (Fookes 1997). Other definitions are based on the ratios of silica (SiO_2) to sesquioxides (Fe_2O_3, Al_2O_3). This has the index k_r. In 'true' laterites the k_r values, are less than 1.33. Those between 1.33 and 2.0 are indicative of lateritic soils, and those greater than 2.0 are indicative of non-lateritic soils (Gidigasu 1976). Bell (1981) gave a comprehensive review on the general characteristics and genesis of laterites and lateritic soil.

Table 1. *Weathering profile according to Pastore (1995)*

Residual or transported soil	I, organic horizon II, lateritic soil horizon
Residual soil	III, saprolitic soil horizon
Transition from soil to rock	IV, saprolite horizon
Rock mass	V, highly weathered rock horizon VI, weathered rock horizon VII, sound rock

Other mineralogical indices

Rocha Filho *et al.* (1985) reviewed many mineralogical indices such as X_d, N_q and N_{qo} (Lumb 1962), and concluded

Fig. 2. A 40 m high cut exposing lateritic (red) and saprolitic (light yellow) soils from gneiss, in Rio de Janeiro State. The weathered rock layer is below the group of people in the centre of the photograph.

that mineralogical weathering indices based upon the content of unstable reference minerals are of limited use, mainly because of the difficulty in their determination. Those workers found that 'referential minerals should be chosen as a function of rock type and climate, and should be used to evaluate the degree of decomposition along the same profile'. Jenny (1941) introduced the 'degree of lixiviation', β, defined as

$$\beta = ba_1(\text{weathered rock})/ba_1(\text{sound rock}) \quad (1)$$

where

$$ba_1 = (K_2O + Na_2O)/Al_2O_3. \quad (2)$$

Fig. 3. Gneiss saprolite, showing relict planar structures of the sliding surfaces rich in mica at right, in shadow.

Fig. 4. Slickensides in the planar layer of mica-rich saprolitic soil.

Bernardes *et al.* (1992) have shown that in a residual soil of leptinitic gneiss β increases with depth, from 0.1 to unity, which reflects a gradual weathering of the parent rock, as seen in Figure 7. They also found that the only parameter affected by β was the effective cohesion intercept for flooded direct shear specimens.

Sandroni (1977) studied gneissic residual soils, and found that the mineralogy of the sand fraction, including the proportions of mica and feldspar, correlates well with the strength of the soil, which tends to be higher as the feldspar content increases and the mica content decreases. Figure 8 shows how the mineralogy of the coarse fraction affects the shear strength of saprolitic gneissic soil.

Structure of saprolitic and lateritic soils

Soils are classified according to grain size: clay, silt, sand, gravel, etc. The mineralogy of each fraction deserves special attention, and generally a description is made of the various clay minerals, as well as of the sand and silt particles (quartz, carbonate, feldspar, etc.). A transported and sedimented soil has a structure in which the grain sizes are sorted during the depositional process, and this is sufficient for their characterization.

However, a simple description in terms of mineralogy and grain size does not suffice when the subject is residual soil. Residual soil has an inherited structure from the parent rock, and also a structure resulting from pedogenetic processes. Saprolitic soils are derived from saprolites, an advanced state of weathered rock. As these soils undergo further weathering the clay content increases, and by exposure to lixiviation plus oxidation of iron and aluminum ions such soils are transformed to lateritic soils. An advanced state of lateritic soils is laterite. Lateritic soils are erosion-resistant, whereas saprolitic soils are not. In short, saprolitic soils are formed in the transition from weathered rock to completely weathered rock, and their clay fraction is small, usually less than 10%, as can be seen in Figure 5. Lateritic soils have a larger clay fraction, up to 60%, but this clay fraction is aggregated with oxides, forming links with the larger grain sizes.

Collins (1985) stated that 'it is clear that a wide variety of microstructural forms and levels of fabric organization will exist in lateritic and saprolitic soils. Complex multi-level pore systems will be found, involving unequal pore sizes'. In our opinion each geotechnical application (stability of natural slopes, compacted soils, pavements, foundations, etc.) should use the most appropriate description of the soil for the purpose in view, including its genesis, climate and so on, and include mechanical tests so as to obtain data that are comparable with those for other soils.

Some attempts have been made in recent decades to find simple methods to characterize residual soils, employing laboratory tests.

Nogami & Villibor (1981) and Nogami *et al.* (1989) made an interesting proposal for the classification of

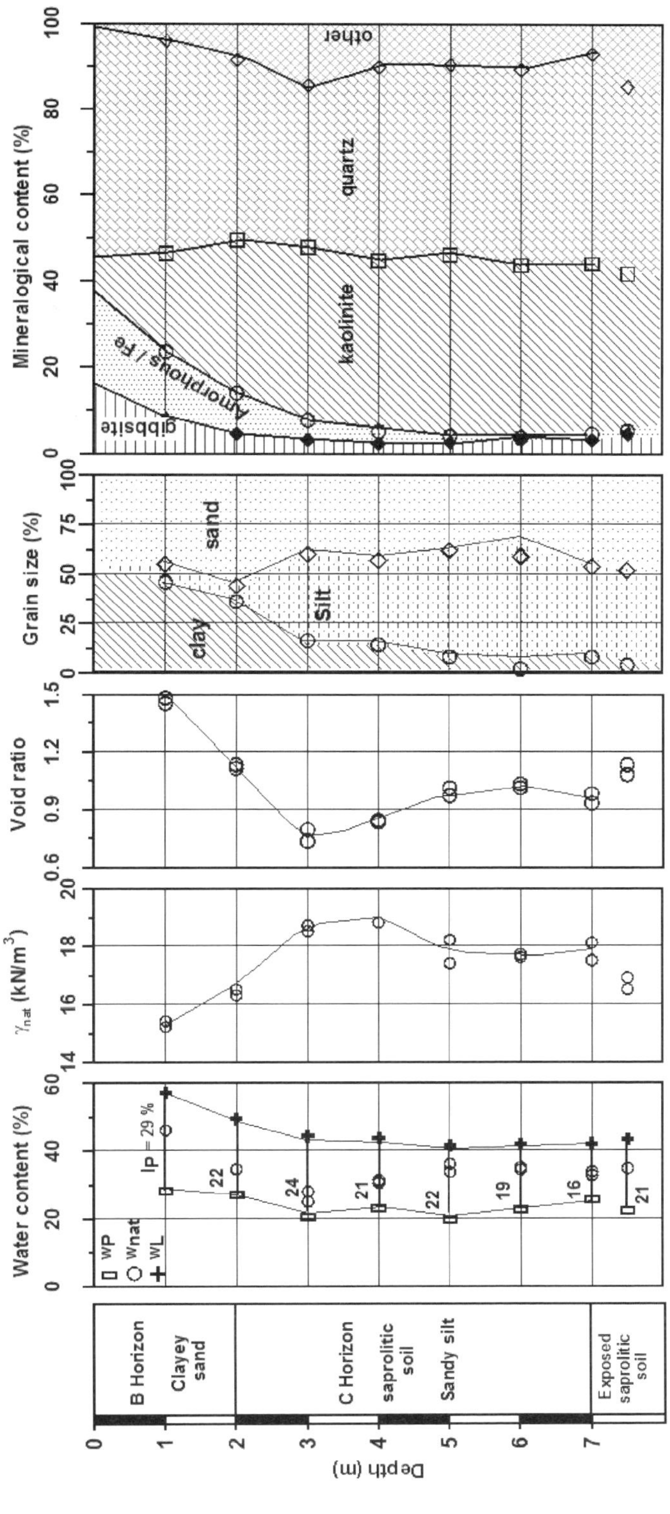

Fig. 5. Profile of the residual gneissic soil at Ouro Preto, Minas Gerais, Brazil (Futai *et al.* 2004).

Fig. 6. Laterite from Goiás State, Brazil.

residual soils, using mechanical compaction tests (mini-MCV, 'moisture condition value') to distinguish soils with a lateritic and non-lateritic behaviour. This proposal started out as an answer to problems encountered in pavement design in tropical regions, where some lateritic soils, by the usual classification systems used in road design, where soil structure is completely destroyed, were rejected as unacceptable, mainly on grounds of undesirable swelling characteristics. Nevertheless, they behaved extremely well in roads that were not 'designed' according to the then prevailing standards. Also in pavement research, De Medina (1989) suggested the use of resilient moduli to differentiate the behaviour of lateritic soil and saprolitic soils.

Lateritic and saprolitic soils show the presence of macro- and micro-pores the size and distribution of which can be investigated by means of the mercury intrusion technique. Futai *et al.* (2004) have performed such tests in a typical tropical soil, whose profile and characterization data can be seen in Figure 5. This is a typical residual soil from gneiss in the State of Minas Gerais, near Ouro Preto, Brazil (Futai *et al.* 2004). Pore-size distribution (PSD) was investigated by Futai *et al.* (2004) means of scanning electron microscopy (SEM) and mercury intrusion porosimetry (MIP). These studies were performed on oven-dried specimens. SEM studies for the soil at 1 m and 5 m depth are shown in Figure 9. The particles of the soil at 1 m depth (Fig. 9a) appear to be aggregated and large pores may be observed, although the clay minerals are not clearly observed. SEM for the soil at 5 m depth (Fig. 9b) showed voids and large parallel plates of kaolin. Porosimetry measurements obtained by Futai *et al.* (2004) using MIP are shown in Figure 10, for incremental intrusion. The soil at 1 m depth showed a clear bimodal distribution of pores, with macro-pores in the range 10–100 μm and micro-pores smaller than 0.1 μm. The soil at 5 m depth had only macro-pores larger than 2 μm, but concentrated in the ranges 2–10 μm and 30–300 μm. Porosimetry measurements appear to confirm the results of the grain-size analysis; that is, horizon B possesses smaller pores and higher clay content than horizon C. The overall analysis of grain size, soil microscopy and porosimetry suggests a metastable structure for the horizon B soil comprising micro- and macro-pores.

The soil water retention curve for this soil reflects the pore-size distribution. The bimodal retention curve for the soil at 1 m depth is typical of tropical highly weathered soils (Carvalho & Leroueil 2000) containing aggregated particles that are uncemented or cemented by iron oxides linked by clay bridges. According to the same researches, these soils have two air entry points, corresponding to macro- and micro-pores. The soil at 5 m depth has a retention curve with two slopes but no plateau, which appears to be consistent with the two ranges of macro-pores at 2–10 and 30–300 μm (Fig. 11).

Porosimetry measurements appear to confirm the results of the grain-size analysis; that is, horizon B possesses smaller pores and higher clay content than horizon C. The

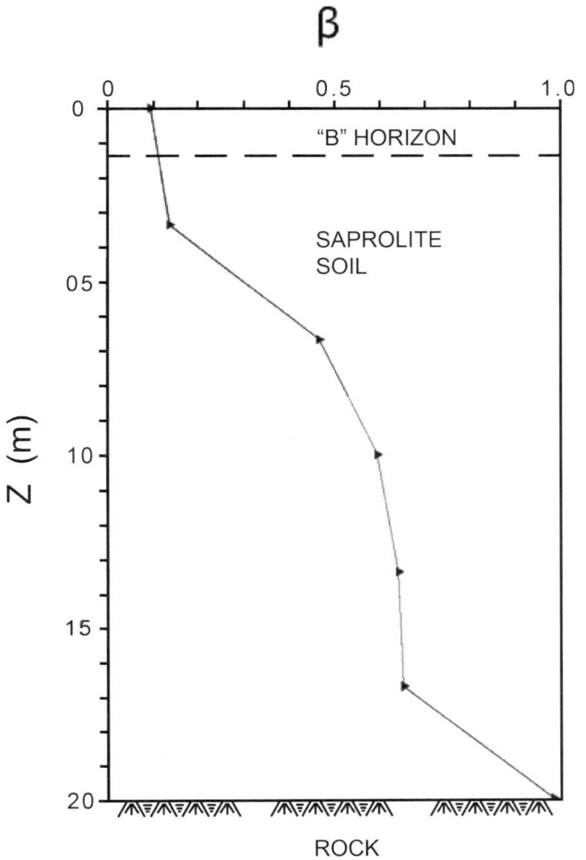

Fig. 7. The beta index v. depth in a gneiss residual soil profile in Rio de Janeiro (Adapted from Bernardes *et al.* 1992).

overall analysis of grain size, soil microscopy and porosimetry suggests a metastable structure for the horizon B soil comprising micro- and macro-pores.

Identification of lateritic soils

For a rapid assessment of whether a soil is lateritic or not the immersion technique (slaking test, or crumb test) can be used. Lateritic soils remain intact even 24 h after immersion, because of their true effective cohesion, whereas saprolitic soils crumble rapidly. Of course, soils in the initial state of laterization may also crumble, but not as completely and as rapidly as a saprolitic soil. Therefore this test is not failproof, and the determination of the k_r index is necessary.

Another way to distinguish these soils is by means of the Soil Conservation Service (SCS) double hydrometer test (Decker & Dunningan 1976). The SCS test was originally aimed at identifying dispersive soils, but can be used to distinguish soils with lateritic and non-lateritic behaviour. The clay content of a lateritic soil determined by a grain-size analysis without any dispersing agent and low mechanical action is practically zero, because the aggregates are not destroyed (Fig. 12). If the analysis is made with mechanical agitation and dispersing agent, the clay particles are loosened from the aggregates and clay contents as high as 60% (or higher) are measured.

In situ void ratios

The natural void ratio varies according to the type of parent rock and degree of weathering. Table 2 shows the range of void ratios found by several researchers in Brazil. It can be seen that there is an increase in void ratio from saprolitic soil to lateritic soil. However, there are instances where the colluvial layer may exhibit an *in situ* void ratio of the same order as that of the underlying residual soil, depending on the genesis of the colluvial layer. If the colluvium is formed by a translational or rotational slide of the residual soil, it may preserve the characteristics of the residual soil. If the colluvium was weathered by laterization after complete destructuring of the residual soil, it will exhibit larger void ratios. It may, therefore, be collapsible or not at low vertical stresses, depending on the degree of laterization.

Shear strength of tropical soils

Some work has been done on this subject. Sandroni (1977, 1991), Maccarini (1988) and Massey *et al.* (1989) called attention to the influence of mineralogy on the shear strength parameters of a saprolitic soil from gneiss, as shown in Figure 8.

The Mohr envelope of shear strength tests is generally curved (Fig. 13). The curved portion is due to structure effects. The test results given in the present study were all obtained in saturated conditions. Gan & Fredlund (1996) also found an initial curved Mohr envelope for two saprolitic soils of Hong Kong, as can be seen in Figure 14. Suction has a great influence on the Mohr envelope, shifting it upwards, as shown by Fredlund & Rahardjo (1993). The influence of suction is to increase the cohesion intercept, with an insignificant influence of the effective friction angle.

The existence of true cohesion

Cohesion is found in rocks, and is lost in the weathering process. Effective cohesion in soils is usually zero. However, true effective cohesion can occur in saturated cemented soils. The oxides of lateritic soils form a weak cement, and true cohesion, although small, exists. This cohesion is not

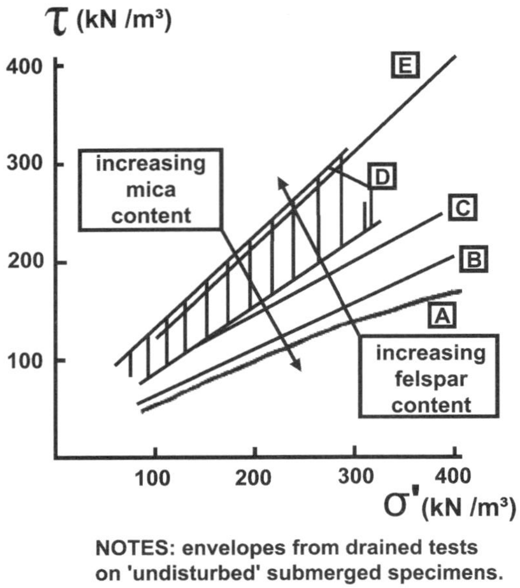

AUTHOR SITE	GRAIN SIZE (%) SAND	SILT	CLAY	VOIDS RATIO	STRENGTH ENVELOPE	COARSE GRAINS MINERALS
CAMPOS (1974) CATINGUEIRO	92	5	3	.76	CURVED A	Q: 30-60 F: – 0 M: 40-70
VARGAS (1951) VILA ANASTÁCIO	40	44	16	1.0	CURVED B	High mica Content Some felspat
SERAPHIM (1974) ESTRADA DE FURNAS	75	19	6	.95	c'=40 kN /m³ ø' = 29.5° C	Q: 48-60 F: 11– 23 M: 22-28
SANDRONI (1973) RIO	50 to 80	15 to 30	5 to 25	.30 to .90	c'=17 to 35 kN /m³ ø' = 34 to 43° D	Low mica content
SERAPHIM (1974) ESTRADA DE FURNAS	83	13	4	.70	c'=25 kN /m³ ø' = 43° E	Q: 48-58 F: 30– 42 M: 6-14

NOTE: Q: quartz F: feldspar M: mica

Fig. 8. Relationship between the mineralogy and the shear strength parameters of residual soils from gneiss in Rio de Janeiro (Sandroni 1977).

lost in a saturated state. Saturated saprolitic soils do not show this effective cohesion, whereas saturated lateritic soils do. However, the determination of this true cohesion poses a challenge in laboratory tests. Standard triaxial tests require that the minimum principal stress be greater than zero, or equal to zero, in the case of the unconfined compression test. The determination of this cohesion is difficult, even using direct shear tests, because of the low confining

Fig. 9. Scanning electron microscopimages of the lateritic (1 m depth) and saprolitic soil (5 m depth) at Ouro Preto (Futai *et al.* 2004).

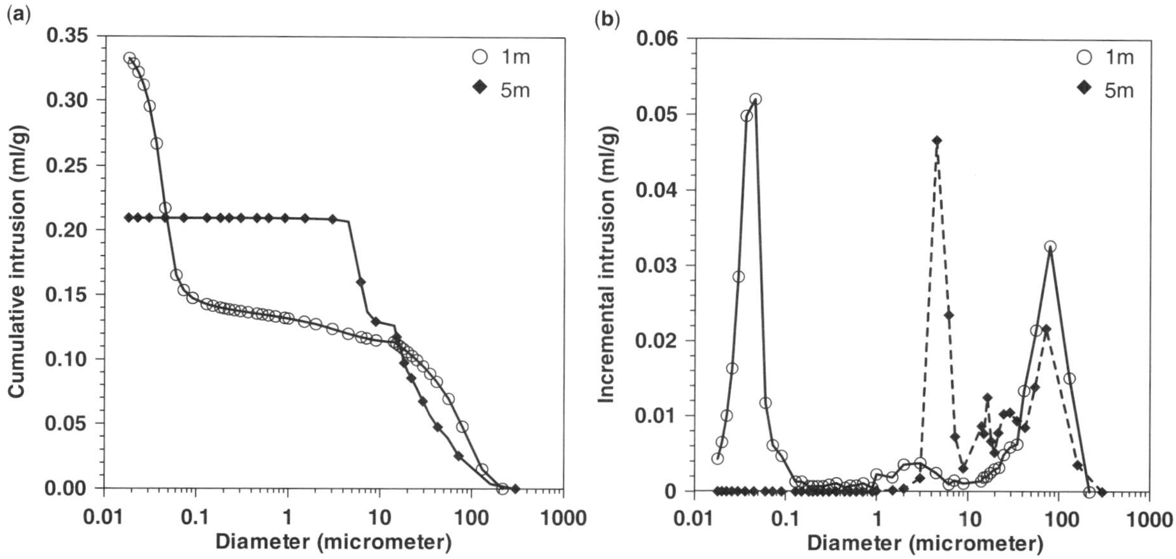

Fig. 10. Pore-size distribution obtained by the mercury intrusion porosimetry technique (Futai *et al.* 2004).

pressures involved. Bishop & Garga (1969) used a triaxial test with a reduced area in the centre of the specimen, but the procedures for this test are cumbersome. Using this technique, Meyer *et al.* (1999) determined the true cohesion of Waitemata clay, an undisturbed residual soil from Auckland, New Zealand. Cohesion intercepts of 13.7 kPa were measured.

A simple way to overcome the experimental difficulty is by means of the 'Brazilian test'. This test is widely used in rock mechanics for the determination of the tensile

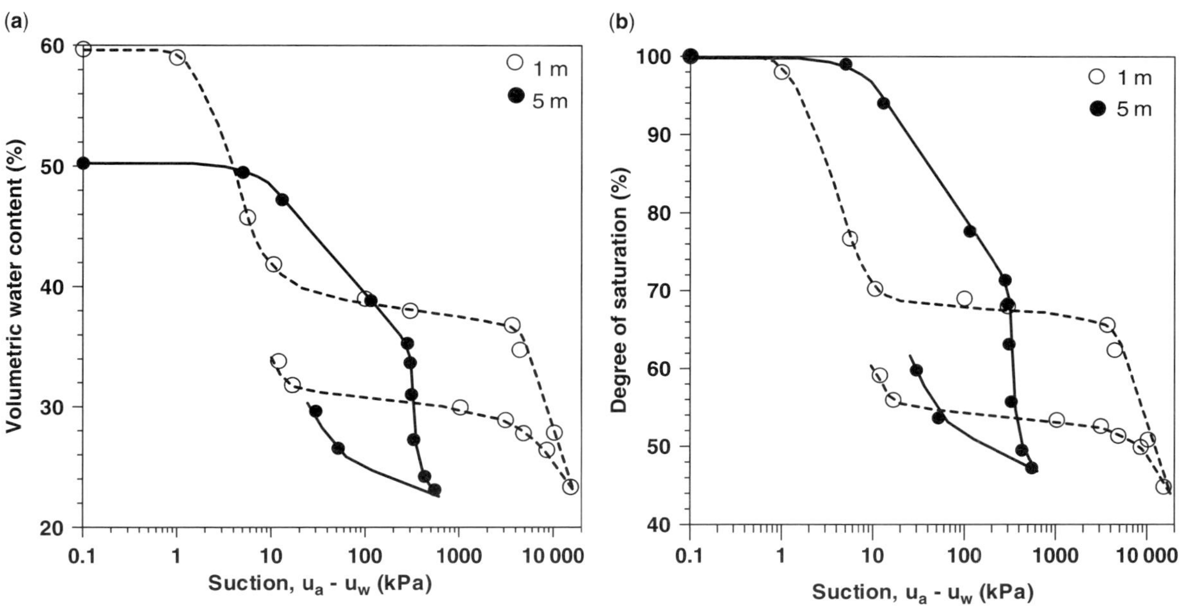

Fig. 11. Soil water retention curve (Futai *et al.* 2004).

Table 2. *Natural void ratios of undisturbed residual or saprolitic soils*

Reference and location	*Soil and rock type*	*Void ratio (e_0)*
Pinto *et al.* (1993), State of São Paulo, BR	Lateritic, migmatite	0.79–1.32
	Lateritic, basalt	1.3–1.6
	Colluvium (lateritic)	1.0–1.76
Clementino & Lacerda (1992), Rio de Janeiro, RJ	Saprolitic soil, granite	0.35–0.65
	Lateritic, granite	0.75–1.0
	Colluvium (lateritic), granite	1.1–1.6
Lacerda & Silveira (1992), Rio de Janeiro, RJ	Saprolitic, granite	1.1–1.2
	Colluvium (lateritic), granite	2.1–2.5
Futai *et al.* (2004), Ouro Preto, State of Minas Gerais	Lateritic, gneiss	1.1–1.5
	Saprolitic, gneiss	0.7–1.0

strength of rock, and was developed to measure the tensile strength of concrete cylinders. It was adapted for testing discs of cylindrical samples cut from undisturbed samples of lateritic soil (Rodriguez 2005). Figure 15 shows the experimental set-up, adapted from rock mechanics laboratory tests. The details of the loading platens can be seen in Figure 16. The soil specimens were discs cut from block samples with diameter (D) of 54 mm and thickness (t) of 27 mm ($D/2$). Loads and vertical displacements were measured by means of a load cell and an displacement transducer. The velocity of axial movement of the loading frame was 0.054 mm min^{-1}, the same as used in direct shear tests on the same soil. The groundwater was at great depth, and could not be sampled; therefore the specimens were immersed in distilled water for 24 h, and tested under water. This procedure was intended to allow all capillary stresses to dissipate. Of course, there would be a difference in the cohesion obtained if the natural groundwater were used, because of the presence of dissolved ions. However, the objective of the tests was to verify the

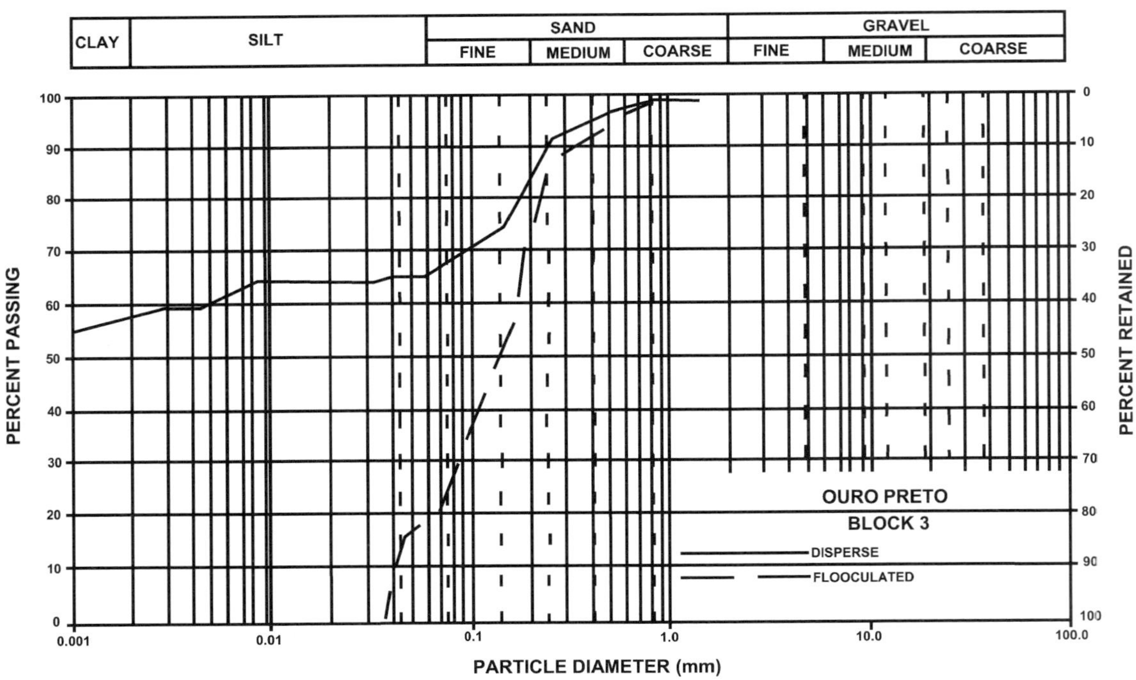

Fig. 12. Grain-size analysis of a lateritic soil without (dashed line) and with (continuous line) dispersing agent (SCS test) (Rodriguez 2005).

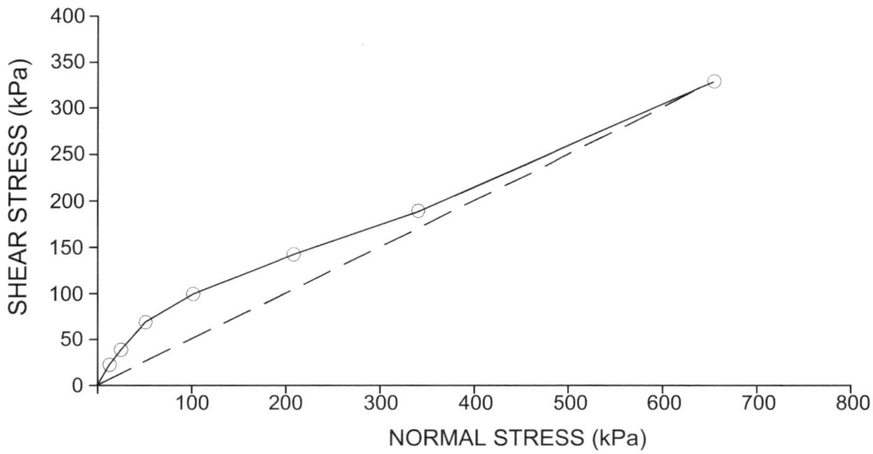

Fig. 13. Mohr envelope of a residual soil from gneiss in saturated direct shear tests (Rodriguez 2005).

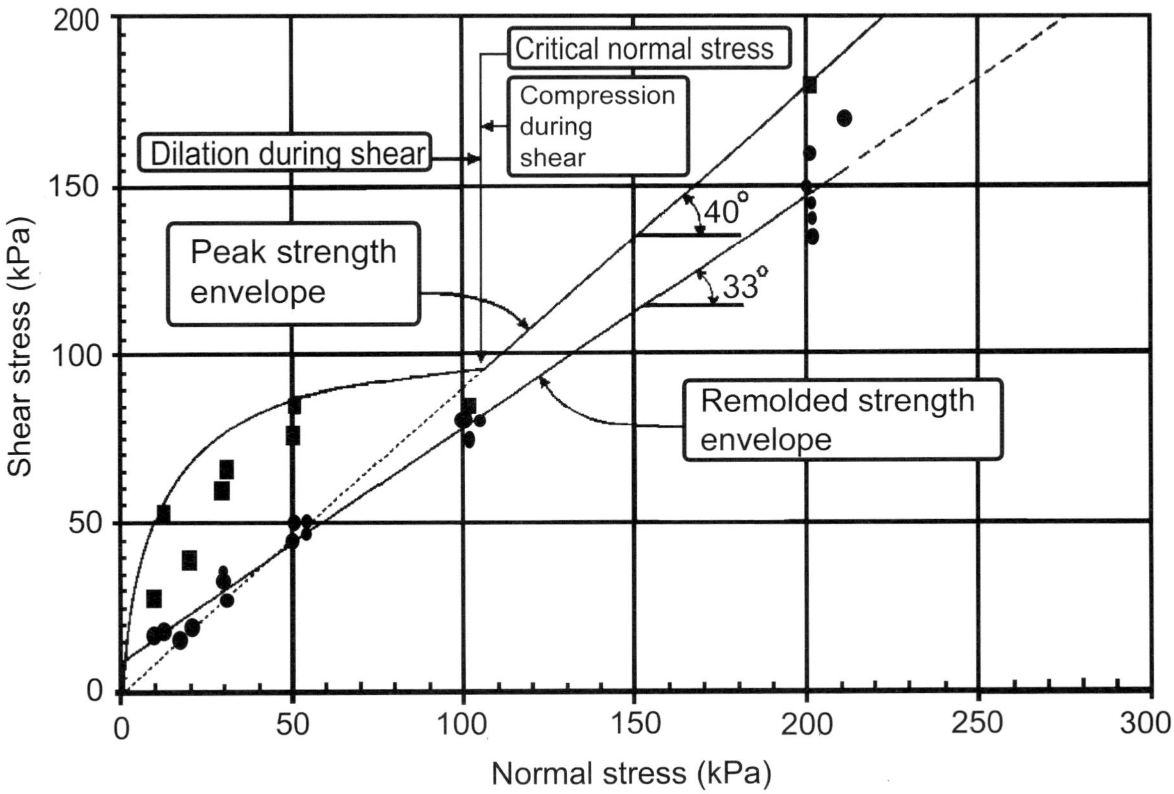

Fig. 14. Bi-segment Mohr–Coulomb failure envelope and remoulded strength envelope for a completely decomposed fine ash tuff in saturated triaxial tests (Gan & Fredlund 1996).

Fig. 15. Experimental set-up of the Brazilian test. Left, test without submersion; right, test with specimen under water (Rodriguez 2005).

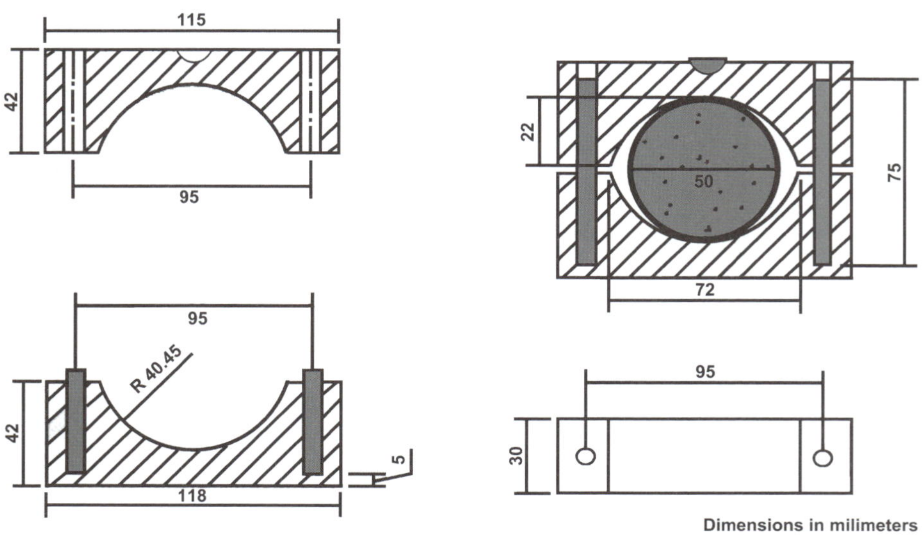

Fig. 16. Loading platens (Rodriguez 2005).

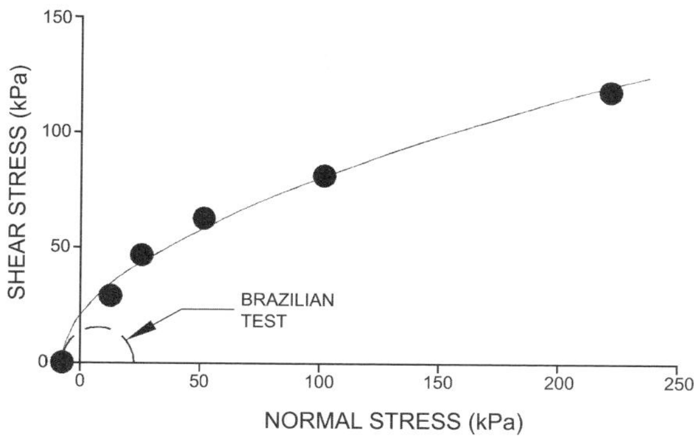

Fig. 17. Mohr envelope of a lateritic soil with true cohesion (Rodriguez 2005).

existence of true cohesion. The load at which the first fissure occurred was identified by a peak in the load v. deformation curve. The tensile strength was calculated by means of the equation

$$\sigma_t = \frac{2P}{\pi D t} \tag{3}$$

where P is the peak force, D is the diameter (in cm) and t is the thickness (in cm).

The curved envelope for soils that do not show cohesion usually is of the form $y = a.x^n$. For soils with true cohesion the equation proposed by Baker (2004) is

$$\tau = 100A\left(\frac{(\sigma + t)}{100}\right)^n \tag{4}$$

where t is the tension intercept on the σ axis, (in kPa), σ is the normal stress (in kPa), τ is the shear stress (in kPa), and A and n are non-dimensional parameters. This equation obeys Mohr's conditions if $A > 0.1$ and $0.5 \leq n \leq 1.0$.

Figure 17 shows the result of triaxial tests and Brazilian tests on a lateritic soil, and Figure 18 shows those on a saprolitic soil from gneiss. The envelopes with cohesion show a similarity to Mohr–Coulomb envelopes of rocks, but the similarity ends here. From rocks to saprolitic soils, cohesion falls to zero. In a further stage, as a result of pedogenetic processes, saprolitic soils turn into lateritic soils, and cohesion is gained through cementation.

Influence of Mohr envelope curves on the stability of shallow lateritic slopes

Steep slopes (angles greater than 30°) are common in granitic–gneissic regions of southern Brazil. The rock is covered by a shallow mantle of residual soil (1–4 m thick). During heavy rainfall the phreatic level may reach

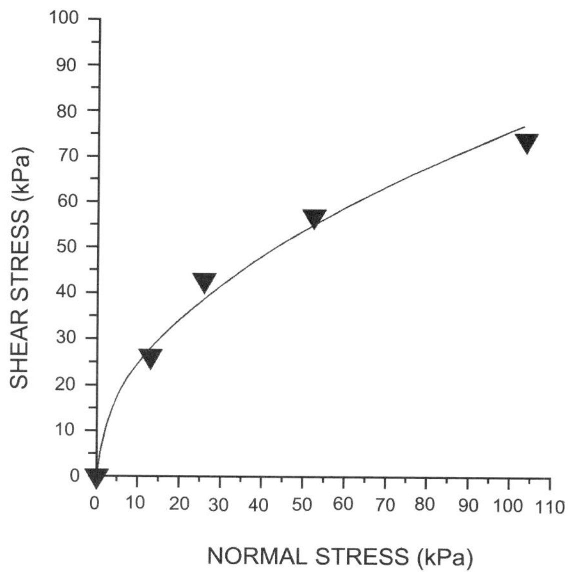

Fig. 18. Mohr envelope of a saprolitic soil without true cohesion (saturated, direct shear tests) (Rodriguez 2005).

depths close to the surface, and as the usual friction angle of these soils is typically in the range 30–40°, the slope may fail. Therefore the existence of a true effective cohesion is crucial in the development of these slides, which are of the so-called 'infinite slope' type and appear as scars on the mountainside when viewed from afar.

Because of the curved Mohr envelopes, the usual linear envelope used in standard stability analysis programs varies depending on the range of depth of the soil profile (this does not happen if the computer program allows a curved strength envelope). Figure 19 shows three Mohr envelopes: one curved, passing through the origin (zero effective cohesion, line A), one curved, with the tension stress considered (line B), and one linear, based on direct shear test results on submerged specimens loaded with normal stresses above 10 kPa (line C).

If one extrapolates linearly the data obtained in the stress range above 20 kPa, the effective cohesion intercept will be 28 kPa and the effective friction angle 29°. If an analysis is made using the infinite slope method with this envelope for very shallow slides, with the phreatic level at the surface and flow parallel to the slope, typically reaching less than 3 m in depth, the result could yield a factor of safety of the order of 1.50 for a natural slope angle of 40°. However, the curved envelope drops sharply below 20 kPa, and if the soil thickness is small, an envelope for the range up to 30 kPa in vertical stress would give, in this case, a cohesion intercept of 10, and an effective friction angle of 53°; a factor of safety of the order of 1.10 would be obtained. Therefore it is recommended to use a curved Mohr envelope in the analysis of shallow slides.

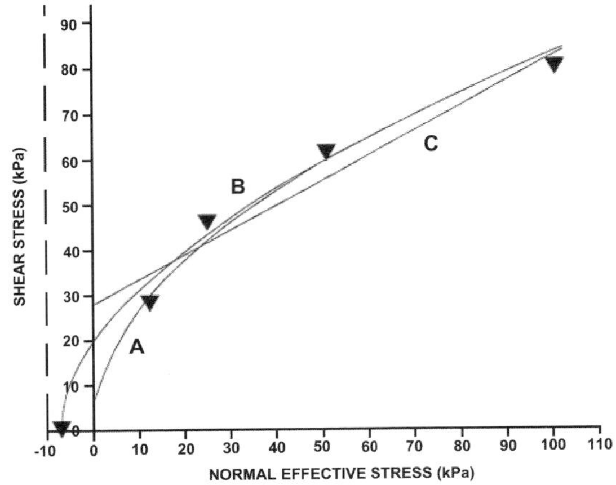

Fig. 19. Mohr envelopes A, B and C in the analysis of an infinite slope (Rodriguez 2005).

Residual friction angle

The movement along a shear surface of a landslide usually is able to reduce the soil shear strength to its residual condition. Leroueil *et al.* (1996) noted that after a first failure, when the peak strength is reached, the reactivation of a landslide can occur if the residual strength is mobilized. Fonseca (2006) analysed two such slides in a residual tropical soil in southern Brazil and showed that this explains the mechanisms of failure observed in the field.

Many researchers have made important contributions to the study of residual strength in recent decades (Lupini *et al.* 1981; Vaughan 1988; Fonseca & Lacerda 2003; Rigo *et al.* 2006). Initially sedimentary clays were studied, and Lupini *et al.* (1981) suggested a band that clays would follow when the residual friction angle is plotted against the clay content of the soil; the higher the clay content, the lower the friction angle. However, since that study a significant amount of data was collected, and Fonseca & Lacerda (2003) and Rigo *et al.* (2006) showed conclusively that lateritic soils behave as a granular soil, and their residual friction angle is independent of the clay content (determined in the usual grain-size tests using dispersing agent). Saprolitic soils have a trend similar to that observed by Lupini *et al.* (1981), but when the mica content of the silt fraction is significant the residual friction angle falls well below the band of Lupini *et al.* This can be seen in Figure 20 (Fonseca & Lacerda 2003).

Conclusions

The study of the properties of tropical soils is relevant particularly in relation to the mechanics of landslides. Regions with high rainfall and subject to severe and frequent rainstorms develop landslides and debris flows that may affect large populations. The initiation of these landslides can be explained when the shear strength characteristics of these soils are better understood. Colluvial slopes in tropical regions have special characteristics, as shown by Lacerda

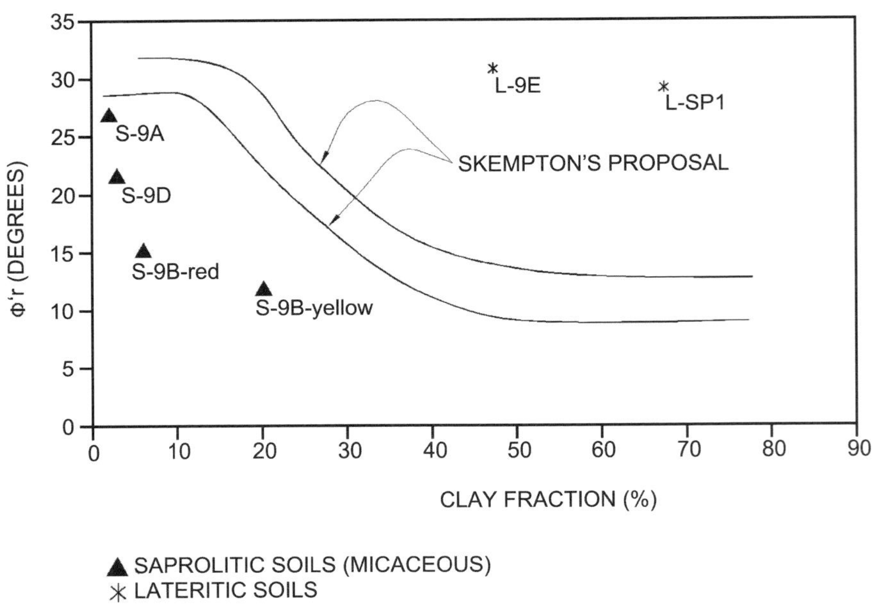

Fig. 20. Residual friction angles of lateritic and saprolitic (micaceous) soils from gneiss (Fonseca & Lacerda 2003).

(2004). Many slides in residual soils are directly related to relict structures inherited from the parent rock, and the residual strength is relevant in these cases. The existence of 'true' cohesion in lateritic soils is particularly important in the initiation of shallow slides.

Acknowledgements. The author thanks the National Research Council of Brazil (CNPq) and the Foundation of Research of the State of Rio de Janeiro (FAPERJ) for their continued support for research on the properties of tropical soils at COPPE/Federal University of Rio de Janeiro, through the PRONEX programmes.

References

ANONYMOUS 1985. *TropicaLS'85 First International Conference on Geomechanics in Tropical Lateritic and Saprolitic Soils, sponsored by the ISSMFE and ABMS (Brazilian Society for Soil Mechanics), Brasilia*. ABMS, Brasilia, Brazil.

AYDIN, A. 2006. Stability of saprolitic slopes: nature and role of field scale heterogeneities. *Natural Hazards and Earth System Sciences*, **6**, 89–96.

BAKER, R. 2004. Nonlinear Mohr envelopes based on triaxial data. *Journal of Geotechnical and Geoenvironmental Engineering*, **130**, 498–506.

BELL, F. G. 1981. *Engineering Properties of Soils and Rocks*. Butterworth, London.

BERNARDES, G. P., ANTUNES, F. S. & MUNECHIKA, V. L. 1992. Geotechnical properties of a leptinite (gneissic) residual soil of the Dona Marta belvedere, Rio de Janeiro. *In*: *Proceedings of 1st Brazilian Conference on Slope Stability, Rio de Janeiro*, Vol. 2, 419–431 [in Portuguese].

BISHOP, A. W. & GARGA, V. K. 1969. Drained tension tests on London clay. *Géotechnique*, **19**, 309–313.

BUCHANAN, F. 1807. *A Journey from Madras through the Countries of Mysore, Canara and Malabar*. T. Cadell and W. Davies, London [reprint: Asian Educational Services, New Delhi].

CARVALHO, J. C. & LEROUEIL, S. 2000. Normalizing models for soil retention curves. *In*: *Proceedings of 32nd Pavements Meeting, Brazilian Pavements Association, Brasília*, 96–106 [in Portuguese].

CARVALHO, J. B. Q., MEDINA, J. & NOGAMI, J. S. 1985. Geotechnical classification. *In*: *Peculiarities of Geotechnical Behavior of Tropical and Lateritic and Saprolitic soils*. Committee on Tropical Soils of the ISSMFE, Brazilian Society for Soil Mechanics, Progress Report 1982–1985, 21–43.

CLEMENTINO, R. V. & LACERDA, W. A. 1992. Shear strength of colluvial and residual granitic soils of Licurgo St. landslide. *In*: *Proceedings of 1st Brazilian Conference on Slope Stability, Rio de Janeiro*, Vol. 2, 463–478 [in Portuguese].

COLLINS, K. 1985. Towards characterization of tropical soil microstructure. *In*: *Proceedings of 1st International Conference on Geomechanics in Tropical Lateritic and Saprolitic Soils, TropicaL'S85, Brasilia*, Vol. 1, 85–96.

DECKER, R. S. & DUNNIGAN, L. P. 1976. Development and use of the Soil Conservation Service Dispersion Test. *In*: SHERARD, J. L. & DECKER, R. S. (eds) *Dispersive Clays, Related Piping and Erosion in Geotechnical Projects*. ASTM Special Technical Publication, **623**, 94–109.

DE MEDINA, J. 1989. Tropical soils in pavement design. *In*: *Proceedings of 12th International Conference on Soil Mechanics and Foundation Engineering, Rio de Janeiro*, Vol. 1, 543–546.

FONSECA, A. P. 2006. *Slide mechanisms in a gully in the Bananal river basin (RJ/SP)*. PhD thesis, COPPE/UFRJ, Rio de Janeiro [in Portuguese].

FONSECA, A. P. & LACERDA, W. A. 2003. Residual strength of colluvial and residual soils. *In*: *Proceedings of Soil and Rock America 2003, 12th Panamerican Conference on Soil Mechanics and Geotechnical Engineering*, Vol. 1. VGE, Essen, 485–488.

FOOKES, P. 1997. *Tropical Residual Soils*. Geological Society, London, Professional Handbook Series.

FREDLUND, D. G. & RAHARDJO, H. 1993. *Soil Mechanics for Unsaturated Soils*. Wiley, New York.

FUTAI, M. M., ALMEIDA, M. S. S. & LACERDA, W. A. 2004. Yield, strength and critical state conditions of a tropical saturated soil. *Journal of Geotechnical and Geoenvironmental Engineering, ASCE*, **130**, 1169–1179.

FUTAI, M. M., ALMEIDA, M. S. S. & LACERDA, W. A. 2006. The laboratory behaviour of a residual tropical soil. *In*: TAN, T. S., PHOON, K. K., HIGHT, D. W. & LEROUEIL, S. (eds) *Characterization and Engineering Properties of Natural Soils*. Taylor & Francis, Singapore, 2477–2505.

GAN, J. J.-M. & FREDLUND, D. G. 1996. Shear strength characteristics of two saprolitic soils. *Canadian Geotechnical Journal*, **33**, 595–609.

GIDIGASU, M. D. 1976. *Laterite Soil Engineering*. Elsevier, Amsterdam.

JENNY, H. 1941. *Factors of Soil Formation*. McGraw Hill, London.

LACERDA, W. A. 2004. The behavior of colluvial slopes in a tropical environment, Keynote Lecture. *In*: LACERDA, W. A., EHRLICH, M., FONTOURA, S. A. B. & SAYÃO, A. F. S. J. (eds) *Landslides – Advances in Evaluation and Stabilization, Proceedings of 9th International Symposium on Landslides, Rio de Janeiro, Brazil*, Vol. 2. Balkema, Rotterdam, 1315–1342.

LACERDA, W. A. 2007. Landslide initiation in saprolite and colluvium in southern Brazil: field and laboratory observations. *Geomorphology*, **87**, 104–119.

LACERDA, W. A. & ALMEIDA, M. S. S. 1995. State of the art report 'Engineering properties of regional soils: volcanic, residual, granular and soft clays'. *In*: *Proceedings of 10th Panamerican Conference on Soil Mechanics and Foundation Engineering, Guadalajara, Mexico*, Vol. 4, 133–188.

LACERDA, W. A. & SILVEIRA, G. C. 1992. Shear strength and compressibility of colluvial and residual soils of Soberbo slope. *In*: *Proceedings of 1st Brazilian Conference on Slope Stability, Rio de Janeiro*, Vol. 2, 445–462. [in Portuguese].

LACERDA, W. A., SANDRONI, S. S., COLLINS, K., DIAS, R. D. & PRUSZA, V. Z. 1985. Compressibility properties of lateritic and saprolitic soils. *In*: *Peculiarities of Geotechnical Behavior of Tropical and Lateritic and Saprolitic Soils*. Committee on Tropical Soils of the ISSMFE, Brazilian Society for Soil Mechanics, Progress Report 1982–1985, 85–113.

LEROUEIL, S., VAUNAT, J., PICARELLI, L., LOCAT, J., FAURE, R. & LEE, H. 1996. A geotechnical characterization of slope movements. *In*: *Proceedings of 7th International Symposium on Landslides, Trondheim, Norway*, Vol. 1, 53–74.

LUMB, P. 1962. The properties of decomposed granite. *Géotechnique*, **13**, 226–243.

LUPINI, J. F., SKINNER, A. E. & VAUGHAN, P. R. 1981. The drained residual strength of cohesive soils. *Géotechnique*, **36**, 269–274.

MACCARINI, M. 1988. Indexing the engineering properties of residual soils. *Quarterly Journal of Mechanics and Applied Mathematics*, **21**, 69–84.

MASSEY, J. B., IRFAN, T. Y. & CIPULLO, A. 1989. The characterization of granitic saprolitic soils. *In*: *Proceedings of 12th International Conference on Soil Mechanics and Foundation Engineering, Rio de Janeiro*, Vol. 1, 533–542.

MEYER, V. M., PENDER, M. J. & WESLEY, L. D. 1999. The very low effective stress behaviour of a residual soil. *In*: *Proceedings of 8th Australia and New Zealand Conference on Geomechanics, Hobart*, 2/877–2/883.

NOGAMI, J. S. & VILLIBOR, D. F. 1981. A new soil classification for road engineering purposes. *In*: *Proceedings of Brazilian Symposium on Tropical Soils, COPPE/UFRJ/ABMS, Rio de Janeiro*, 30–40.

NOGAMI, J. S., BANERJI, P. K., NETTERBERG, F. & SCHELLMANN, W. 1985. Characterization, identification and classification of tropical lateritic and saprolitic soils for geotechnical purposes. *In*: *Peculiarities of Geotechnical Behavior of Tropical and Lateritic and Saprolitic Soils*. Committee on Tropical Soils of the ISSMFE, Brazilian Society for Soil Mechanics, Progress Report 1982–1985, 3–20.

NOGAMI, J. S., COZZOLINO, V. M. N. & VILLIBOR, D. F. 1989. Meaning of coefficients and index of MCT soil classification for tropical soils. *In*: *Proceedings of 12th International Conference on Soil Mechanics and Foundation Engineering, Rio de Janeiro*, Vol. 1, 547–550.

PASTORE, E. L. 1995. Weathering profiles. *In*: *Proceedings of 10th Panamerican Conference on Soil Mechanics and Foundation Engineering, Guadalajara, Mexico*, Vol. 1, 353–364.

PINTO, C. S., GOBARA, W., PERES, J. E. E. & NADER, J. J. 1993. Properties of residual soils. *In*: *Solos do interior de São Paulo*, ABMS, DGEE São Carlos, São Carlos, 95–142 [in Portuguese].

PIRES, F. R. M. 1998. Arcabouço Geológico. *In*: CUNHA, S. B. & GUERRA, A. J. T. (eds) *Geomorfologia do Brasil*. Bertrand, Rio de Janeiro, 17–70.

RIGO, M. L., PINHEIRO, R. J. B., BRESSANI, L. A., BICA, A. V. D. & SILVEIRA, R. M. 2006. The residual shear strength of tropical soils. *Canadian Geotechnical Journal*, **43**, 431–447.

ROCHA FILHO, P., ANTUNES, F. S. & FALCÃO, M. F. G. 1985. Qualitative influence of the weathering degree upon the mechanical properties of a young gneiss residual soil. *In*: *Proceedings of 1st International Conference on Geomechanics in Tropical Lateritic and Saprolitic Soils, TropicaL'S85, Brasilia*, Vol. 1, 281–294.

RODRIGUEZ, T. T. 2005. *Colluvium classification: a geotechnical approach*. PhD thesis, COPPE/UFRJ, Rio de Janeiro [in Portuguese].

SANDRONI, S. S. 1977. Micromorphology and the interpretation of a landslide in gneissic residual soil. *In*: *Proceedings of International Symposium on the Geotechnics of Structurally Complex Formations, Associazione Geotecnica Italiana, Capri, Italy*, Vol. 1, 423–431.

SANDRONI, S. S. 1981. Residual soils – researches performed at PUC-RJ. *In*: *Proceedings of Brazilian Symposium on Tropical Soils in Engineering, COPPE-UFRJ, Rio de Janeiro*, Vol. 2, 30–65. [in Portuguese].

SANDRONI, S. S. 1991. Young metamorphic residual soil, Special Report. *In*: *Proceedings of 9th Pan-American Conference on Soil Mechanics and Foundation Engineering, Valparaiso, Chilean Society of Soil Mechanics and Geotechnical Engineering*, Vol. 1, 1771–1788.

VARGAS, M. 1953. Some engineering properties of residual clay soils occurring in southern Brazil. *In*: *Proceedings of 3rd International Conference on Soil Mechanics and Foundation Engineering, Switzerland*, Vol. 1, 84–90.

VARGAS, M. 1974. Engineering properties of residual soils derived from igneous and metamorphic rocks. *In*: *Proceedings of 2nd International Congress of the IAEG, São Paulo, Brazil*, Vol. 1, IV-pc1–IV-pc26.

VARGAS, M. 1985. The concept of tropical soils. *In*: *Proceedings of 1st International Conference on Geomechanics in Tropical Lateritic and Saprolitic Soils, TropicaLS'85, Brasilia*, Vol. 3, 101–134.

VAUGHAN, P. R. 1988. Characterizing the mechanical properties of in-situ residual soils, Keynote Lecture. *In*: *Proceedings of 2nd International Conference on Geomechanics of Tropical Soils, Singapore*, Vol. 2. Balkema, Rotterdam, 469–487.

Recurrent landslides predisposed by fault-induced weathering of flysch in the Western Carpathians

T. Pánek*, J. Hradecký, J. Minár & K. Šilhán

Department of Physical Geography and Geoecology, Faculty of Science, University of Ostrava, Chittussiho 10, Ostrava 710 00, Czech Republic

**Corresponding author (e-mail: tomas.panek@osu.cz)*

Abstract: The interrelationship between slope deformation and fault-induced weathering as a predisposing factor for the development of sliding is analysed through several case studies from the Western Carpathians in the Czech Republic. The study area comprises flysch nappes with alternating sandstone and shale of different permeability. These lithological structures are affected by systems of faults. Recurring slope instability is found associated with zones of deep weathering in tectonically weakened areas. Climatic variability of landslide activity can be identified during the Holocene by means of radiocarbon dating and pollen analysis. Areas affected by recurring landsliding suggest gradual and cyclic landslide frequency.

The Flysch Carpathians is the most landslide-affected region of the Czech Republic (Fig. 1). Both shallow and deep-seated landslides are found in tectonically deformed Mesozoic (dominant) and Tertiary units of sandstones, claystones and shales comprising highly permeable beds alternating with less permeable ones (Krejčí *et al.* 2002*a*, *b*; Baroň *et al.* 2004; Hradecký *et al.* 2007). There has been considerable interest in landslide analysis (preparatory and triggering factors and related hazard mitigation) since a heavy rainfall event in July 1997, when several hundred landslides occurred in the eastern part of the Czech Republic. Another important recent landslide phase occurred after rapid snow melt in March–April 2006, when more than 80 mostly shallow landslides occurred on clay-rich substratum in the vicinity of the towns of Zlín and Uherské Hradiště (Bíl & Müller 2007). Recently, at least 3700 (both ancient and active) landslides have been documented in the Czech part of the Flysch Carpathians (Krejčí *et al.* 2002*a*). They comprise mainly deep-seated lateral spreads of mountain ridges, rotational and translational landslides, debris slides and earthflows. On the other hand, evidence of catastrophic slope failures such as debris flows and rock avalanches is less abundant (Rybář & Stemberk 2000).

Geomorphological analysis of slope deformations (Baroň *et al.* 2004), landslide susceptibility mapping (Klimeš 2002) and detailed multidisciplinary study of selected slope failures (Baroň *et al.* 2004; Margielewski 2006*a*, *b*; Hradecký *et al.* 2007) have revealed complex behaviour of slope deformation. The deformations are usually multigenerational with multiple shear surfaces predisposed by an anisotropic flysch substratum (Margielewski 2006*b*). The main triggering factors include increased pore-water pressures resulting from long-lasting heavy rainfall (e.g. the July 1997 event), rapid snowmelt (e.g. the April 2006 event) or flash floods (e.g. 114 mm of precipitation measured on 24 August 2005 in the highest part of the Czech Flysch Carpathians). Other triggering mechanisms (such as earthquakes) are of lesser importance but cannot be ignored (Bober & Wojcik 1977).

There has been considerable climatic variation in Central Europe and this has influenced weathering during the Quaternary. These changes have resulted in frost weathering processes during Pleistocene stadials and chemical weathering during more humid and warmer interglacials. These variations in weathering processes have played a predisposing role in relation to landslide genesis. Periods of intensive weathering can also be identified during the Holocene when periods of relative landslide inactivity were replaced by increased landslide activity induced by one or more triggering factors. Phases with more pronounced landslide activity can be identified within the Carpathians in Central–Eastern Europe during the Late Pleistocene and Holocene (Fig. 2).

Hypothesis and methods

This study concerns weathering-related recurring slope phenomena that lead to enhanced denudation of tectonically weakened zones in selected parts of the Czech Flysch Carpathians (Figs 1, 3 & 4). This precondition has led to the occurrence of recurring slope instability (at least during the Holocene) as verified by both radiocarbon dating and the analysis of historical sources (aerial

From: CALCATERRA, D. & PARISE, M. (eds) *Weathering as a Predisposing Factor to Slope Movements*.
Geological Society, London, Engineering Geology Special Publications, **23**, 183–199.
DOI: 10.1144/EGSP23.11 0267-9914/10/$15.00

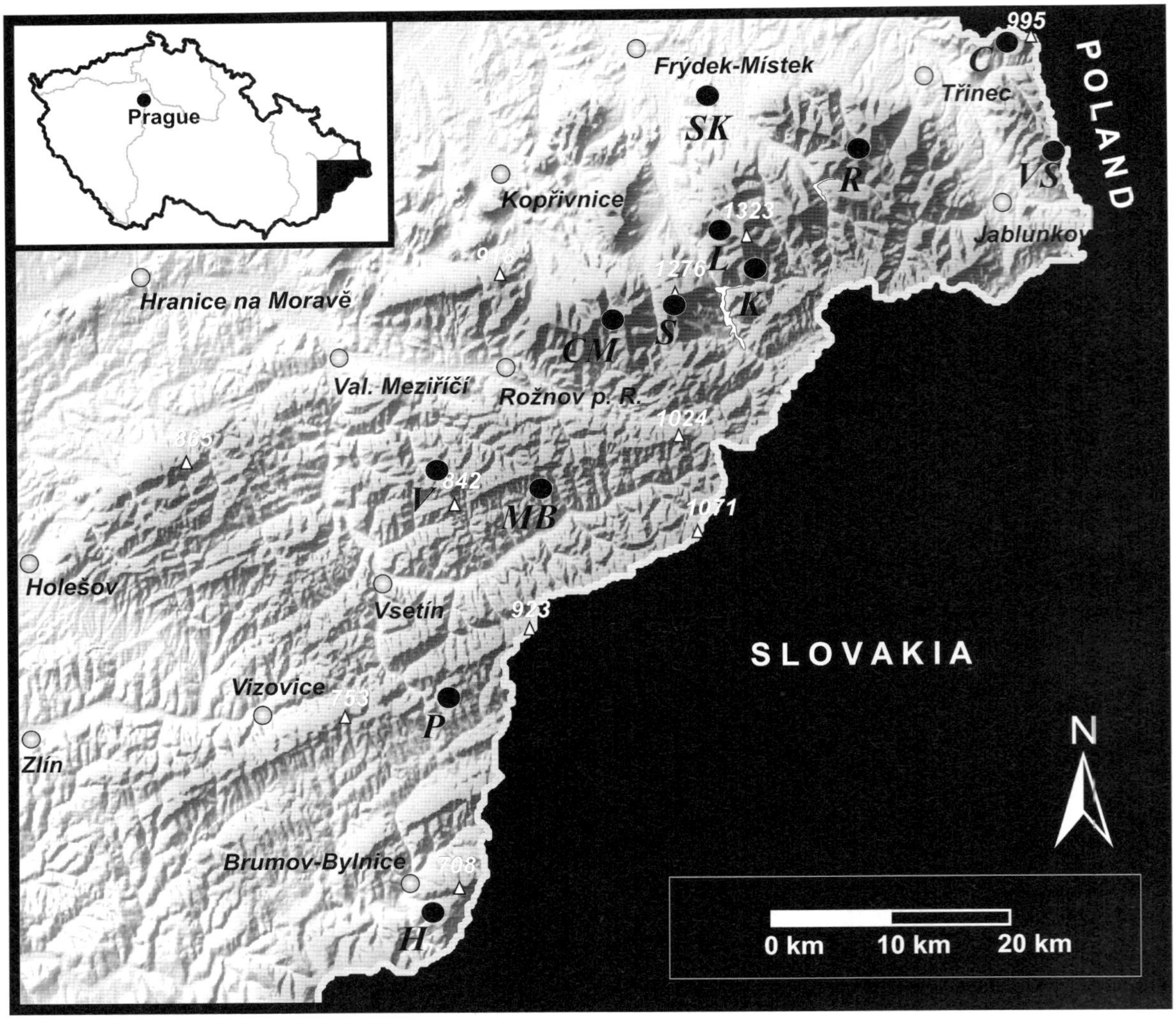

Fig. 1. Study area with locations of landsliding (black circles) referred to in the text. C, Velká Čantoryje; CM, Čertův Mlýn; H, Hluboče; K, Kobylík; L, Lukšinec; MB, Malá Brodská; P, Pulčínské skály rocks; R, Ropice; S, Smrk; SK, Skalice; V, Vaculov; VS, Velký Stožek.

photographs, old maps and eyewitness accounts). Landslide occurrence depends not only on triggering factors but also on the nature and duration of predisposing factors, of which weathering seems to be one of the most important (Fig. 3). In this paper dating of ancient landslides is compared with Holocene climatic reconstruction to determine the possible influence of weathering intensity on landslide frequency. This study of weathered and gravitationally disturbed bedrock was accomplished by means of electrical resistivity tomography (ERT) and ground penetrating radar (GPR) sounding; these techniques are used in tectonic (Suzuki *et al.* 2000), weathering (Beauvais *et al.* 2007) and landslide (Sass *et al.* 2008) studies worldwide.

Fault-induced weathering and recurring slope instabilities

Fault zones are often the focus for groundwater infiltration, geochemical alternation, frost weathering and/or activation of shrink–swell activity of smectite-rich clays. Landslides have been reported within weathered fault zones both from metamorphic or igneous domains (Ietto *et al.* 2007) and sedimentary sequences (Seijmonsbergen *et al.* 2005).

Massive occurrence of recurring slope instabilities along the tectonically weakened and weathered Alpine Fault zone of New Zealand was discussed by Korup (2004).

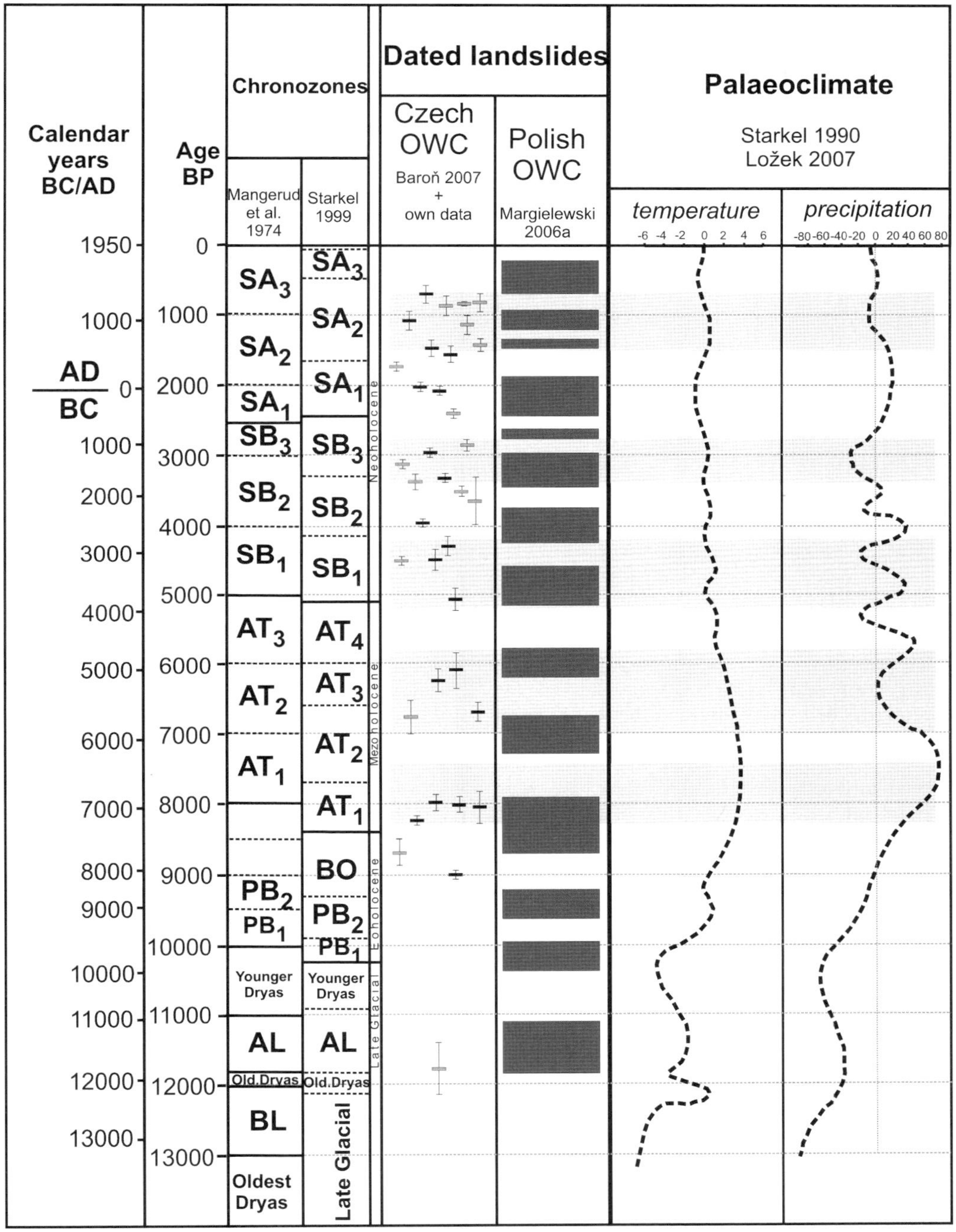

Fig. 2. Correlation of dated landslides (in both Czech and Polish parts of the Flysch Carpathians) with palaeoclimate. The scheme is based on Margielewski (2006*a*), with the time-span of chronozones after Mangerud *et al.* (1974) and Starkel (1999). Dated landslides in the Czech part of the Outer Western Carpathians (OWC) are after Baroň (2007), Baroň & Vařilová (2007) (20 cases, black boxes) and the authors of the present study (15 cases, grey boxes). Dated landslides (landslide phases derived from the dating of 67 landslides) in the western part of the Polish OWC are after Margielewski (2006*a*). Palaeotemperature and palaeoprecipitation are after Starkel (1990). Despite the fact that landslides occurred in the Czech part of the OWC throughout the entire Late Glacial and Holocene, significant clusters of landslides (horizontal grey bars) correlate with periods of high precipitation–low temperature. Chronozones: BL, Bølling interstadial; AL, Allerød interstadial; PB, Preboreal; BO, Boreal; AT, Atlantic; SB, Subboreal; SA, Subatlantic.

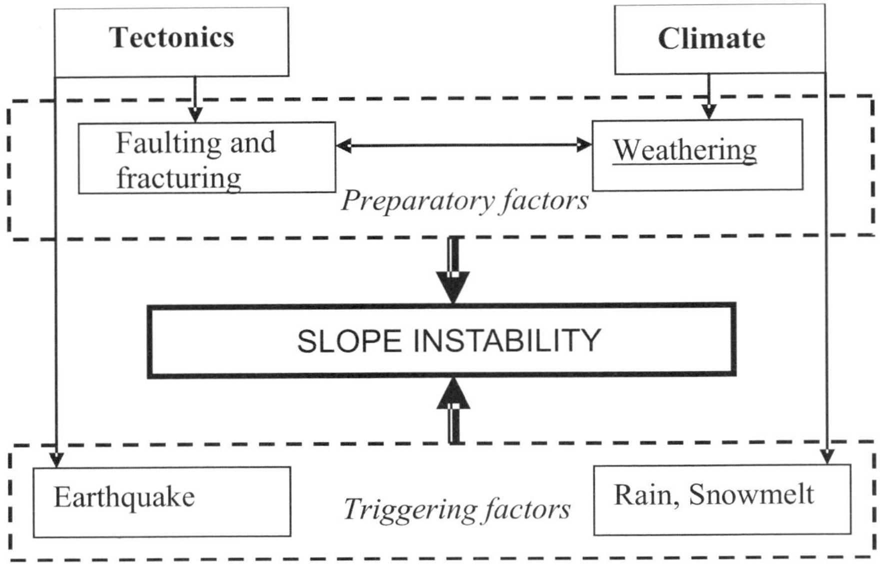

Fig. 3. Conceptual scheme of the relationships between weathering related to zones of faulting and landsliding.

Clusters of landslides and active gullies along this weakened zone are important denudational agents, sources of fluvial deposits and geomorphological hazard hotspots (Korup 2004). Kellogg (2001) analysed the importance of weathered rocks along reverse faults in relation to the development of huge slope failures in the Williams Fork Mountains (Colorado, USA). Many examples of hazardous slope failures related to weathered fractured zones have been described in mountainous environments. Strouth *et al.* (2006) stated that an important predisposing factor for the development of a huge rockslide near Newhalem (Washington, USA) was a highly fractured zone passing through the main scarp of the landslide. Similar observations were made by Boultbee *et al.* (2006), Brideau *et al.* (2006) and Geertsema *et al.* (2006) in mountainous areas of British Columbia. Likewise, a catastrophic rockslide–debris avalanche (17 February 2006) at St. Bernard (Leyte, Philippines) occurred within an older landslide area situated on volcaniclastic rocks affected by the intersection of at least three major discontinuity sets (Catane *et al.* 2007).

The relationship between deeply weathered tectonically weakened zones and the occurrence of landslides in flysch in the area of the Czech Western Carpathians has already been noted (see, e.g. Fig. 4). Baroň *et al.* (2004) attributed the development of one of the largest active landslides within the Czech Republic (Vaculov-Sedlo locality, Vsetínské vrchy Mts) to a 250 m wide, deeply weathered (perhaps to a depth of *c.* 100 m) sinistral fault zone. This huge translational–rotational slope failure, which originated before 6100 $\pm$ 250 ^{14}C BP (Atlantic), was reactivated at roughly 1550 $\pm$ 250 ^{14}C BP (Subatlantic) and again most recently during an event in July 1997 (Baroň *et al.* 2004). Other evidence of very large slope instabilities related to deeply weathered fault zones concerns deep-seated landslides that have not been reactivated during the past few decades, but that geomorphological and geophysical evidence suggests are repeatedly reactivated. Krejčí *et al.* (2002*b*) described a gravitationally collapsed anticlinal ridge of Hradisko (Pulčínské skály rocks, Javorníky Mts), where sandstones weathered to a depth of nearly 40 m are prone to toppling, the development of crevice-type caves, and earthflows in the frontal part of the slope failure.

Four main types of relationship can be distinguished regarding the positions of tectonically weakened zones, deep weathering and landslides, in the flysch bedrock of the Czech part of the Western Carpathians (Fig. 5). Fault zones passing along or through highly elevated ridges comprising flysch sequences of thick (usually several tens of metres) sandstones and predominantly claystone intercalations allow water penetration through the weakened sandstone beds and saturation of claystone layers (Fig. 5, type A). Spreading and subsequent sliding of the sandstone beds led to the development of 'double ridges' often accompanied by other (often active) mass movements (Hradecký & Pánek 2007). A similar consequence can be observed in connection with tectonically induced weathering of inclined flysch sequences with nearly slope-parallel bedding (Fig. 5, type B). Translational landslides in such settings are usually prone to repeated activation and they are one of the most common mass movements in the study region. Very large slope failures also occur in sandstone-dominated (thickly bedded flysch) formations with a negligible content of claystones where strong sandstones are significantly tectonized and altered by weathering (Fig. 5,

Fig. 4. Examples of weathered flysch rocks in the main scarp areas of landslides. (**a**) Detached sandstone blocks along a fault line in the ridge-top depression (Čertův Mlýn); (**b**) blocky disruption of thickly bedded flysch in a strongly tectonized zone of the Ropice slope failure; (**c**) weakened coarse-grained sandstones and conglomerates along a normal fault in the main scarp area of the Velký Stožek landslide; (**d**) thrust and weathered bedrock in the Kobylík slope failure; (**e**) main scarp of the Malá Brodská earthflow, which originated in thinly bedded weathered flysch bedrock.

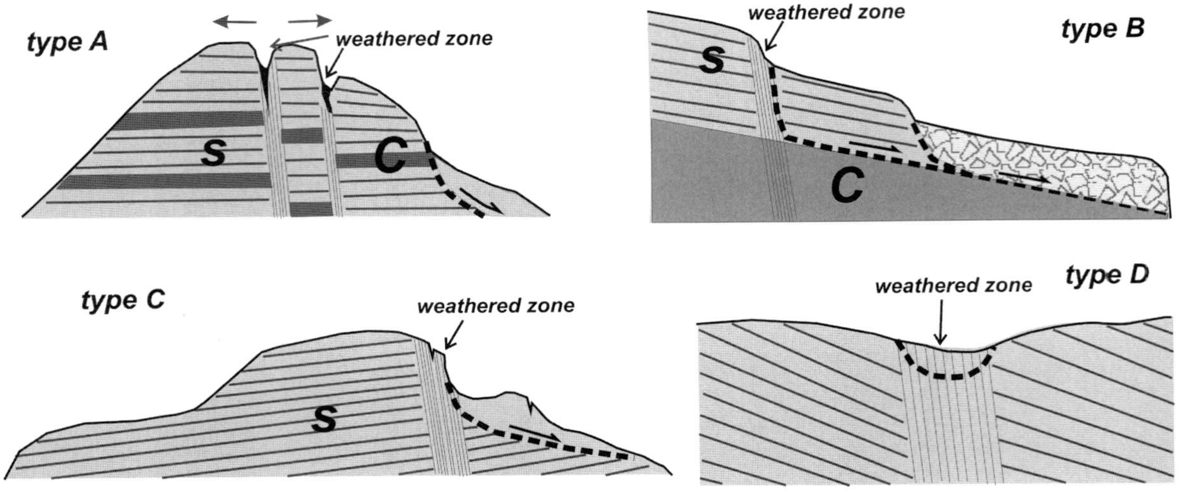

Fig. 5. Sketch of possible fault–weathering relationships within the landslide development; S, sandstones; C, claystones.

type C). Deep-seated (often rotational) landslides can occur as a result of tectonically and weathering induced 'homogenization' (*sensu* Margielewski 2006*b*) of anisotropic bedrock. In such cases, densely jointed bedrock physically resembles homogeneous material and tends to develop rotational shear surfaces (Margielewski 2006*b*). Specific conditions arise when dominantly flow-like landslides run parallel to weakened fault zones. This leads to gradual evacuation of weathered bedrock from shallow hillslope hollows (Fig. 5, type D).

Deep-seated spreading of mountain ridges

Documentation of several tens of gravitationally spread mountain ridges in the Czech Flysch Carpathians has revealed the important influence of tectonically weakened and deeply weathered fault zones (Fig. 6). The primary factor in the evolution of very large ridge collapses is gravitational disequilibrium caused by the position of massive rigid sandstones overlying weak, plastic claystones and shales (Krejčí *et al.* 2004; Hradecký & Pánek 2007). However, the location of fault zones determines the location of extensional features (ridge rents, double ridges, etc.), which focus the gravitational disruption of ridges (Figs 5 & 6). Prominent landforms associated with lateral spreading of flysch mountain ridges are tension cracks (sometimes occurring in an en echelon arrangement), which usually elongate in the direction of (Fig. 6a), and rarely also diagonally (Fig. 6b) or perpendicular (Fig. 6c) to, the main mountain ridge. In many localities, the limited depth extent of bedrock disruption beneath the ridge depressions was confirmed by speleological research on crevice-type caves. According to Wagner *et al.* (1990) most caves are less than 20 m deep, and rarely up to 50 m deep (e.g. Kněhyňská jeskyně cave, 57.5 m). Geophysical investigation of several double ridges indirectly verified the speleological research in places where open cracks are absent. For example, the combination of seismic refraction and resistivity measurements has revealed 40 m deep disruption under a diagonal trench crossing Mt. Smrk (1276 m; for location see Fig. 6b).

The Čertův Mlýn Mt. (1206 m) in the central part of the Moravskoslezské Beskydy Mts is an impressive example of fault-induced spreading of a mountain ridge (Figs 6a & 7, type A according to Fig. 5). Gently inclined glauconitic sandstone units of the Godula Formation (locally comprising an up to 6 km thickness of Cretaceous flysch complex) are intersected by at least three major faults in the vicinity of the top of the ridge (Krejčí *et al.* 2004). Morphotectonic analysis has revealed further failures in the near vicinity of the mountain ridge (Fig. 6). Gravitational spreading is associated with NNW–SSE and north–south faults. One of these is associated with a 100 m long and 4 m deep trench passing directly along the ridge top (Fig. 5, type A). Another discontinuity influenced development of a 330 m long depression with several crevices extending to a depth of up to 30 m (Wagner *et al.* 1990; Fig. 7). Several ERT cross-sections of the disrupted ridge indicate complicated spreading of highly fragmented bedrock (Fig. 8). The main ridge-top trench continues as a wedge-like, high-resistivity (>4000 Ωm) structure to a depth of 30 m. ERT sounding revealed highly fragmented bedrock even outside the morphological expression of the trench, in its northern prolongation. This indicates that gravitational deformation has extended further along the fragmented fault zone. The resistivity structure of the eastern undisturbed slope (with

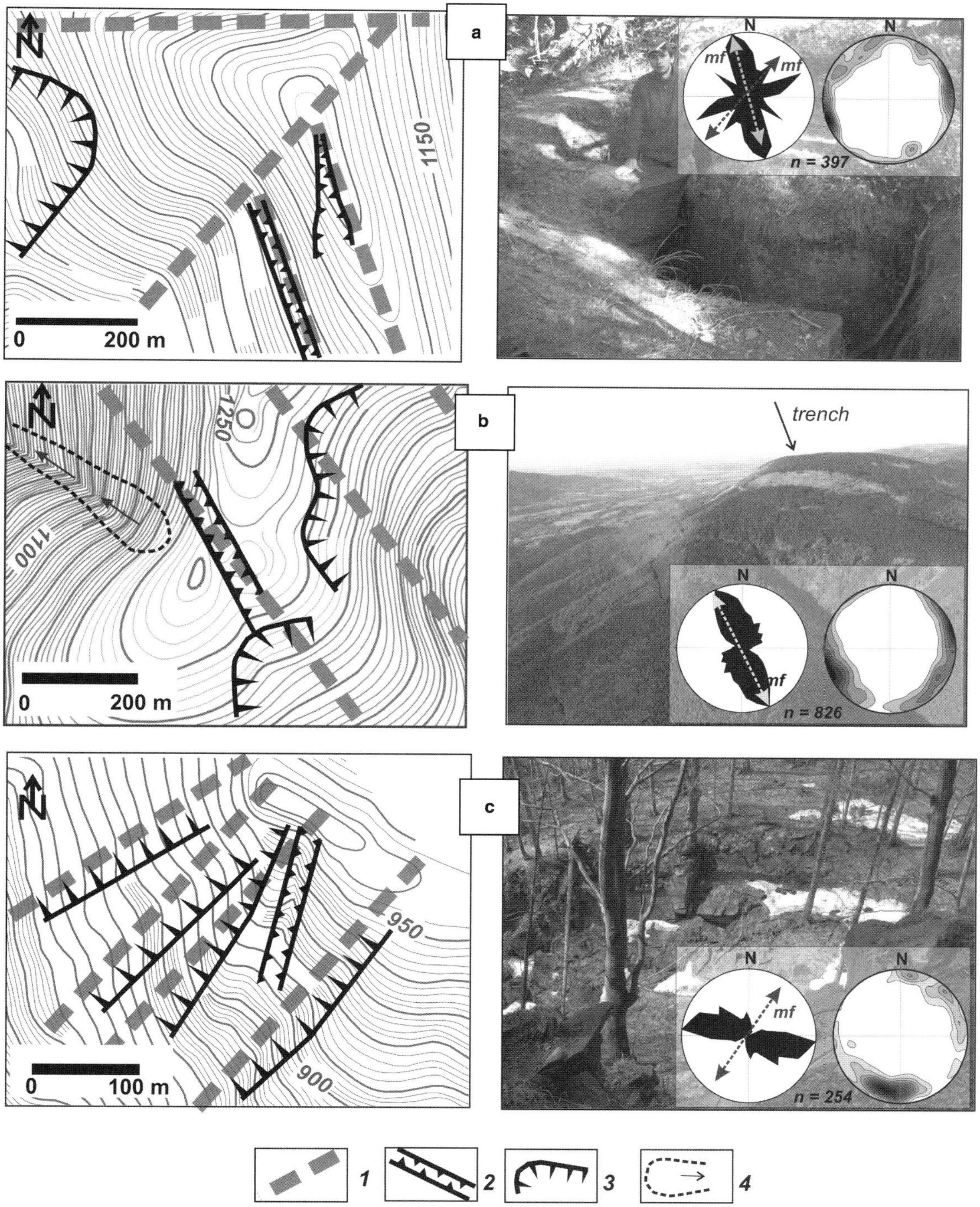

Fig. 6. Types of deep disintegration of mountain ridges. (**a**) Ridge-parallel trench predisposed by faults; (**b**) ridge-diagonal trench predisposed by faults; (**c**) trench perpendicular to the ridge. Legend: 1, fault; 2, trench; 3, landslide main scarp; 4, debris-flow basin. Stereoplots show the orientation of main faults (mf) and tectonic conditions in the vicinity of the trenches.

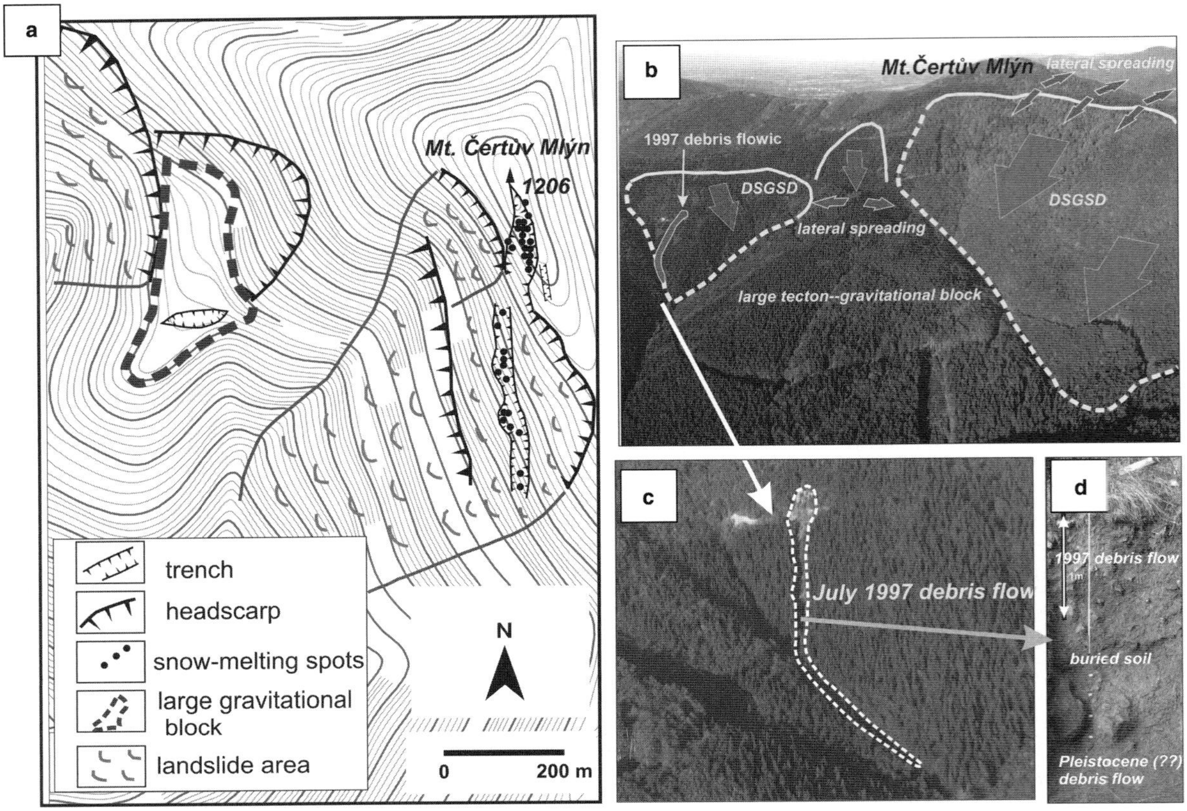

Fig. 7. (**a**) Geomorphological map of the gravitationally disrupted Čertův Mlýn Mt. (**b**) Main landforms of the Čertův Mlýn Mt, shown in an oblique aerial photograph (DSGSD, deep-seated gravitational slope deformation). (**c**) Detailed view of the July 1997 debris-flow path. (**d**) Outcrop showing the July 1997 debris-flow deposits overlying older, strongly weathered debris-flow sediments, demonstrating the multigenerational nature of the landslide area.

slope-parallel resistivity zones) strongly contrasts with the western, highly fragmented and faulted slope. The depth of the gravitationally spread bedrock in the western sector of the ridge is at least 60 m. Besides lateral spreading and toppling, landslides (mainly translational) and sliding of a huge block, with dimensions of 300 m × 150 m, led to the simultaneous development of further gravitationally derived landforms (Fig. 7). This huge tectonic–gravitational block situated in the lower part of the landslide area was disrupted by gravitational spreading that led to the development of a peat bog with a 6 m thick sequence of peat and colluvial deposits. The base of the 3.6 m thick peat deposit was dated at 3410 ± 1201 ^{14}C BP, which, along with pollen analysis, indicates a late Subboreal age for the block slide. The main sliding activity predated the inferred Subboreal age, but a fresh tension crack on the main ridge (see Fig. 6a) indicates continuing lateral spreading. The repetitive nature of the strongly tectonized and weathered bedrock is consistent with the location of recent reactivations of some parts of the disrupted ridge of the Čertův Mlýn Mt. (e.g. the debris flow with a volume of *c.* 1.5×10^3 m that occurred during the July 1997 event).

Recurring landslides have also been described under the gravitationally disrupted ridge of the Velká Čantoryje Mt. (995 m) (Hradecký & Pánek 2007) (see Fig. 6c). A highly fragmented zone of the overthrust sandstone-dominated nappe, which lies on a weathered claystone-dominated sequence, contains numerous multiple rotational slides. Basal peat bog deposits situated in the near main scarp depressions of these landslides show minimal ages of 3540 ± 80 ^{14}C BP to 3680 ± 350 ^{14}C BP (late Subboreal) and a further reactivation at approximately 2400 ± 70 ^{14}C BP to 2890 ± 90 ^{14}C BP (Subatlantic). Recent (rather minor) mass movements have occurred sporadically throughout the last 100 years (e.g. during the July 1997 event).

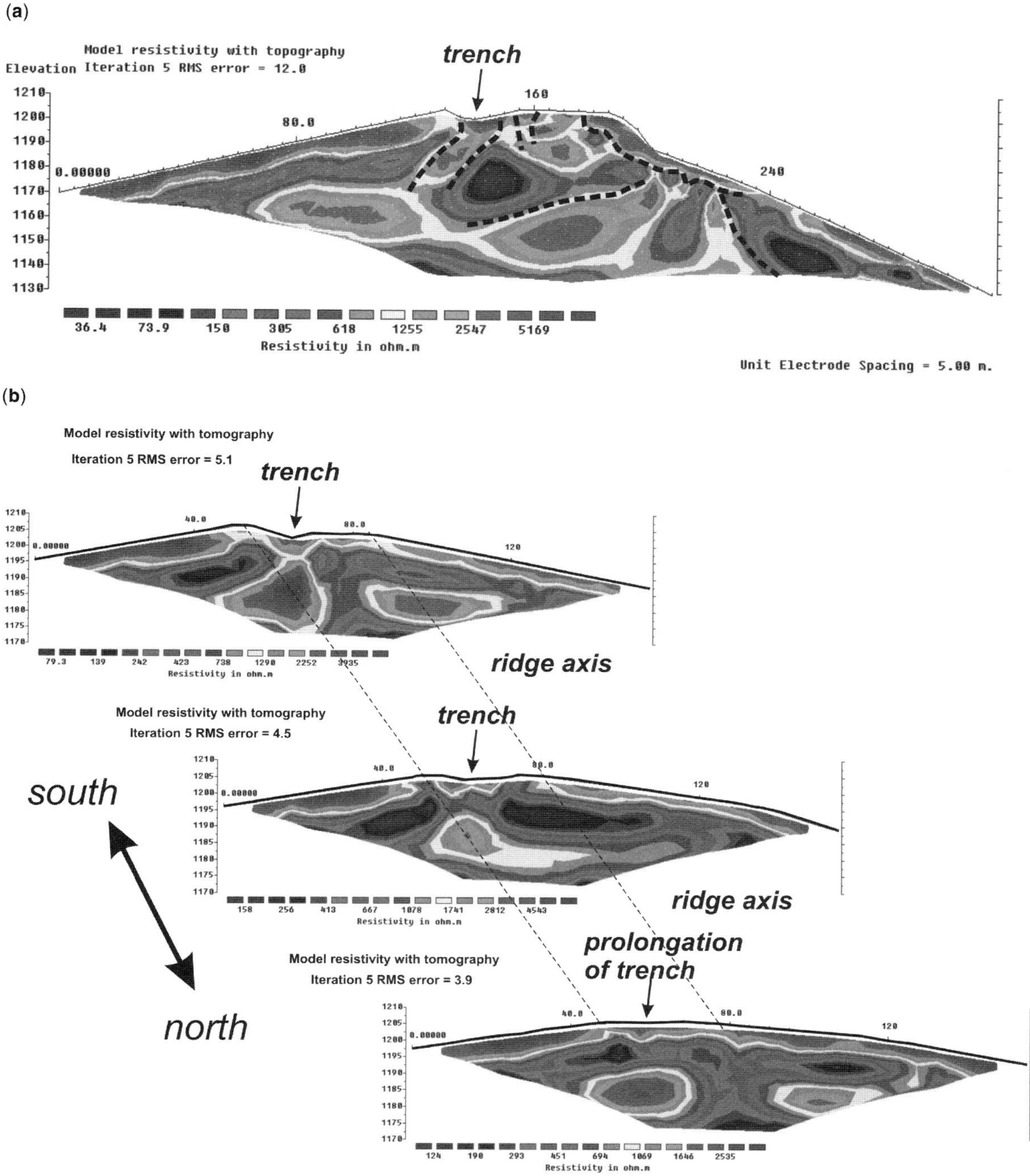

Fig. 8. Electrical resistivity tomography (ERT) in selected cross-sections at Čertův Mlýn Mt. (**a**) Deep geoelectrical structure (to 60 m depth) of the mountain ridge; (**b**) shallower structure (to 30 m depth) in selected cross-sections through the spread ridge with a wedge-like tensional fracture up to 30 m deep, which is evident also in the section without prominent morphological expression (lowermost cross-section).

Massive bedrock landslides

Several deep-seated landslides involve massive sandstone without significant intercalations of claystones. Such examples occur where tectonic fragmentation and subsequent weathering have allowed the development of zones up to several hundred metres wide of weakened bedrock (type C according to Fig. 5; see also Fig. 4b).

The best example of this type of slope deformation occurred on the northern slope of the Ropice Mountain (1082 m) in the eastern part of the Moravskoslezské Beskydy Mountains (Fig. 1). Gently dipping (up to 5°) sandstones of the Godula Formation (single beds 1–5 m thick) form a massive ridge with pronounced structural flats on the watersheds. The Ropice massif is situated at the intersection of three fault structures, the most important of which is an east–west-trending fault zone north of the mountain top (Fig. 9). The fault zone comprises a highly fragmented belt more than 400 m wide. This has influenced the development of several deep-seated landslides conspicuously arranged an the east–west direction. Outcrop observations have shown that the weathered sandstones have been fragmented into blocks less than 20 cm wide. This is in contrast to unaltered Godula sandstones, which typically have fracture spacings of >100 cm (Fig. 4b).

Several rotational landslides, toppled blocks and very large flow-like landslides evolved as a cascade on a 1300 m long and 800 m wide segment of the slope. The most prominent failure is a rotational slide with a 60 m high rocky main scarp, tilted blocks several tens of metres wide, and extensional trenches up to 5 m deep. Older, denuded scars of the rotational slide are situated above the main scarp and there is evidence downslope of advancing and multiple rotational behaviour of slope deformation (*sensu* Dikau *et al.* 1996). GPR measurements in the depletion zone of the slope failure revealed a dense system of wedge-like structures reaching a depth of more than 40 m. Despite the fact that the main shear surface was not identified because of the limited penetration of the GPR, it is assumed that its depth is more than 50 m (Fig. 10). A part of the very large rotated block collapsed and transformed into a 1000 m long flow-like landslide that blocked an adjacent river valley. At least two ancient flow-like landslides can be inferred from the morphology of accumulation lobes and radiocarbon dating of backwater deposits above the landslide dam. The first event occurred at 1450 ± 100 ^{14}C BP (Subatlantic age of a basal sediment layer of a former landslide-dammed lake). Another smaller flow-slide (identified by a change in sedimentation within the impoundment) probably postdated 310 ± 60 ^{14}C BP. Geomorphological evidence points to even more landslide events in fragmented bedrock in the northern part of Ropice Mt. Failures before *c.* 1.5 ka BP are indicated by the GPR record in the lower part of the slope failure, stratified diamictons in the distal part of the accumulation lobe, and a large alluvial fan in the main valley downslope from the failure (Pánek *et al.* 2009).

Recent catastrophic failures

Some of the catastrophic failures that have occurred within the last 10 years clearly demonstrate a relationship with fault-related weathered zones. Some of the largest active slope deformations, for example, the Vaculov-Sedlo failure according to Baroň *et al.* (2004) or the Malá Brodská flow-like landslide (see Fig. 2e) described by Rybář & Stemberk (2000), can be associated with deeply weathered flysch complexes.

A clear relationship to a weathered tectonic zone is revealed by the largest active landslide that has occurred since the July 1997 event in the Czech Republic. This flow-like landslide, with a length of 770 m and maximum width of 110 m, occurred between 3 and 4 April 2006 in the vicinity of the town of Brumov-Bylnice (Hluboče site) after a rapid snowmelt and prolonged antecedent rainfall. Slope failure affected the clay-rich Svodnice Formation of Paleocene age (part of the Magura Nappe, Bílé Karpaty Mts). The landslide has two distinct depletion zones (western and eastern), each occupied by translational–rotational landslides, which coalesce in the central part of the slope and continue as an earthflow for 350 m, as far as the valley bottom (Fig. 11). ERT sounding has detected possible shear surfaces at a depth of 5–10 m in both of the landslide depletion zones. Geomorphological mapping, morphotectonic observation and digital elevation model analysis show that the Hluboče landslide is located in a north–south fault zone (Fig. 11). This weakened belt passes through the western depletion zone of the landslide. The landslide is classified as type D of Figure 5. Drilling has confirmed that bedrock in this zone is severely weathered to a minimum depth of 6 m. Tectonic structures striking near vertically are evident in the ERT record of the landslide's western depletion zone (Fig. 11). A significant low-resistivity zone is situated directly under the western branch of the landslide.

Analyses of aerial photographs and eyewitness accounts show gradual, recurring landslides in the western branch since the 1950s. The April 2006 event was simply a culmination of gradual creep processes and the long-term evolution of shear surfaces. Geomorphological mapping has identified several historical reactivations of earthflows (old accumulation lobes and hillslope hollows) in the vicinity of the April 2006 failure. The evidence of landforms and aerial photography clearly display a repetitive nature of the slope failure.

One of the most hazardous landslides in the studied region is situated in the hilly landscape on the eastern slope of the Skalická Strážnice Hill (438 m), close to the town of Frýdek-Místek. The slope failure occurred within predominantly Lower Cretaceous thinly bedded flysch (intercalated with

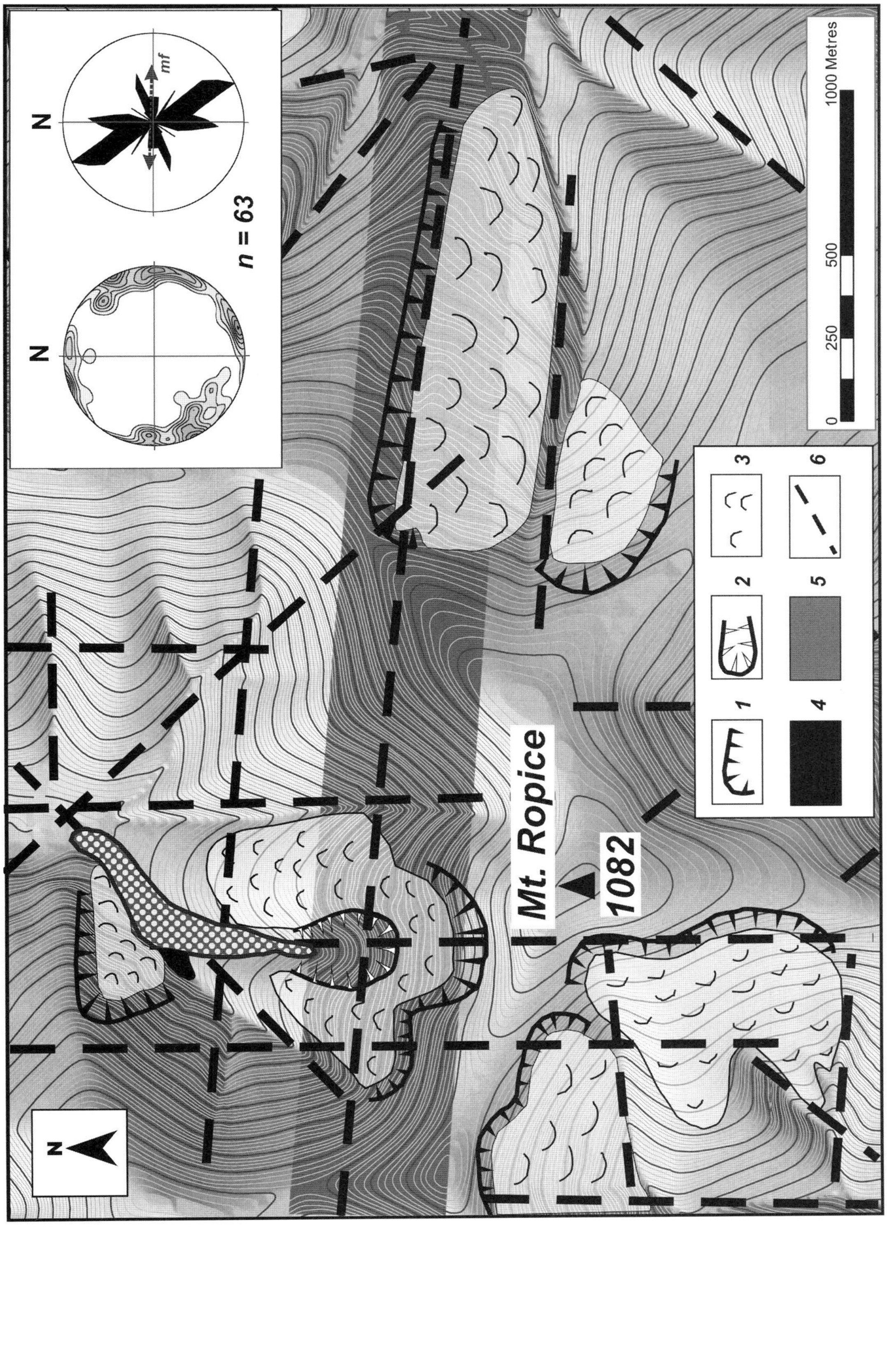

Fig. 9. Morphotectonic sketch and distribution of deep-seated landslides in the vicinity of Ropice Mt. Legend: 1, landslide main scarp; 2, flow-like landslide main scarp (likely rock avalanche); 3, landslide area; 4, valley blocked by landslide; 5, assumed weathered zone; 6, topolineaments.

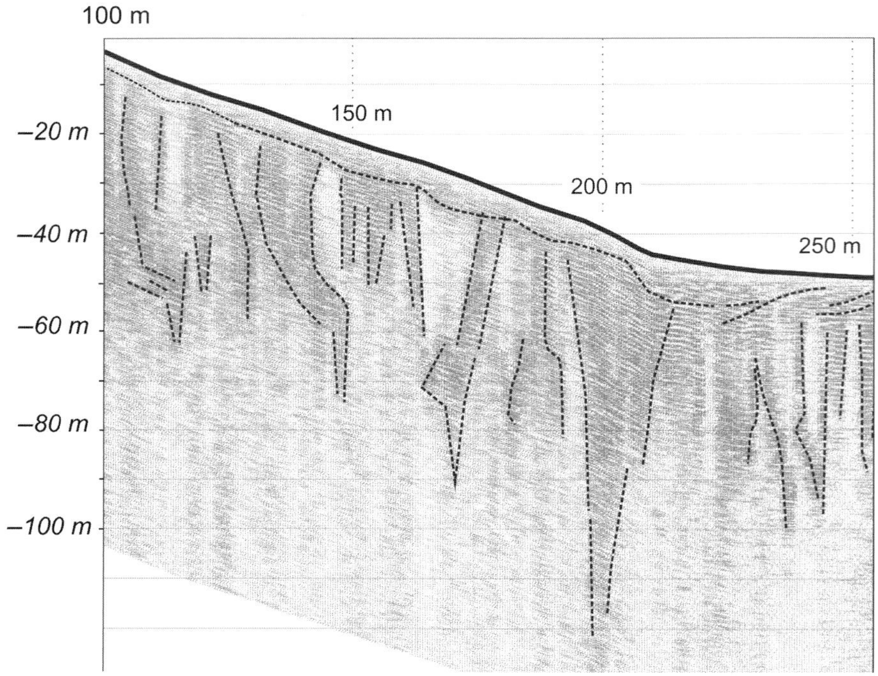

Fig. 10. GPR structure of the zone of depletion of the Ropice failure with conspicuous wedge-like tensional fractures that developed within strongly fragmented bedrock (after Pánek *et al.* 2009).

limestone layers up to several metres thick). A landslide area, more than 2 km wide, developed as a result of prolonged lateral erosion by the Morávka River (Fig. 12). An original landslide, perhaps of Pleistocene age, was reactivated several times during the 20th century. The main landslide events in August 1972 and July 1997 destroyed an important regional road and several cottages, and they still endanger a local baroque church (which is a protected site) and a graveyard. Kovář *et al.* (2006) observed 1.5–15 mm of horizontal movements on shear surfaces (at depths of 3.5–7 m) within the most active part of the landslide by inclinometer measurements between 20 December 2004 and 13 April 2006.

Morphotectonic mapping has revealed the presence of NE–SW- and north–south-striking topolineaments passing through the wider region of the landslide area (Fig. 12). These discontinuity sets influenced the development of main scarps. In addition to the prevailing orientation of bedding, these discontinuities should be considered as an important predisposing factor in the development of the recurring landsliding (type B of Fig. 5). ERT sounding confirmed the structural influence on the main scarp of the most active failure, in the northern part of the landslide area (Figs 12 & 13). Weathered thinly bedded flysch of Early Cretaceous age can be observed in conspicuous tension cracks and crevice-type caves above the main landslide body (Fig. 12). Boreholes within and directly above the main landslide body (for location see Fig. 12) have revealed weathered bedrock to a depth of 5–11 m (Kovář *et al.* 2006). Future progressive up-slope deformation can be expected within the weakened fragmented zone above the landslide main scarp (Fig. 13b, d).

Conclusion

Geomorphological analysis in the Czech part of the Flysch Carpathians confirms the relationship between highly fragmented and weathered fault-related zones and recurring slope instabilities during the Holocene. Although slope failures in the Flysch Carpathians are widespread and associated with other predisposing factors (e.g. lithological boundaries, gradual undercutting of slopes, etc.), weathered fault zones influenced the development of some of the largest and most active slope failures in the studied region. In comparison with geomorphologically active high-mountain environments with dense occurrence of catastrophic geomorphological processes, the mid-mountain conditions of the Flysch Carpathians are characterized by separate, location-related occurrences of high-energy geomorphological systems. These systems are predisposed particularly by fault zones and associated deep weathering. Despite our still poor knowledge of the physical and geochemical

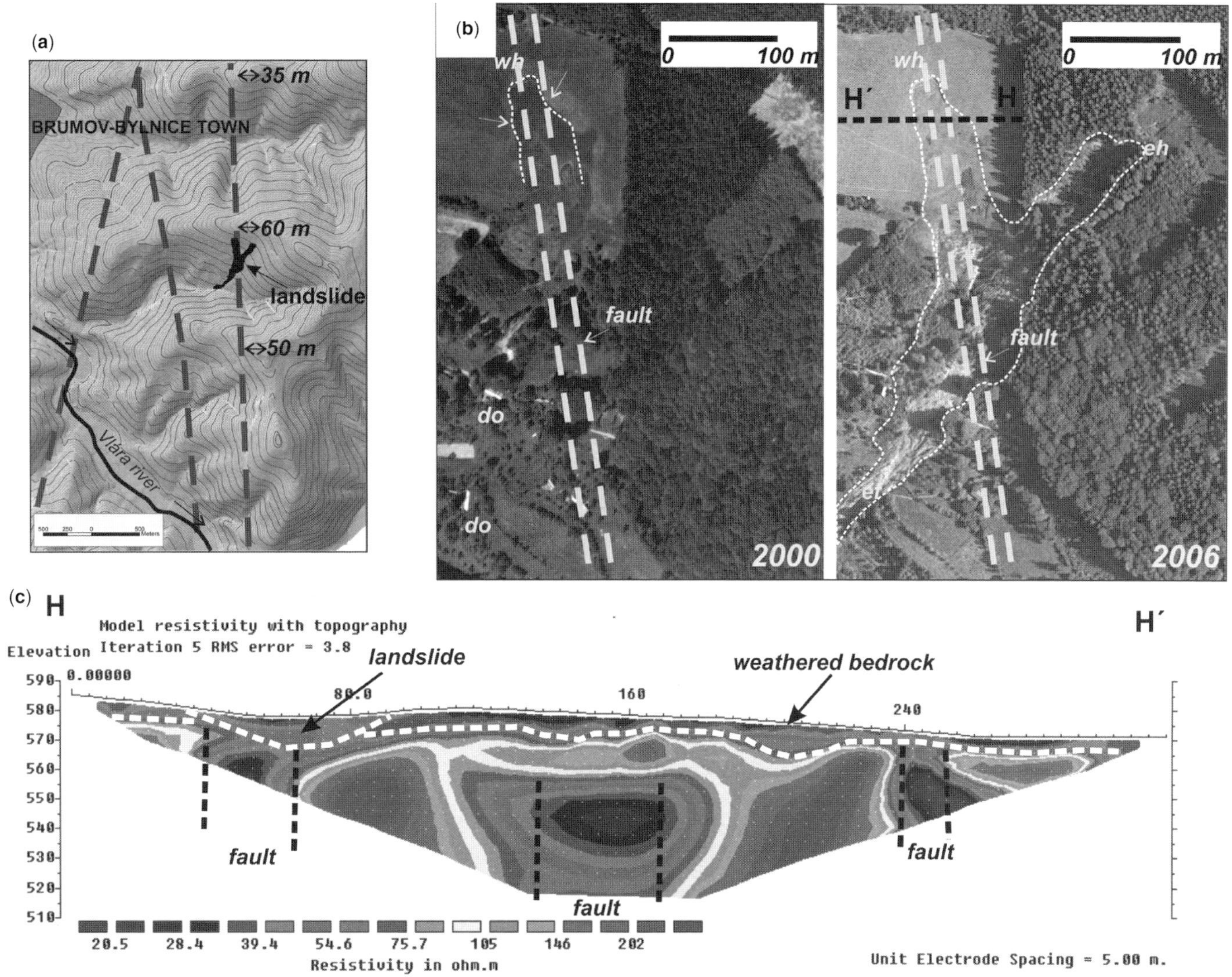

Fig. 11. (**a**) Location and morphotectonic context of the April 2006 Hluboče landslide (faults are marked by grey dashed lines; numbers along the fault line indicate assumed neotectonic ridge dislocation). (**b**) Comparison of pre-landslide (year 2000) and post-landslide (year 2006) topography, showing the western (wh) and eastern (eh) depletion zones, fault traces (double white dashed line) and destroyed houses (do). The dark area (juvenile stage of the slope deformation) in the western depletion zone represents a wet, fault-related preferentially weathered zone, which was the subject of the April 2006 catastrophic slope failure. (**c**) ERT cross-section in the western main scarp area [H–H′ in (**b**)].

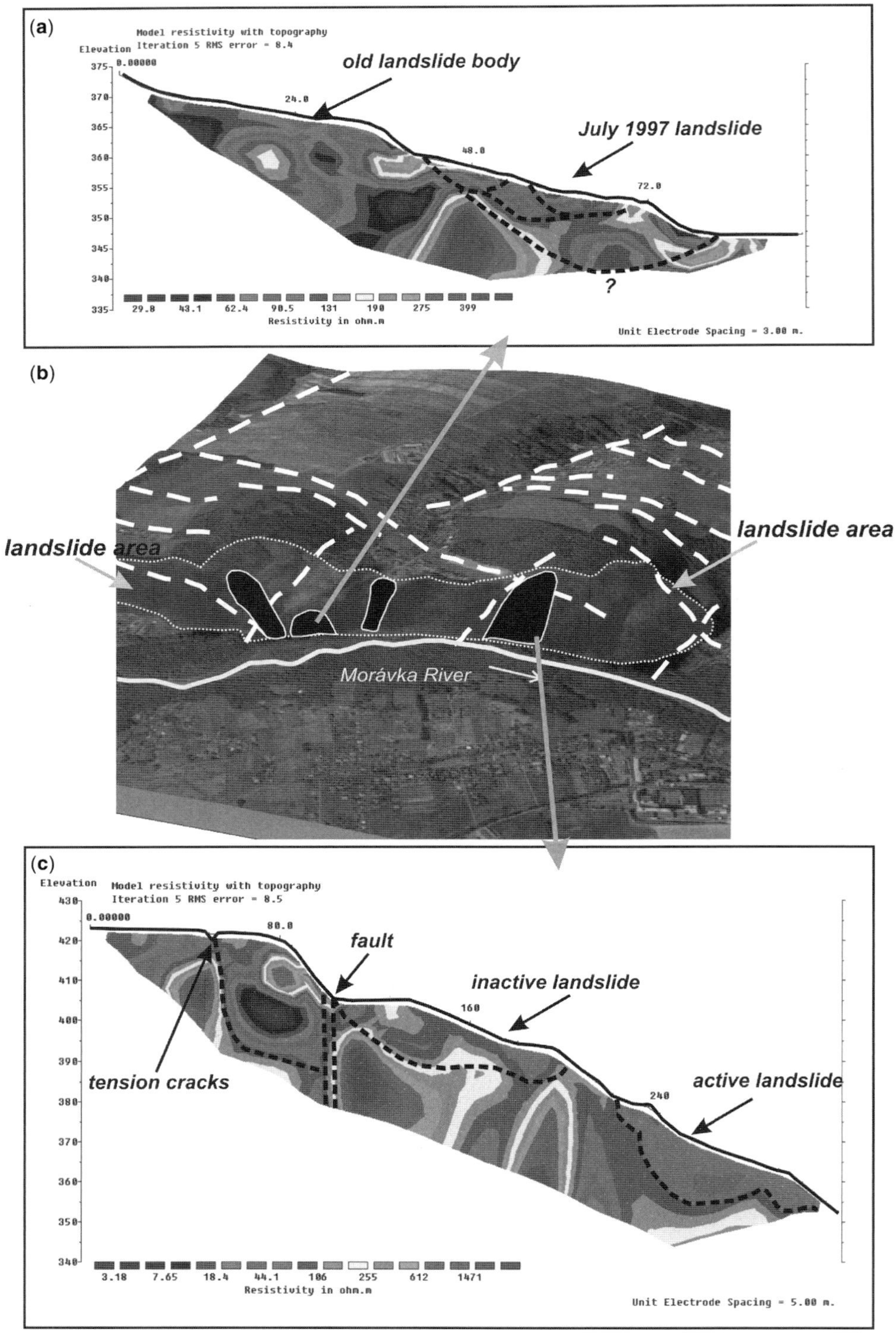

Fig. 12. The Skalice landslide area. (**a**, **c**) ERT sections through the most active, fault-related landslides (black dashed lines indicate probable shear surfaces. (**b**) Morphotectonic sketch. Assumed faults are marked by white dashed lines; black areas denote landslides active since 1972.

Fig. 13. (**a**) The most destructive July 1997 landslide in the Skalice area; lateral part of landslide main scarp. (**b**) Tension cracks above the main landslide body indicate active retrogression. (**c**) Intensively weathered flysch rocks in the main scarp. (**d**) Strongly fragmented thinly bedded flysch in a crevice-type cave predisposed by tension cracks.

nature of the weathered flysch bedrock in tectonic zones, we can reach the following conclusions on the weathering–landslide relationships in the Czech Flysch Carpathians.

(1) Weathering mainly in association with fault-related fracturing of the rocks has a key predisposing influence on landslide development, character and frequency.

(2) Four basic types of relationship exist between fault-weathering systems and landslides in the Czech Flysch Carpathians. All show evidence of repeatedly recurring instabilities, which are probably connected with gradual (cyclic) occurrence of instability in time.

(3) With the highest probability, processes of deep weathering are connected with more humid chronozones during the Holocene. Therefore, the issues of landslide frequency and climatic changes should be studied further.

Acknowledgements. This research was supported by a project grant from the Grant Agency of the Czech Republic, No. P209/10/0309: 'Effect of historical climatic and hydrometeorological extremes on slope and fluvial processes in the Western Beskydy Mts. and their forefield'. The authors thank F. Hubatka for GPR measurement, and T. Kosačík and V. Smolková for their assistance during fieldwork. We also extend our sincere gratitude for pollen analysis to V. Zernitskaya from the Institute of Geological Sciences, Belarussian Academy of Sciences in Minsk, and V. Skripkin and N. Kovalyuk from the Kiev Radiocarbon Laboratory of the Institute of Environmental Geochemistry (Ukraine).

References

BAROŇ, I. 2007. Výsledky datování hlubokých svahových deformací v oblasti Vsetínska a Frýdeckomístecka. *Geologické výzkumy na Moravě a ve Slezsku v roce 2006*, **14**, 12.

BAROŇ, I. & VAŘILOVÁ, Z. 2007. Hrazené pánve V NP Česke Švýcarsko jako doklad starých skalnúch řícení. *Zprávy o geologických výzkumech v roce 2006. Ceská geologická sluzba*, Praha, 53–57.

BAROŇ, I., CÍLEK, V., KREJČÍ, O., MELICHAR, R. & HUBATKA, F. 2004. Structure and dynamics of deep seated slope failures in the Magura Flysch Nappe, Outer Western Carpathians (Czech Republic). *Natural Hazards and Earth System Science*, **4**, 549–562.

BEAUVAIS, A., PARISOT, J. C. & SAVIN, C. 2007. Ultramafic rock weathering and slope erosion processes in a South West Pacific tropical environment. *Geomorphology*, **83**, 1–13.

BÍL, M. & MÜLLER, I. 2007. Příčiny vzniku mělkých sesuvů na Moravě na jaře 2006. *In*: HRADECKÝ, J. & PÁNEK, T. (eds) *Geomorfologický sborník 6*. Ostravská univerzita v Ostravě & Ceská asociace geomorfologu, Ostrava, 10.

BOBER, L. & WÓJCIK, A. 1977. Structural landslides in the region of the Prusów Ridge (Beskid Żiwiecki Mts.). *Studia Geomorphologica Carpatho-Balcanica, Kraków*, **11**, 155–167.

BOULTBEE, N., STEAD, D., SCHWAB, J. & GEERTSEMA, M. 2006. The Zymoetz River rock avalanche, June 2002, British Columbia, Canada. *Engineering Geology*, **83**, 76–93.

BRIDEAU, M. A., STEAD, D. & COUTURE, R. 2006. Structural and engineering geology of the East Gate Landslide, Purcell Mountains, British Columbia, Canada. *Engineering Geology*, **84**, 183–206.

CATANE, S. G., CABRIA, H. B., TOMARONG, C. P., SATURAY, R. M., ZARCO, M. A. H. & PIOQUINTO, W. C. 2007. Catastrophic rockslide–debris avalanche at St. Bernard, Southern Leyte, Philippines. *Landslides*, **4**, 85–90.

DIKAU, R., BRUNSDEN, D., SCHROTT, D. & IBSEN, M. L. 1996. *Landslide Recognition: Identification, Movement and Cause*. Wiley, Chichester.

GEERTSEMA, M., HUNGR, O., SCHWAB, J. W. & EVANS, S. G. 2006. Large rockslide–debris avalanche in cohesive soil at Pink Mountain, northeastern British Columbia, Canada. *Engineering Geology*, **83**, 64–75.

HRADECKÝ, J. & PÁNEK, T. 2007. Deep-seated gravitational slope deformations and their influence on consequent mass movements (case studies from the highest part of the Czech Carpathians). *Natural Hazards*, **45**, 235–253.

HRADECKÝ, J., PÁNEK, T. & KLIMOVÁ, R. 2007. Landslide complex in the northern part of the Silesian Beskydy Mountains (Czech Republic). *Landslides*, **4**, 53–62.

IETTO, F., DONATO, F. F. & IETTO, A. 2007. Recent reverse faults and landslides in granitoid weathered profiles, Serre Mountains (southern Calabria, Italy). *Geomorphology*, **87**, 196–206.

KELLOGG, K. S. 2001. Tectonic controls on a large landslide complex: Williams Fork Mountains near Dillon, Colorado. *Geomorphology*, **41**, 355–368.

KLIMEŠ, J. 2002. Analýza faktorů podmiňujících vznik sesuvů na okrese Vsetín. *Geografie-sborník ČGS*, **107**, 40–49.

KORUP, O. 2004. Geomorphic implications of fault zone weakening: slope instability along the Alpine fault, South Westland to Fiordland. *New Zealand Journal of Geology and Geophysics*, **47**, 257–267.

KOVÁŘ, L., MARSCHALKO, M. & LIBERDA, M. 2006. Významná svahová deformace "U hřbitova" ve Skalici u Frýdku-Místku. *In*: BAROŇ, I., KLIMEŠ, J. & KAŠPERÁKOVÁ, D. (eds) *Svahové deformace a pseudokras*. Sborník referátů a prezentací z odborného semináře. Ceská geologická služba & Ustav struktury a mechaniky hornin AV CR, v.v.i., Vsetín, 4.

KREJČÍ, O., BAROŇ, I., BÍL, M., HUBATKA, F., JUROVÁ, Z. & KIRCHNER, K. 2002*a*. Slope movements in the Flysch Carpathians of Eastern Czech Republic triggered by extreme rainfalls in 1997: a case study. *Physics and Chemistry of the Earth*, **27**, 1567–1576.

KREJČÍ, O., BAROŇ, I. ET AL. 2002*b*. Some examples of deep-seated landslides in the Flysch belt of the Western Carpathians. *In*: RYBÁŘ, J., STEMBERK, J. & WAGNER, P. (eds) *Landslides – Proceedings of the 1st European Conference on landslides, Prague, Czech Republic, 24–26 June 2002*, A. A. Balkema, Lisse, 373–380.

KREJČÍ, O., HUBATKA, F. & ŠVANCARA, J. 2004. Gravitational spreading of the elevated mountain ridges in the Moravian–Silesian Beskids. *Acta Geodynamica et Geomaterialia*, **3**, 1–13.

LOŽEK, V. 2007. *Zrcadlo minulosti*. Česká a slovenská krajina v kvartéru. Dokořán, 1–198.

MANGERUD, J., ANDERSEN, S. T., BERGLUND, B. & DONNER, J. J. 1974. Quaternary stratigraphy of Norden, a proposal for terminology and classification. *Boreas*, **3**, 109–126.

MARGIELEWSKI, W. 2006*a*. Records of the Late Glacial–Holocene palaeoenvironmental changes in landslide forms and deposits of the Beskid Makowski and Beskid Wyspowy Mts. Area (Polish Outer Carpathians). *Folia Quaternaria*, **76**, 1–149.

MARGIELEWSKI, W. 2006*b*. Structural control and types of movements of rock mass in anisotropic rocks: case studies in the Polish Flysch Carpathians. *Geomorphology*, **77**, 47–68.

PÁNEK, T., HRADECKÝ, J., MINÁR, J., HUNGR, O. & DUŠEK, R. 2009. Late Holocene catastrophic slope collapse affected by deep-seated gravitational deformation in flysch: Ropice Mountain, Czech Republic. *Geomorphology*, **103**, 414–429.

RYBÁŘ, J. & STEMBERK, J. 2000. Avalanche-like occurrences of slope deformations in the Czech Republic and coping with their consequences. *Landslide News*, **13**, 28–33.

SASS, O., BELL, R. & GLADE, T. 2008. Comparison of GPR, 2D-resistivity and traditional techniques for the subsurface exploration of the Öschingen landslide, Swabian Alb (Germany). *Geomorphology*, **93**, 89–103.

SEIJMONSBERGEN, A. C., WONING, M. P., VERHOEF, P. N. W. & DE GRAFF, L. W. S. 2005. The failure mechanism of a Late Glacial Sturzstrom in the Subalpine Molasse (Leckner Valley, Vorarlberg, Austria). *Geomorphology*, **66**, 277–286.

STARKEL, L. 1990. Holocene as interglacial – problems of stratigraphy. *Przegląd Geologiczny*, **38**, 13–16 [in Polish with English summary].

STARKEL, L. 1999. Chronostratigraphy of Late Vitulian and Holocene in Poland. *In*: PAZDUR, A., BLUSZC, A., STANKOWSKI, W. & STARKEL, L. (eds) *Geochronologia górnego czwartorzedu Polski, w świetle datowania radiowęglowego i luminiscencynejnego*. Uniwersytet Wroclawski, Wroclaw, 280–283 [in Polish with English summary].

STROUTH, A., BURK, R. L. & EBERHARDT, E. 2006. The Afternoon Creek rockslide near Newhalem, Washington. *Landslides*, **3**, 175–179.

SUZUKI, K., TODA, S., KUSUNOKI, K., FUJIMITSU, Y., MOGI, T. & JOMORI, A. 2000. Case studies of electrical and electromagnetic methods applied to mapping active faults beneath the thick Quaternary. *Engineering Geology*, **56**, 2–45.

WAGNER, J., DEMEK, J. & STRÁNÍK, Z. 1990. *Caves of the Moravskoslezské Beskydy Mountains*. Knihovna České speleologické společnosti, Praha [in Czech].

Visual observation of erosion processes on the Black Marl badlands in the southern Alps, France

T. Yamakoshi[1]*, N. Mathys[2] & S. Klotz[3]

[1] Public Works Research Institute, Erosion and Sediment Control Research Group, 1–6 Minamihara, Tsukuba 305-8516, Japan

[2] Cemagref, Snow, Avalanches Engineering and Torrent Control Research Unit, BP 76, 38402 Saint Martin d'Hères, France

[3] Cemagref, Snow, Avalanches Engineering and Torrent Control Research Unit, Maison de l'Erosion, 04420 Draix, France

**Corresponding author (e-mail: yamak226@pwri.go.jp)*

Abstract: To clarify the erosion processes on a marly bare slope in the southern Alps, the erosion processes on a steep and erodible slope composed of Black Marls formations were observed using a time-lapse video camera. The observations revealed that miniature debris flows (MDFs) occurred at the time of the rainfall–runoff event during which the most severe erosion took place over the 3 month observation period. Analysing the camera images, we show some characteristics of the MDFs and discuss them in the context of real rainfall–runoff phenomena observed at the outlet of the small experimental basin that contained the visually observed slope. The following results were obtained. (1) It is roughly estimated that the total amount of sediment discharged by the MDFs was not quantitatively negligible in comparison with the total sediment discharge from the entire experimental basin. (2) The MDFs occurred only during the rising limb of the hydrograph, which lasted 6 min. The visual observation also provided information on triggering conditions. Compared with the precipitation data, it was found that a rainfall intensity of 1 mm min^{-1} triggered MDFs and an antecedent cumulative rainfall of 10.4 mm was also enough to trigger an MDF. This is consistent with a previous study in Draix. (3) Although the Darcy–Weisbach friction factor and Manning's *n* of the MDFs observed differ considerably, they are consistent with results in the literature reporting erosion phenomena on very steep slopes up to 36°. (4) Based on this observation and a review of the literature, on very steep and highly erodible slopes, MDFs or similar phenomena might play an important role in erosion and transport processes. It is important to consider this possibility when examining erosion models for catchments composed of steep and erodible slopes.

In the southern Alps, the Black Marls formation, or 'Terres Noires', covers a large area. This formation is very susceptible to weathering and erosion (Antoine *et al.* 1995). The mean annual soil loss in marly badlands is about 7–10 mm a^{-1} (Descroix & Olivry 2002). In the Mediterranean and mountainous climate, which produces frosts in winter and high-intensity rainfall events, in summer, this results in a badlands morphology and high solid transport, causing heavily loaded floods downstream. The suspended sediment concentration reaches very high values during flood events, frequently over 300 g l^{-1}, and up to 800 g l^{-1} (Mathys *et al.* 2003). Therefore, the erosion rates of the Durance basin, which includes this Black Marls formation area devoid of vegetation, are among the highest in the world (Walling 1988).

To quantify this phenomenon and the effect of restoration strategies, erosion processes are intensively monitored and studied in the Draix experimental and representative basins (Draix ERBs; e.g. Mathys *et al.* 2003). The sediment discharge from the DRAIX ERBs has been modelled by physically based distributed water flow and sediment transport models, the ETC model (Brochot & Meunier 1996; Mathys *et al.* 2003), and SHETRAN (Bathurst *et al.* 1998). These studies have demonstrated fairly good results for the calculated sediment discharge at the outlet point of the Laval basin (86 ha), which is poorly vegetated and mainly composed of the Black Marls formation. However, this success was achieved because the researchers calibrated the parameters in their model using actual sediment discharge data obtained on a smaller sub-basin (1330 m^2) nearby. This suggests that their validity to a large extent depends on the reference data and it means that their model can be applied only if there is sufficient comparability with the reference sub-basin. To make the existing models more universally applicable, erosion processes that are smaller than the sub-basin scale, namely those at a unit slope scale, should be clarified.

Generally speaking, erosion processes at a small scale are simpler and their description involves less uncertainty than those at a larger scale. In Draix, the erosion processes at a unit slope scale have not yet been clarified. Mathys *et al.* (2007) showed that there is no general relationship between

From: CALCATERRA, D. & PARISE, M. (eds) *Weathering as a Predisposing Factor to Slope Movements*. Geological Society, London, Engineering Geology Special Publications, **23**, 201–212.
DOI: 10.1144/EGSP23.12 0267-9914/10/$15.00

sediment concentration and discharge even at a unit slope scale. For a given discharge, the sediment concentration may vary in a ratio of 1–20 or more. This suggests that sediment discharge is controlled not only by the sediment transport capacity but also by the sediment yield processes.

This study shows how the erosion processes occurred during real rainfall–runoff phenomena in the southern French Alps using visual observations. A time-lapse camera was installed to obtain close-up pictures of the actual erosion processes occurring on the slope surface during rainfall events. Based on the visual observations, the data at the sub-basin's point of outlet were interpreted.

The Draix experimental basins

The Draix ERBs make up a field laboratory for mountain erosion studies. They are located near Digne-les-Bains in the southern French Alps (Fig. 1). They were established in mountainous watersheds whose dominant geology is Black Marls formations, a very erodible outcrop. On steep slopes devoid of vegetation, erosion and gullying are intense and result in badland topography. Floods generate high levels of solid transport (Mathys *et al.* 2003). Five basins with a variety of vegetation covers have been equipped since 1983 to study and quantify runoff and erosion processes, taking into account the basin size and the extent of vegetation cover. Only four are still being monitored today: three are in denuded areas [Roubine (1330 m^2), Moulin (0.09 km^2) and Laval (0.86 km^2)] and the last one [Brusquet (1.08 km^2)] was afforested at the end of the 19th century. The vegetation land cover ranges from 21% (Roubine) to 87% (Brusquet) (Fig. 1). Rainfall is measured in the upper and lower parts of each basin. The flow level, used to calculate flow discharge, is measured in calibrated gauging flumes or weirs using various types of level recorders. Upstream of the gauging station, a sediment trap

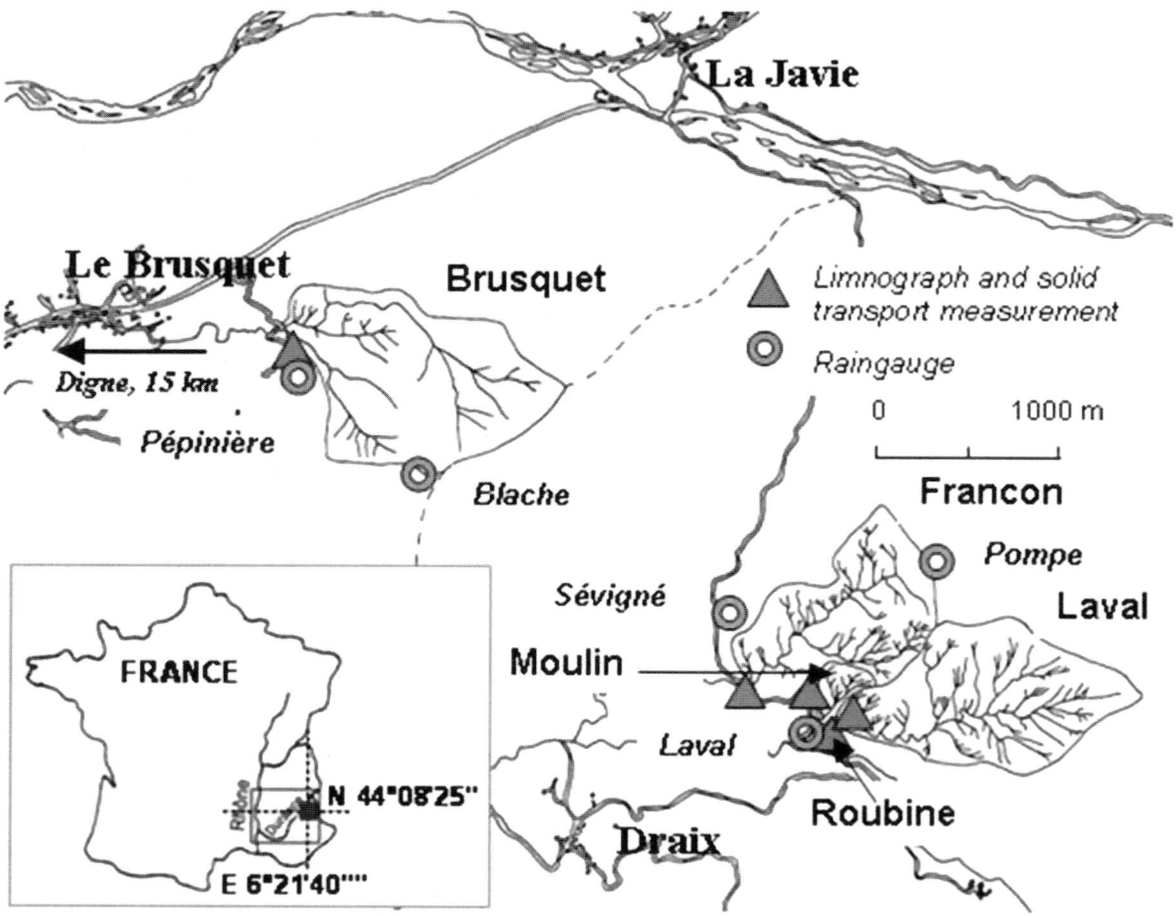

Fig. 1. Location map of the experimental catchments.

retains the coarse material, transported as bedload. The volume of deposited sediments is measured using topographic methods after each flood. In the gauging station, an automatic sampler takes samples during floods with a programme referring to both the level of water and the time lag between two samples: this allows the calculation of the fine sediment fluxes during floods. Since 1993, optical fibre sensors, using the backscattering of light by the particles, have been used to take continuous measurements of sediment concentration (Mathys *et al.* 2007). The smallest basin, the Roubine experimental basin, hereafter simply referred to as Roubine (0.0013 km^2), is a small, steep, bare basin typical of this area (Fig. 2). The slope gradient of the main channel is greater than 35% and can be considered an elementary unit for observing erosion phenomena. Here, continuous observation data on precipitation as well as water and sediment discharge are available.

The shallow layer of the bare slope is heavily weathered, as shown in Figure 2, because the climatic conditions are very favourable for weathering. There are many freeze–thaw cycles (almost 100 cycles in a typical winter), and many wet–dry cycles in summer, which result in the fragmentation of black marls on the surface. Depending on the weathering process, there is a characteristic variety in the regolith layers. Intact bedrock is found in the layer 20–30 cm deep. Above this bedrock layer, there are some regolith layers, which typically include a fine-particle-rich crustal layer, a platelet layer, and the upper regolith layer, as shown in Figure 3.

The weathering and erosion processes vary with the geomorphological settings of the basins, especially in terms of the relationship between the dip angle of the stratification in the Black Marls formation and the slope angle of the surface topography (Fig. 4). A slope whose orientation is nearly perpendicular to the dip angle is more susceptible to weathering because water can infiltrate into deeper layers more easily, resulting in more severe weathering through freeze–thaw cycles. The slope visually monitored in this study approximately corresponds to a slope whose orientation is nearly perpendicular to the dip angle; that is, a slope that is highly susceptible to weathering and has a thick cover of marl platelets.

Methods

Observation with a time-lapse video camera

An *in situ* visual observation system was installed inside Roubine, where water and sediment discharge are continuously measured at the outlet of the experimental basin. The system installed consists of a time-lapse camera (Figs 5 & 6), a rain gauge, a light, and a data logger. The camera takes moving pictures for 1 s every 10 s if the rainfall intensity threshold of 0.5 mm within 10 min is exceeded. It continues to work as long as the rainfall intensity remains above this threshold, which was determined from analysis of past rainfall–runoff data obtained in Roubine. Twenty-five frames are included within a 1 s film. The video camera used in this system was a commercial digital video camera (Canon DM-IXY DV2) but special external circuitry was added. This circuit emits signals to a controlling port in the video camera, which tells it when to start an observation, when to take a 1 s film every 10 s, and when to terminate a running observation. The camera and the light are controlled by the datalogger (CR10X, Campbell Sci. Inc.). The camera's field of view was set to cover almost 2 m × 1.5 m. All the instruments were installed in early September 2006.

Visual interpretation of hydraulic parameters

There are three main rills in the camera's view, as shown in Figure 6. As described below, the video camera observed that the flow inside the rills sometimes formed intermittent surges identical to debris flow in natural torrents. Within the camera's field of view, the depth, the width, and the longitudinal slope angle of the rills are nearly uniform, at 10–15 cm, 30–40 cm, and 30°, respectively. The velocities of the surges, hereafter referred to as miniature debris flows (MDFs), in the rills were determined on the film obtained because the 1 s, 25-frame film allowed us to read the displacement frame by frame. The heights of the MDFs were very roughly estimated. Compared with the depth of the rills (10–15 cm) the height of each MDF was read and classified into two levels; namely, the nearly full-bank or the middle level. Assuming that the heights of the MDFs for the nearly full-bank and the middle levels are 10 cm and 5 cm, respectively, the apparent Darcy–Weisbach friction factors and Manning's roughness coefficients can be roughly determined for each MDF as

$$ff = \frac{8gHS}{V^2} \tag{1}$$

where ff is the Darcy–Weisbach friction factor, g is acceleration due to gravity, H is flow height, S is the longitudinal gradient of the rill, and V is the velocity of the flow; and

$$n = \frac{H^{2/3}S^{1/2}}{V} \tag{2}$$

where n is Manning's roughness coefficient.

The total amount of sediment discharged by each MDF is difficult to determine quantitatively, because no direct measurement was taken in the field except for the video images. The measurement of large-scale debris flows is often estimated only from video images (e.g. Suwa 1989). The flow velocity, height, width, and number of debris flows roughly estimate the volume of the debris flow. The following is the method used to estimate the total sediment

Fig. 2. Photographs of the experimental catchments in Draix. (**a**) General view; (**b**) close-up view and basin limit of the Roubine experimental station.

volume of each MDF after the method was actually used to measure a largescale debris flow.

The height and width of MDFs had already been measured and the ranges of their values are known and shown in Table 1. Consequently, the length and sediment concentration of the surges are needed to estimate the total volume of the sediment discharged. Ooustwoud Wijdenes & Ergenzinger (1998) measured the sediment concentration of the surge in their field experiments in Draix. They conducted a series of artificial rainfall experiments and sampled muddy water using a small hand-operated container. They found occurrences of MDFs during their

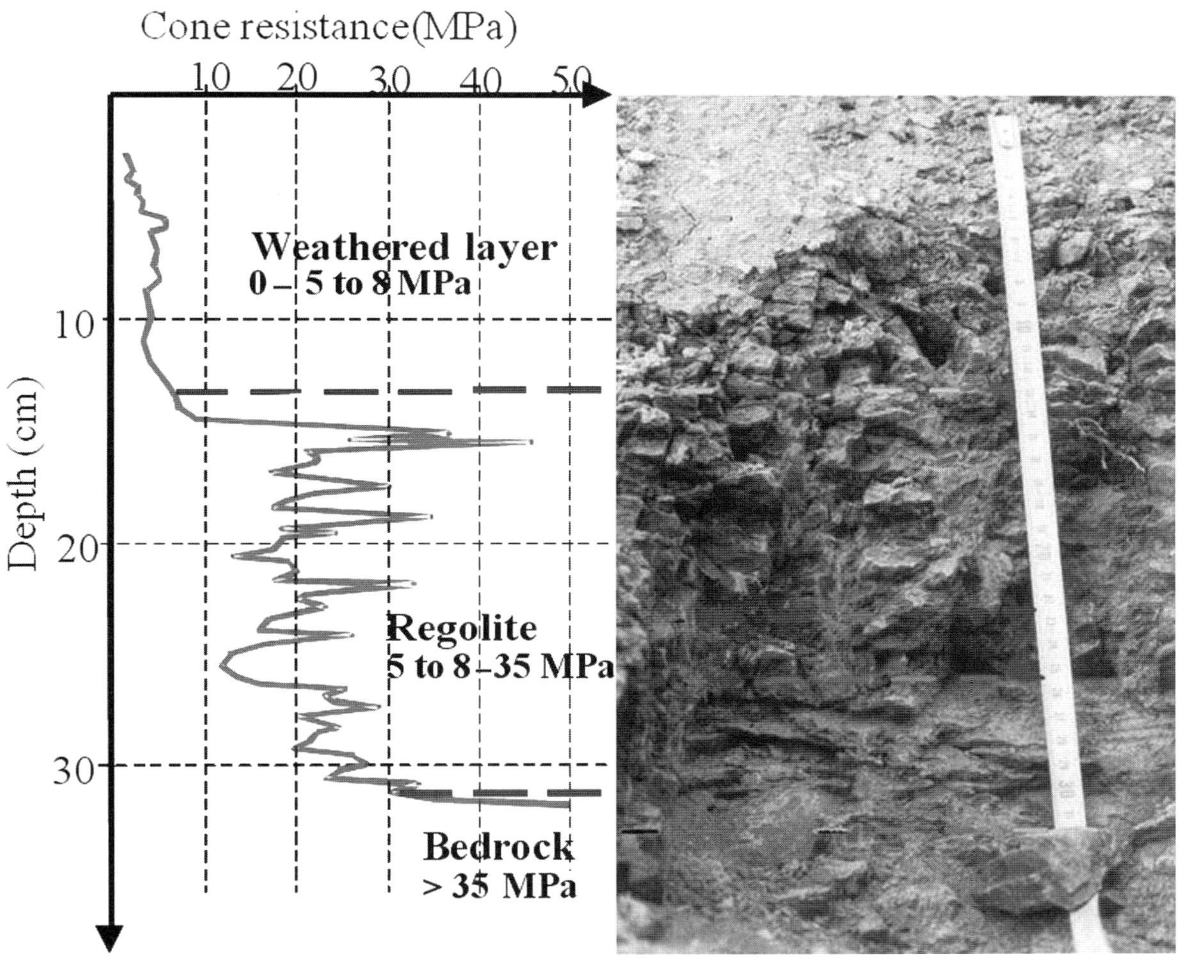

Fig. 3. Photograph and cone-resistance profile of weathered black marls.

experiments and succeeded in measuring the sediment concentration, which was in the range of 550–1440 g l^{-1}.

On the other hand, it is very difficult to determine the length of the MDF because the entire length of many of the MDFs could not be seen during the 1 s image sequences. However, the lengths of some MDFs were visible. The extent of the MDF was defined as a section where the texture of the surface of the flow was different from that of the upper or lower section in the same rill, around 1 m. Therefore, this value is regarded as a representative length for all the MDFs observed. As for a large-scale debris flow, the highest peak of an MDF is at its tip and the height decreases gradually toward the tail. Therefore, if we regard it as a triangular pyramid whose top and bottom correspond to the tail and the tip of the MDF, respectively, the volume of each MDF can be calculated.

Other observations and measurements

In addition to this newly installed equipment, water runoff and sediment discharge are also measured at Roubine (Mathys *et al.* 2003). There is a 90° gauging weir at the outlet where the water stage is measured both mechanically and electronically. In addition, there is also an automatic sampler that can collect the muddy water in the weir; this means that the suspended sediment concentration of the water running out can be measured. This complements the total sediment discharge data by capturing the loss of wash load, providing the temporal tendency of the sediment discharge from the experimental basin. The dried weight of the sampled muddy water and the deposit inside the weir and the bed load sediment trap are also measured every time a major rainfall–runoff event occurs. The observation

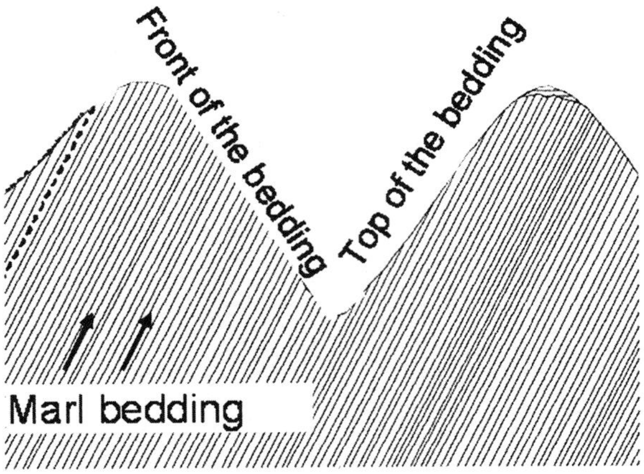

Fig. 4. Geomorphological setting of black marl bedding and surface topography.

was conducted during the rainy season in autumn, from 4 September to 8 December 2006. Photographs with the same angle were taken and are compared to show how the rill evolves.

Results

Data at the outlet of the Roubine experimental basin

During the observation period, from 4 September to 8 December 2006, the total rainfall observed was 343 mm. As a result of this rainfall, 4721 kg of sediment was discharged from the experimental basin. Most of the sediment discharge was observed in September. The greatest intensity of rainfall was measured on 8 September and the largest amount of rainfall was on 14–15 September. Figure 7 shows a hyeto-hydrograph and the temporal change in

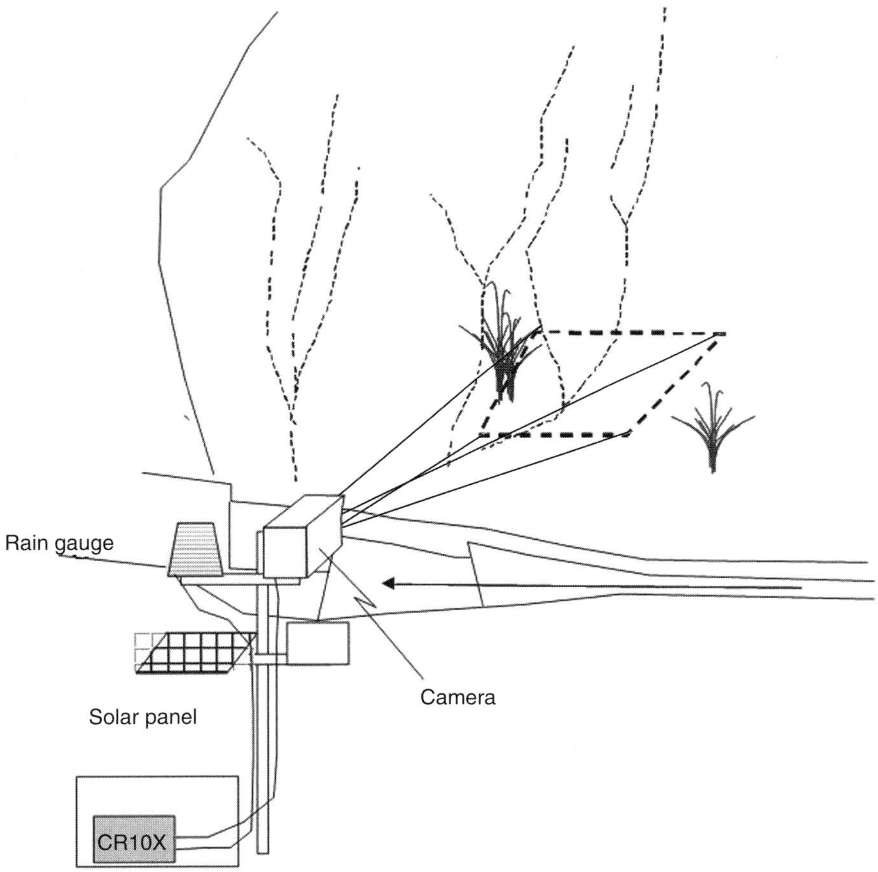

Fig. 5. The observation system.

Fig. 6. Top right: photograph from the automatic camera. The dashed white line shows the camera's field of view. Top left, camera set-up; below, view of the slope of interest in Roubine.

Table 1. *Height and velocity of surges of the observed miniature debris flows*

Number	*Time of occurrence*	*Height of surge (H)* (cm)	*Velocity (V)* (m s^{-1})	*Darcy–Weisbach friction factor*	*Manning's roughness coefficient* ($m^{-1/3}$ s)
1	14:15	10	0.3	50.3	0.5
2	14:15	5	0.5	9.1	0.2
3	14:17	10	0.4	28.3	0.4
4	14:19	10	0.4	28.3	0.4
5	14:19	10	0.4	28.3	0.4
6	14:19	5	0.6	6.3	0.2
7*	14:19	10	–	–	–

*This MDF was not sufficiently visible to measure the velocity.

suspended sediment concentration, which was measured by the automatic sampler of the rainfall–runoff event measured in the Roubine experimental basin on 8 September 2006.

On 8 September 2006, there was a brief, heavy rainfall that produced 37.8 mm of the total rainfall, with a maximum intensity of 2.6 mm min^{-1}, lasting nearly 1 h. It resulted in the rapid response of the water and sediment runoff. The amount of suspended sediment discharged was measured and calculated to be 1330 kg, and adding the amount of the sediment deposited in the weir and the sediment trap, the total amount was 4278 kg in dried weight.

On the other hand, on 15 September, there was a long-lasting, but relatively low-intensity rainfall, producing 57.8 mm of the total rainfall, with a maximum intensity of 1.0 mm min^{-1} lasting almost 1 day. During this event, water runoff was measured and showed several low peaks corresponding to the peaks in the rainfall intensity, but the sediment discharge was small, a total of 376 kg (Fig. 8). The sediment discharge during the other rainfall events, totalling 247.4 mm, was negligible.

Visual data of miniature debris flow

The camera observed the slope of interest every time rain fell. Notable phenomena were observed at the time of the rainfall–runoff phenomena on 8 September 2006. According to the images obtained by the camera, the following was observed.

At 13:09 UT, water flow appeared in the bottoms of the rills. Then, at 13:11, thin overland flow appeared on the interrill slope. During the period from 13:14 to 13:17 UT, runoff disappeared both in the rills and on the interrill slope because of a short interruption of the intense rainfall. At 13:18 UT, the rainfall intensity began to increase and the intensity of the overland flow increased. At the same time, intermittent surges began to be observed. Oostwoud Wijdenes & Ergenzinger (1998) conducted artificial rainfall experiments in Draix and found that similar phenomena occurred during their experiments. They called them miniature debris flows (MDFs). Within the camera's field of view, seven MDFs were observed in the rills on the slope of

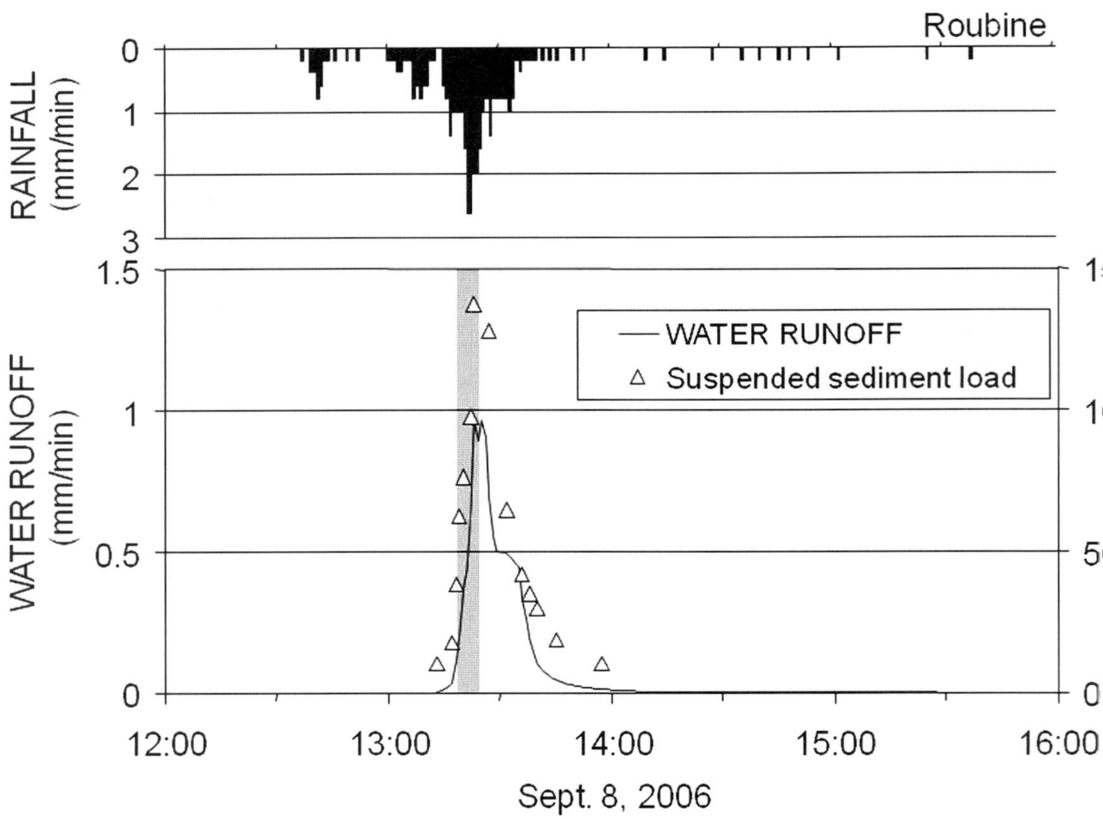

Fig. 7. Hyeto-hydrograph and the suspended sediment discharge from the storm measured in the Roubine experimental basin on 8 September 2006. The grey zone shows the period during which the miniature debris flows occurred.

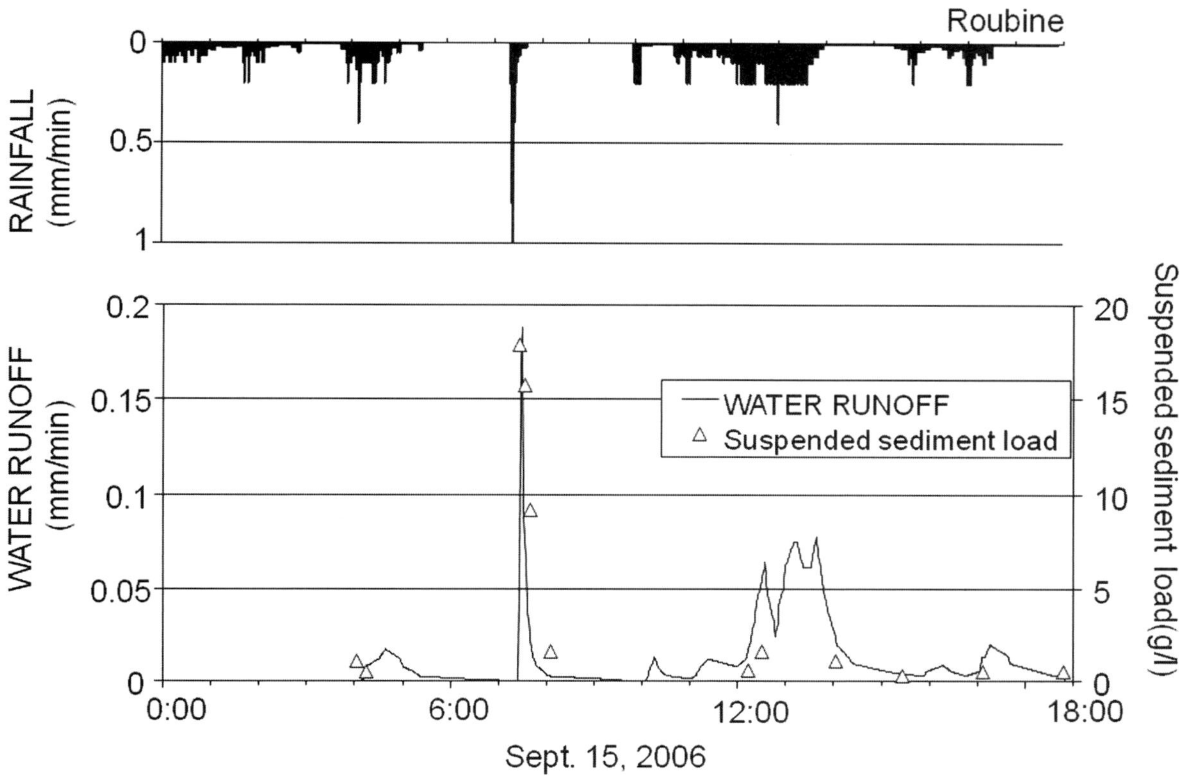

Fig. 8. Hyeto-hydrograph and the suspended sediment discharge from the storm measured in the Roubine catchment on 15 September 2006.

interest. Figure 9 shows two successive photographs taken 1 s apart. The snout of an MDF can be seen in the circle in the right-hand photograph. The MDFs occurred during the 13:18–13:24 UT period, when the instantaneous rainfall intensity was between 1.0 and 2.6 mm min^{-1}. On the other hand, no MDF was observed even though the continuous sediment-laden flow continued up to the end of the runoff event in spite of the high rainfall intensities between 0.8 and 1.6 mm min^{-1} (Fig. 7). After the passages of the MDFs, temporary deposits were occasionally observed at the foot of the slope. However, most of them were eroded away by the successive flows.

The MDFs were seen to occur two or three times in each rill. The video interpretation roughly estimated the velocities to be 0.3–0.6 m s^{-1}. The Darcy–Weisbach friction factors and Manning's roughness coefficients were roughly determined for each MDF and are shown in Table 1. The Darcy–Weisbach friction factor and the Manning's roughness coefficient estimated in this study are 6.3–50.3 and 0.2–0.5 $m^{-1/3}$ s, respectively.

Discussion

Erosion processes on the slope of interest in the Roubine experimental basin

During the entire observation period, most of the sediment discharged from the Roubine experimental basin was generated during the rainfall–runoff event on 8 September 2006. Throughout this event, the occurrence of MDFs was observed. MDFs are known in Draix from occasional eyewitness accounts during rainstorms, from the deposits seen after a rainstorm and as observed in small artificial rainfall experiments (Oostwoud Wijdenes & Ergenzinger 1998). Therefore, it is inferred that this kind of sediment transport could be common in the Draix Black Marls formation.

MDFs and other similar phenomena have also been reported on slopes of decomposed granite in Japan (Kimoto *et al.* 2003), marly badlands in Spain (Bouma & Imeson 2000), badlands in Israel (Yair *et al.* 1980), a post-fire

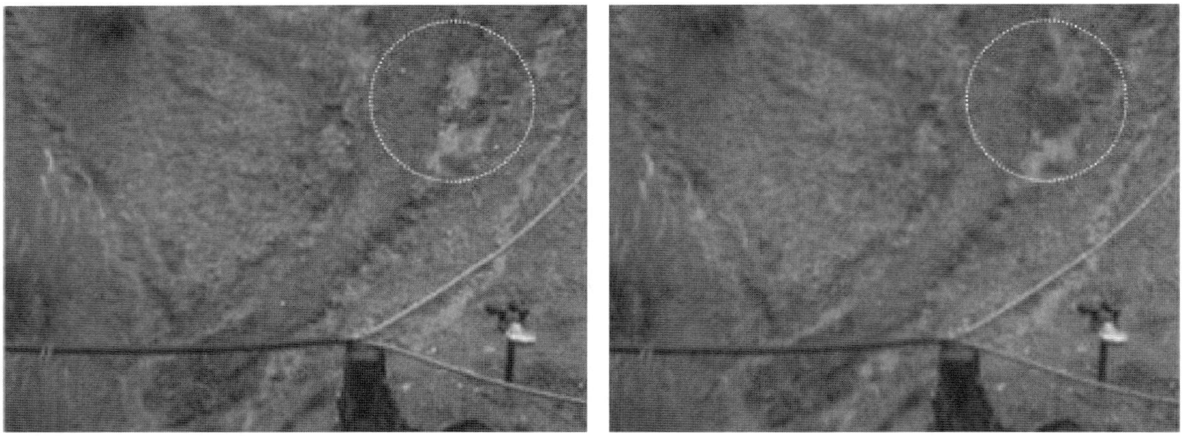

Fig. 9. A miniature debris flow observed by the camera (see inside the circle in the upper right corner of the images). The right-hand image was taken 1 s after the left one.

hillslope in the USA (e.g. Gabet 2003), and a post-eruption tephra-covered hillslope in Japan (T. Chiba, pers. comm.). Table 2 shows that the slope angles in all the reported cases are greater than 25°, except on the tephra-covered slope because of the lack of quantitative data. In addition, they are all bare land and susceptible to erosion, such as badlands with severely weathered marls, or hillslopes burned by wild fire or covered with volcanic tephra. The situation in this study is very similar to these cases. The slope observed in this study is 30°. The Black Marls formation is highly susceptible to weathering and erosion. This implies that MDF, are not specific to Draix but are common on very steep, bare and very susceptible slopes throughout the world.

The impact of sediment input from the MDF on the Roubine experimental basin sediment discharge

The total amount of sediment discharged by the MDFs is difficult to determine quantitatively. Here, we have attempted to estimate the impact of sediment input from the MDF on the sediment discharge from the Roubine experimental basin. The dimensions of the MDFs observed in the event on 8 September are summarized in Table 3.

Assuming that the conditions that led to MDFs in the three rills observed would be the same for the entire Roubine experimental basin, the total amount of sediment discharged by the MDFs from the entire basin can be estimated (Table 3). Compared with the total amount of discharged sediment, 4278 kg, the proportion of sediment estimated to be discharged as MDFs is calculated to be 5–36% in weight. Quantitatively, it is difficult to say that it is dominant, but it is not negligible.

In previous field experiments, the abrupt rise in sediment concentration in the runoff water caused by MDF occurrence has been reported (Oostwoud Wijdenes & Ergenzinger 1998; Kimoto *et al.* 2003). Throughout the temporal change in suspended sediment concentration, no spike in concentration but instead a moderately high concentration

Table 2. *Miniature debris flows reported in the literature*

Reference	*Slope (°)*	*Name used in the reference*	*Description on the slope condition*
Oostwoud Wijdenes & Ergenzinger (1998)	36–44	Miniature debris flow	Black marl, badlands
Kimoto *et al.* (2003)	33.6	Small-scale debris flow	Decomposed granite, scarps of the surface failure
Bouma & Imeson (2000)	37–52	Mass movement	Marl, badlands
Gabet (2003)	25–30	Thin debris flow	Paso Robles Formation burned by wild fire
Yair *et al.* (1980)	30	Mudflow	Taquiya Formation, badlands
T. Chiba (pers. comm.)	–	–	Volcanic ash, disturbed by a recent eruption
This study	30	Miniature debris flow	Black mari, badlands

Table 3. *Estimation of the total weight of the sediment discharged by the MDFs in the Roubine experimental basin within the runoff phenomenon on 8 September 2006*

Term	*Value*	*Remark*
Bareland in the Roubine experimental basin (A)	1050 m^2	Measured in the field
Drainage area of the observed three rills (A')	45 m^2	Measured in the field
Length of an MDF (L)	100 cm	Observed in the image
Height of an MDF (H)	5–10 cm	Observed in the image
Width of the tip of an MDF (B)	30–40 cm	Measured in the field
Volume of one MDF (V)	2.5–6.7 l	Calculated ($= 1/2 \times H \times B \times L \times 1/3$)
Sediment concentration of an MDF (Cs)	550–1440 g l^{-1}	Estimated from the literature*
Weight of the sediment of one MDF (W)	1.4–9.6 kg	Calcualted ($= V \times Cs$)
Number of MDFs in the rills observed (N')	7	Observed in the image
Number of MDFs in the whole basin (N)	163	Calculated ($= N' \times A/A'$)
Total weight of the discharged sediment	230–1560 kg	Calculated ($= W \times N$)

*Ooustwoud Wijdenes & Ergenzinger (1998).

of suspended sediment was observed at the Roubine outlet during the period that the MDFs occurred. This is because the temporal change in the suspended sediment concentration could be attenuated to some extent by a large sand pocket in the upper stream of the gauging weir.

Conditions that trigger MDFs

Ooustwoud Wijdenes & Ergenzinger (1998) showed the initiation conditions for the MDFs observed in their experiments on the marly slope in Draix. They found that the infiltration of 8 mm of rainfall was capable of liquidizing the upper 3 cm of surface soil and triggering MDFs. In the rainfall on 8 September, the cumulative rainfall before the onset of the MDFs was 10.4 mm. As the material composing these slopes is considered to be similar, the initiation conditions found in this study seem to be highly consistent with the results described by Ooustwoud Wijdenes & Ergenzinger (1998).

In addition, it seems that the MDFs occurred only in the rising limb of the hydrograph (see Fig. 7). The MDFs began to occur at 13:18 UT when the rainfall intensity became heavier than 1 mm min^{-1}. They continued to occur sporadically until 13:24 UT when the rainfall intensity began to decrease. Given that no MDFs were observed during other rainfall–runoff phenomena whose intensities were below 1 mm min^{-1}, even where cumulative rainfall was far greater than 8 mm, high-intensity rainfall seems to be necessary to generate MDFs on the slope of interest. On the other hand, the MDFs stopped occurring at 13:24 UT After this time, the rainfall intensity began to decrease, although it remained high, 0.8–1.6 mm min^{-1}, for several minutes. MDFs never occurred in the other rainfall events during the observation period, as shown in Figure 8. In the rainfall event on 15 September, rainfall that was more intensive than 1 mm min^{-1} was observed but did not trigger any MDFs. These facts suggest that the erodible sediment had already been eroded away by the preceding erosion.

Hydraulic characteristics of the MDFs

The slope angle of the slope of interest in this study is 33°. There are few studies that have shown the hydraulic characteristics of the flow on such a steep slope.

Ooustwoud Wijdenes & Ergenzinger (1998) conducted an experimental study in Draix. They measured the depth and the velocity of the sediment-laden flow in the rills of the steep marly slopes (36° and 38°) and reported that the Darcy–Weisbach friction factor was in the range of 1.1–44.46. Hassel *et al.* (2003) conducted their field experiments on slopes in croplands with a wide range of gradients, between 3.5° and 32°, and obtained a Manning roughness coefficient within the range of 0.05–0.2 $m^{-1/3}$ s.

The Darcy–Weisbach friction factor and Manning's n estimated in this study are 6.3–50.3 and 0.2–0.5 $m^{-1/3}$ s, respectively. The ranges seem to be extraordinarily wide but correspond to those found in the studies shown above. On the other hand, the values vary within a very wide range. To pursue the discussion, the precision of the measurement needs to be improved.

Conclusion

In this study, the erosion processes on the steep marly slope were successfully observed using the time-lapse video camera. It is also shown that visual observation using a time-lapse video camera can provide not only qualitative information, but also quantitative data. According to the visual observation, the MDFs were found to occur at the moment of the most severe erosion during the observation period. A rough estimation of the total amount of sediment discharged by the MDFs showed that it was not negligible quantitatively. The observation also showed that the MDFs occurred only during the rising limb of the hydrograph, lasting 6 min. The visual observation also provided information on triggering conditions. Compared with the precipitation data, it was found that a rainfall intensity of 1 mm min^{-1} triggered MDFs and an antecedent cumulative rainfall of 10.4 mm was enough to trigger an MDF. This is consistent with the MDFs observed by Ooustwoud Wijdenes & Ergenzinger (1998). It is also inferred that not only cumulative rainfall but also rainfall intensity seems to be necessary to generate MDFs.

According to the results of this observation and a review of the literature, in very steep and highly erodible slopes, such as the severely weathered bare slope of the Black Marls formations, MDFs or similar phenomena may play an important role in the erosion and transport processes. Consequently, it is important to take this into account in considering erosion models for catchments composed of steep and erodible slopes.

References

ANTOINE, P., GIRAUD, A., MEUNIER, M. & VAN ASCH, T. 1995. Geological and geotechnical properties of the 'Terres Noires' in southern France: weathering, erosion, solid transport and instability. *Engineering Geology*, **40**, 223–234.

BATHURST, J. C., LUKEY, B., SHEFFIELD, J., HILEY, R. A. & MATHYS, N. 1998. Modelling badlands erosion with SHETRAN at Draix, southeast France. *In*: SUMMER, W., KLAGHOFER, E. & ZHANG, W. (eds) *Modeling Soil Erosion, Sediment Transport and Closely Related Hydrological Processes*. IAHS Publication, **249**, 129–136.

BOUMA, N. A. & IMESON, A. C. 2000. Investigation of relationships between measured field indicators and erosion processes on badland surfaces at Peter, Spain. *Catena*, **40**, 147–171.

BROCHOT, S. & MEUNIER, M. 1996. *Un modèle d'érosion destorrents en crue (ETC)*. Cemagref Editions, Ingénieries-EAT, 9–18.

DESCROIX, L. & OLIVRY, J. C. 2002. Spatial and temporal factors of erosion by water of black marls in the badlands of the French southern Alps. *Hydrological Science Journal*, **47**, 227–242.

GABET, E. J. 2003. Post-fire thin debris flows: sediment transport and numerical modelling. *Earth Surface Processes and Landforms*, **28**, 1341–1348.

HASSEL, R., JETTEN, V. & ZHANG, G. 2003. Estimating Manning's *n* for steep slopes. *Catena*, **54**, 77–91.

KIMOTO, A., UCHIDA, T., MIZUYAMA, T., KOSUGI, K. & TSUTSUMI, D. 2003. Surface runoff and sediment movement on a bare slope in the Tanakami Mountains: implementation of *in situ* artificial rainfall experiment. *Journal of Japan Society of Erosion Control Engineering*, **55–6**, 52–58 [in Japanese with English abstract].

MATHYS, N., BROCHOT, S., MEUNIER, M. & RICHARD, D. 2003. Erosion quantification in the small marly experimental catchments of Draix (Alpes de Haute Provence, France). Calibration of the ETC rainfall–runoff–erosion model. *Catena*, **50**, 527–548.

MATHYS, N., KLOTZ, S., ESTEVES, M. & GRÉSILLON, J. M. 2007. Erosion processes at the gully scale: observations, quantification and interpretation of field data from the Draix laboratory. *In*: CASALI, J. & GIMENEZ, R. (eds) *IV International Symposium on Gully Erosion*. Universitad Publica de Navarra, Pamplona, 90–93.

OOSTWOUD WIJDENES, D. J. & ERGENZINGER, P. 1998. Erosion and sediment transport on steep marly hillslopes, Draix, Haute-Provence, France. An experimental field study. *Catena*, **33**, 179–200.

SUWA, H. 1989. Field observation of debris flow. *In*: *Proceedings of Japan–China (Taipei) Joint Seminar on Natural Disaster Mitigation*, 343–352.

WALLING, D. E. 1988. Measuring sediment yield from river basins. *In*: LAL, R. (ed.) *Soil Erosion Research Methods*. Soil and Water Conservation Society, Iowa, 39–73.

YAIR, A., LAVEE, H., BRYAN, R. B. & ADAR, E. 1980. Runoff and erosion processes and rates in the Zin valley badlands, Northern Negev, Israel. *Earth Surface Processes*, **5**, 205–225.

Relationship between the residual shear strength and the methylene blue value in weathered clay soils

C. Meisina

Department of Earth Sciences, University of Pavia, Via Ferrata 1, 27100 Pavia, Italy
(e-mail: claudia.meisina@unipv.it)

Abstract: A new correlation is proposed between the residual shear strength and the methylene blue value ('value of blue'; VB) for weathered clay soils on argillaceous bedrock and on alluvial soils in the Oltrepo Pavese area (Northern Italy), which can be considered as geologically representative of a large part of the Italian Apennines. Shallow landslides occur periodically in these soils as a result of high-intensity rainfall events. A number of soils were tested. Trench pits were used for sampling and for the soil profile description (lithology, structure, grade of weathering, thickness). Field surveys were integrated with some standard geotechnical laboratory tests. The methylene blue dye adsorption (VB) was determined in accordance with the French AFNOR standards. The residual strength friction angle was measured with direct shear tests; the procedure employed for the measure involved inserting the soil at the liquid limit in the direct shear ring, applying consolidation in stages and then shearing (Kanji method). The applicability of some existing correlations between the residual friction angle and index properties were tested. The correlations are characterized by a large amount of scatter in the data and tend to overestimate the residual friction angle. The VB test is easy and rapid to perform and appears to be a good indicator of the residual strength friction angle for the tested soils.

Residual strength is the minimum constant value attained (at slow rates of shear) at large displacements (Skempton 1985). Skempton (1964, 1985) and Lupini *et al.* (1981) suggested that in clay soils the major cause of the decrease in strength to a residual value is the increased parallel orientation of platy clay particles located along the slip surface. The strength of a clay will be at or close to the residual value on slip surfaces of old landslides, in bedding shears in folded strata, in sheared joints or in faults (Mesri & Shahien 2003). The residual shear strength also plays an important role in slope stability analysis in clay slopes containing pre-existing shear surfaces, along which the available strength is less than that of the intact material and coincides with the residual strength. Such slopes are widespread in the Italian Apennines; they are formed of highly tectonized varicoloured clays, flysch formations (alternating layers of clay shales with calcareous arenaceous marls), their weathering products (clays with diffused mineral or rock fragments of different size) and weathered clay soils within old alluvial terraces (cycles of drying and wetting produce shear surfaces, e.g. slickensides). All these soils are often the location of shallow landslides.

The main factors affecting the residual shear strength are the following (Lo Presti & Froio 2004).

(1) Mineralogy (Stark & Eid 1994): this is the most important factor.

(2) Particle shape: Lupini *et al.* (1981) reached the conclusion that the grain shape controls the sliding mechanism; when round sand particles predominate a turbulent shear (rolling shear) occurs without particle reorientation; in contrast, when flat clay particles are predominant, there is a sliding shear, with complete particle reorientation. A transition zone exists between the two mechanisms.

(3) Normal effective stress: several researchers have shown the non-linearity of the residual strength envelope for various type of clays (Skempton 1985; Stark & Eid 1994).

(4) Displacement rate: the residual shear strength decreases a few per cent as the displacement rate increases one order of magnitude (Leroueil 2001).

(5) Pore solution composition: the residual shear strength angle of various clays reconstituted in the laboratory increases when using saturated NaCl solution instead of distilled water (Di Maio 1996; Calvello *et al.* 2005).

The residual shear strength has been studied extensively in the laboratory and various attempts have been made to correlate this parameter with soil index properties (clay fraction, plasticity index, liquid limit), which are supposed to represent the soil mineralogy. Practical correlations between residual friction angle and the index properties, such as the clay fraction (Skempton 1964, 1985; Borowicka 1965; Binnie *et al.* 1967; Blondeau & Josseaume 1976; Cancelli 1977; Lupini *et al.* 1981; Collotta *et al.* 1989; Yin 2002), the plasticity index (Fleischer and Bucher, both cited by Lupini *et al.* 1981; Voight 1973; Kanji 1974; Seycek 1978; Rouiguia 1996; Wesley 2003) and the liquid limit (Jamiolkowski & Pasqualini 1976; Mesri & Cepeda-Diaz 1986) have been widely reported in the literature. These correlations have been developed for sedimentary soils, but in recent years they have also been demonstrated for tropical soils (Wesley 2003) and residual soils (Wan & Kwong 2002; Kaya & Kwong 2007).

From: Calcaterra, D. & Parise, M. (eds) *Weathering as a Predisposing Factor to Slope Movements.* Geological Society, London, Engineering Geology Special Publications, **23**, 213–233.
DOI: 10.1144/EGSP23.13 0267-9914/10/$15.00

It is well known that the residual shear strength is strongly controlled by clay mineralogy. The 'value of blue' (VB; the quantity of blue methylene consumed per 100 g of soil) is an indicator of the amount of clay and of the type of clay mineral; it is measured with the methylene blue stain test, which is based on the adsorption of the organic dye methylene blue by negatively charged soil particles when in solution, forming a monomolecular layer (Lautrin 1989; Magnan & Youssefian 1989; Shaeffner 1989; Tourenq & Tran Ngoc Lan 1989; Chiappone & Scavia 1999; Chiappone *et al.* 2004). Thus, a correlation between the residual friction angle and the VB is expected because they have the mineralogy factor in common.

The aims of the paper are: (1) to test the applicability of some existing correlations between the residual friction angle and index properties for clay soils derived from the weathering of argillaceous bedrock and alluvial soils of the Oltrepo Pavese area (Northern Italy), which are often involved in shallow landslides; these soils can be considered as geologically representative of a large part of the Italian Apennines; (2) to establish whether any correlation exists between the residual friction angle and the VB for these soils.

Geological setting of the study area

The Oltrepo Pavese is located in Northern Italy. Its southern part corresponds to the northwestern sector of the Apennines. The geology is dominated by sedimentary formations, with a dominant clay component (Beatrizzotti *et al.* 1969; Fig. 1). Clay shales, referred to as Argille Varicolori (Varicoloured Clays), or Complesso Caotico (Chaotic Complex), crop out throughout the Oltrepo Pavese area, whereas calcareous flysch, made up of alternating marls, calcareous marls, and tectonically deformed shales, are mainly located in the eastern part. Processes of chemical alteration and physical degradation result in the weathering of the bedrock and the production of thick sequences of weathered materials,

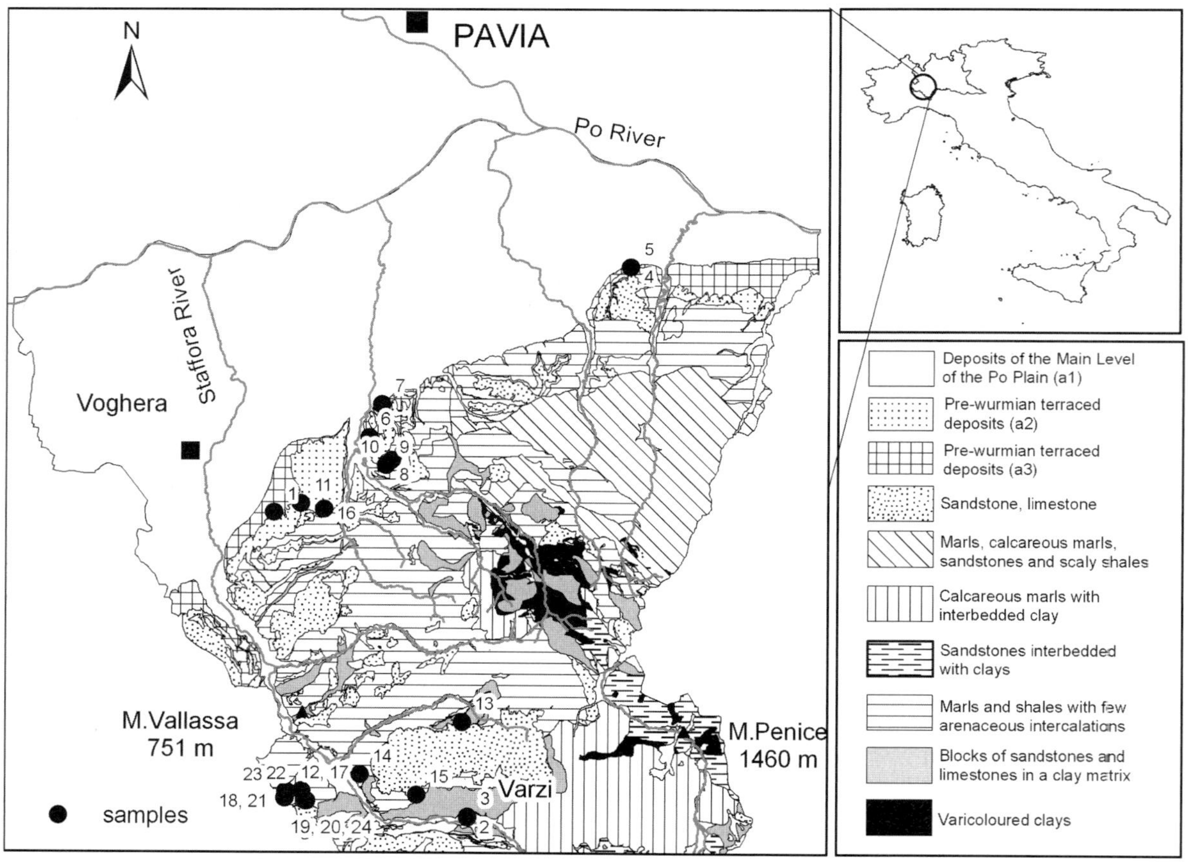

Fig. 1. Lithological map sof the Oltrepo Pavese area with the soil sample location.

generally consisting of clays, including diffused mineral or work fragments of different size, and in the reduction of the soil strength properties. Slopes in the argillaceous and marly successions are mantled by soils derived from downslope transportation of weathered bedrock (hereafter referred to as colluvial soils) varying between 1 and 6 m in thickness. The colluvial soils are often the location of numerous shallow slope instabilities and foundation problems.

The northern sector of Oltrepo Pavese is part of the Po River Plain. The Quaternary cover of the Po River Plain consists of alternating impermeable sediments (silt and clay) and more permeable horizons (gravels and sandy clays of the Apennine river alluvial fans and the Po River sediments). The Quaternary sediment thickness decreases from west to east. Geomorphologically, fluvial terraces and piedmont alluvial fans characterize this zone. All along the piedmont margin, which separates the foothills of the Apennines from the alluvial deposits of the Po Plain, three orders of fluvial terraced deposits are present (a1, a2 and a3). The upper part of the older alluvial deposits is strongly weathered and includes a large amount of clay (Fig. 1).

The soils tested in this study belong to the colluvial soils on the Antognola Marls (marls and shales with few arenaceous intercalations), the Monte Piano Marls (marls and shales with few arenaceous intercalations), the Zebedassi Limestone (calcareous marls with interbedded clays), the Chaotic Complex (blocks of sandstones and limestones in a clay matrix) and to the weathering products of the alluvial deposits. These soils are often affected by shallow landslides.

The landslides

Shallow landslides affecting the colluvial soils are a recurrent problem in many areas of the Oltrepo Pavese. The shallow landslides are generally 20–50 m wide, 10–100 m long, and 1–2 m deep (Fig. 2a). The main sliding factors characterizing the slopes are the following: (1) the colluvium lithology; (2) the slope angle, which varies between 8° and 13°; (3) the presence of a temporary perched water table in colluvial soils; (4) the permeability contrasts within the bedrock, which often contains a water table in artesian conditions; (5) the land use (presence of ploughed areas).

The identification and mapping of the shallow landslides is difficult because of their small size and the intensive farming activity. Field observations in 2000, 2002 and 2009, after heavy rainfall events, indicated that most of the failure surfaces are located along the contact between the colluvium and the underlying partially weathered or unweathered bedrock. The most frequent types are complex (translational slides associated with earthflows; Cruden & Varnes 1996) and they develop on the Zebedassi Limestone. A minor number of shallow slides are represented by earth slides (Monte Piano Marls, Zebedassi Limestone and Chaotic Complex).

Fig. 2. Photographs showing some examples of landslides. (**a**) Shallow landslides (complex earth slide–earthflow) affecting colluvial soils on Zebeddassi Limestone; (**b**, **c**) shallow landslides affecting the soil on alluvial terraces and their effects on roads.

Shallow landslides are seasonal, are mostly related to the reactivation of pre-existing landslides and involve clayey colluvial soils during or after heavy rainfall events, such as occurred in October 1993, October–November 2000, November 2002 and April 2009. The cumulative rainfall for a 60 day period before the events were respectively

450, 350, 300 mm and 340 mm. Some of these heavy rainfall events followed intense drought periods in March 1989–August 1993, May 1998–September 2000 and June 2001–October 2002 (Meisina 2006).

Shallow landslides in weathered clays on alluvial deposits have similar characteristics and are responsible for considerable damage to buildings and roads (Fig. 2b, c). They are complex movements (earth rotational slides evolving into earthflows) and are controlled by a combination of factors, such as the presence in the weathered soils of a temporary perched water table, heavy rainfall, and excavation and cuts in the slope.

Method

Some zones representative of areas in the Oltrepo Pavese where shallow landslides occur (see Fig. 1 for the site locations) were selected for this study and 24 natural soils that include colluvial deposits of different geotechnical properties were chosen. The sites were selected on the basis of landslide frequency of occurrence and type and thickness of colluvial or alluvial soils.

Site investigations were made to characterize the weathered soil profile. A detailed stratigraphical description was carried out, also related to landslide bodies. The soil profile was established on the basis of landslide scars and trench pits with a mean depth of 5–6 m. A limited number of boreholes allowed the stratigraphical description of the alluvial soils to 20 m depth. During the site investigation, samples for laboratory tests were also collected. The samples were taken vertically in weathered and unweathered material. They consisted of hand-prepared undisturbed blocks and disturbed bag samples, obtained from shallow depth. In some cases in the Po River Plain it was possible to take samples from boreholes.

A series of laboratory tests was conducted to define the main geotechnical characteristics of the weathered soils. The laboratory test methods used can be grouped into the following categories: (1) index tests to classify the soils: Atterberg limits (liquid, plastic and shrinkage limit), grain-size distribution (hydrometer tests) and carbonate content; natural water content, unit weight and degree of saturation were also measured to determine the *in situ* state of the material; (2) methylene blue stain adsorption tests to determine the VB; (3) direct shear tests to measure the residual shear strength. Most of the tests were carried out following the American Society for Testing and Materials standards (ASTM 1994). The VB was determined in accordance with the French AFNOR standards (AFNOR 1998).

The methylene blue stain test

The methylene blue stain test was developed in France, mainly for characterizing soils involved in highway construction. It quantifies the ionic absorption capacity of a soil by measuring the quantity of methylene blue necessary to cover the total (external and internal) surface of the clay particles contained in the soil. The basis of the technique is the chemical reaction triggered by an excess in negative electric charges in clay particles and/or the ionic exchange phenomena taking place between the easily exchangeable cations of the clay and the methylene blue cations released by methylene blue during its decomposition in water (Chiappone *et al.* 2004). Methylene blue dye in aqueous solution dissociates into the chloride anion and methylene blue cation. This cation displaces Na^+, Ca^{2+}, K^+ and Mg^{2+} cations, allowing the determination of the specific surfaces of these minerals (Sa) (Magnan & Youssefian 1989; Tourenq & Tran Ngoc Lan 1989). The particles with the largest specific surface and the greatest negative charge are the clay particles. Conversely, the absorption capacity of a soil of given clay content is a function of the type of clay mineral contained in the soil. Indeed, the absorption capacity increases as a function of clay particle specific surface and charge. The AFNOR standards use a 'blue value of soil' index (VB), which provides a semiquantitative evaluation of the activity of a soil and qualitative indications of the type of clay minerals contained in the soil.

The quantity of material to be tested is determined as a function of clay content. The material passing the N° 10 ASTM sieve is crushed using a pestle and mortar and air-dried. The specimen is placed in a container with distilled water for 24 h. During the test the specimen is maintained in suspension by means of a stirrer. With a burette, unit doses (10 ml) of the methylene blue solution, at a concentration of $10\ g\ l^{-1}$, are added to the suspension. The unit dose value was chosen as a function of the material characteristics (clay soils). After adding each unit dose, a small quantity of the suspension is collected with a rod and a small drop is made to fall onto standard filter paper; this produces a dark blue stain, surrounded by a colourless wet area. Further unit doses of the solution are then added and more drops are deposited on the filter paper until a light blue halo is seen to form inside the wet area and around the blue stain. This reflects the presence of an excess quantity of methylene blue that is no longer absorbed by the clay mineral and remains in suspension. At this point, no more methylene blue is added and the suspension is checked at 1 min intervals to determine the stability of the light blue halo; if the halo persists after 5 min, a sufficient quantity of methylene blue has been added. Otherwise, the titration process is continued until the halo becomes stable. Testing time ranges from about 15 to 60 min as a function of the content of the clay minerals contained in the sample. There is a certain degree of subjectivity in deciding whether the titration process has come to an end, with an error of ± 1 unit dose (Chiappone *et al.* 2004).

The index derived by this test method is the blue value of the soil (VB, in g per 100 g of soil) and is given by

$$\mathrm{VB} = V \times 0.01 \times 100/W \qquad (1)$$

where V is the volume (in cm^3) of methylene blue solution used, 0.01 is the concentration ($g\ cm^{-3}$) of methylene blue solution used, and W is the dry weight (in g) of the specimen. The blue value, which represents the quantity of methylene blue adsorbed by the clay minerals contained in the aqueous suspension, ranges between 1 and 16 g per 100 g of soil: the greater the quantity of clay minerals contained in the soil, the greater the quantity of methylene blue adsorbed and the higher the blue value. The blue value is also indicative of the types of minerals present; in particular, it is affected by the percentage of expansive clay minerals, such as smectite. The blue value increases with an increase in the content of this mineral.

The blue value makes it possible to determine the total active (internal + external) specific surface (Sa, in $m^2\ g^{-1}$) of the particles that can be reached by water: having determined the blue value and the concentration of a solution ($0.01\ g\ cm^{-3}$), and knowing that a molecule of methylene blue has a surface area of 130 $\mathring{A}^2$ ($1\ \mathring{A} = 0.1$ nm), the following formula can be applied:

$$\mathrm{Sa} = (\mathrm{VB}/100) \times (N/W_{\mathrm{MB}}) \times (130 \times 10^{-20}) = 21\mathrm{VB} \quad (2)$$

where N is Avogadro's number (6.02×10^{23}), and W_{MB} is the molecular weight of methylene blue. The total active surface ranges from 20 to 800 $m^2\ g^{-1}$ for clay minerals and from 1 to 4 $m^2\ g^{-1}$ for inert minerals.

For the fraction made up of particles smaller than 2 μm, the test allows for the determination of the index of activity of the clay fraction in soil (ACB):

$$\mathrm{ACB} = 100\mathrm{VB}/\mathrm{C} \quad (3)$$

where ACB is the index of activity (in grams of methylene blue in a 100 g clay fraction), VB is the blue value for the sample (in grams of methylene blue in 100 g of dry soil), and C is the clay content (particles smaller than 2 μm). The index of activity increases with increase in the smectite content, its value being low for slightly active or inert clay fractions. Hence, the index of activity is an excellent indicator of the capability of a soil to retain water. Lautrin (1989) proposed a classification of soils as a function of their index of activity: a low value of VB is believed to represent 'stable soil' (non-swelling, with high resistance to shear), and high values indicate that the properties of the soil may change in the presence of water. The methylene blue adsorption test is also a quick and easy method to assess the expansive potential of clay soils (Chiappone & Scavia 1999; Fityus *et al.* 2000; Meisina 2000; Erguler & Ulusay 2003).

The residual shear strength test

The residual strength was evaluated in direct shear apparatus (a Casagrande-type shear box was used) by the Kanji method (Kanji 1974). In this method the soil slides on soil–rock interfaces; the minimum friction angles attained for soil sliding on polished rock are shown to correspond to the residual strength of the soils. The direct shear test by the Kanji method has the advantage that it shears the specimen continuously in one direction; this allows clay particles to be oriented parallel to the direction of shear and a residual strength condition to develop. Direct shear tests were performed on specimen 5 of 60 mm diameter and 22 mm height. Residual strength is independent of initial soil structure so that remoulded samples were tested. These were obtained by air drying a representative sample of each soil, crushing it with a mortar and pestle and processing it through a N°. 10 ASTM sieve. Distilled water was added to the processed soil until a liquidity limit was obtained. The sample was then allowed to rehydrate at least overnight and then consolidation in stages was applied. A spatula was used to place the remoulded soil paste into the specimen container. Each specimen was consolidated to a selected vertical stress, ranging between 100 and 300 kPa. A consolidation stress of 300 kPa was chosen to represent the maximum effective stress that could be encountered in a slope. During consolidation, the specimen container was secured in a water bath filled with deionized water. For each load increment the samples were consolidated until the time–consolidation curve showed more than 95% consolidation. All specimens were sheared at a displacement rate of 2 $\mu m\ min^{-1}$ and one cycle of shearing was carried out. During the shearing process, readings of time, vertical and shear displacements and shear force were recorded until the shear stress became constant. The residual friction angle $\phi' r$ was evaluated using a best-fit linear envelope on three points over the range of vertical effective stresses involved.

Soil profiles and properties

Colluvial soils

The soil profiles were divided into three zones based on the amount of weathering that had occurred (Fig. 3).

(1) Colluvium. The colluvial material extends to more than 4 m depth; it is composed of totally weathered bedrock, in which the original structures have been obliterated, and of the material derived from downslope transportation. The weathered material can be distinguished into horizons C1 and C2. The top 1–1.5 m of the colluvial deposits (hereafter referred to as the upper colluvial soil, C1) consists typically of hard to firm fissured clay and silty clay. Where slope instabilities are present the horizons C2 indicate landslide material of different ages. Often within the clay matrix there are stony clasts (marl, limestone, sandstone) with sharp edges. The upper zone has been significantly disturbed by shrink–swell phenomena, with fissures and slickensides frequently present as the result of past desiccation.

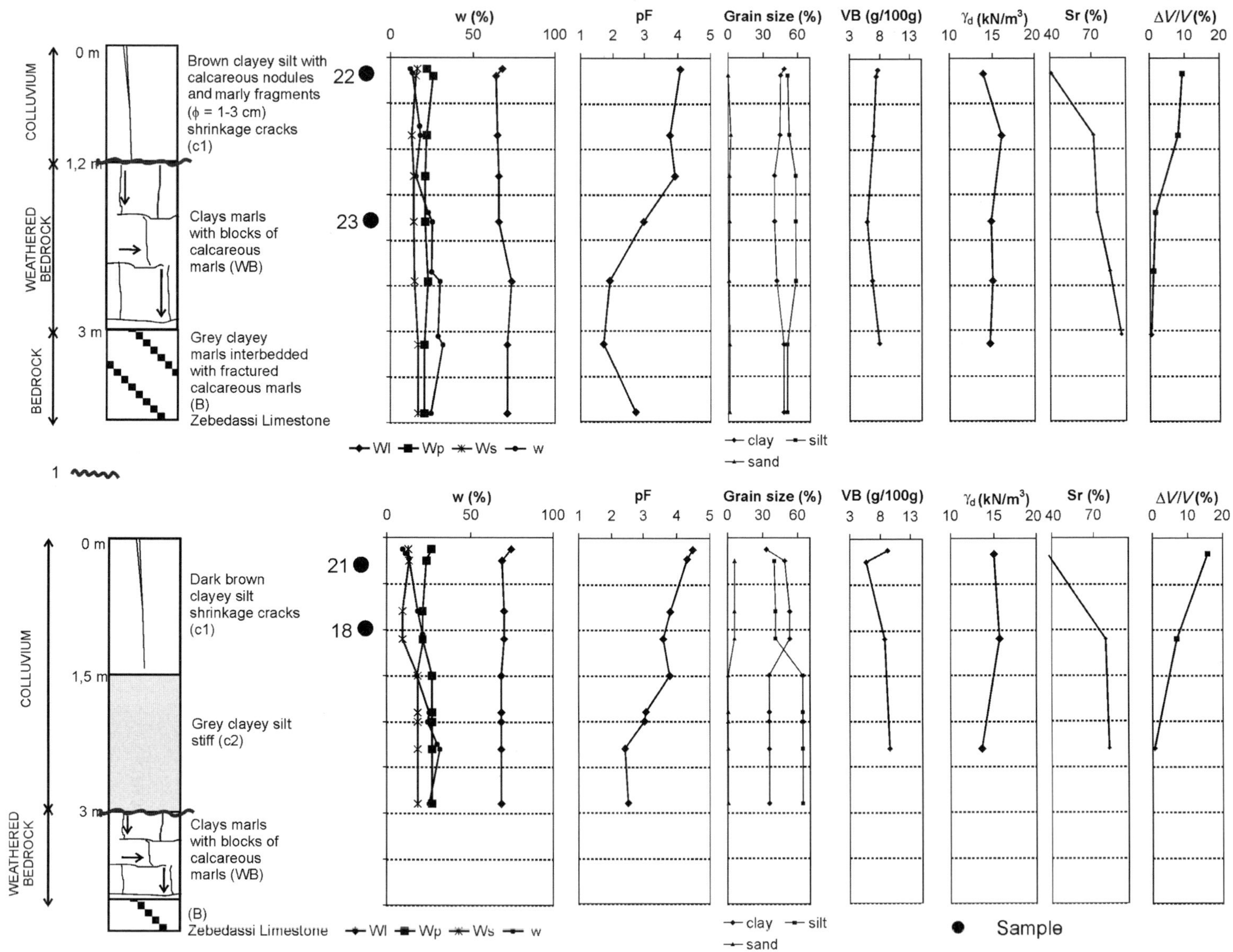

Fig. 3. Lithological and geotechnical profiles of some colluvial soils on Zebedassi Limestones. 1, sliding surface; pF, soil suction; w, natural water content; Wl, liquid limit; Wp, plastic limit; Ws, shrinkage limit; VB, value of blue; γ_d, dry density; Sr, degree of saturation; $\Delta V/V$, swell strain.

The failure surfaces are located along the contact between the colluvium and the underlying weathered bedrock.

(2) Partially weathered bedrock (WB).

(3) bedrock (B). This is less disturbed by the shrink–swell phenomena, more uniform, harder and duller in appearance, and shows fewer slickensides. It is composed of fractured calcareous marls with interbedded shales (Zebedassi Limestone) or consists of hard, high-plasticity clay with marly fragments (Chaotic Complex) or of marls and shales with few arenaceous intercalations (Monte Piano Marls and Monte Antognola Marls).

The geotechnical properties of the upper 3 m are summarized in the geotechnical profiles of Figure 3 and in Table 1. The geotechnical soil profiles (Fig. 3) do not highlight significant geotechnical differences with depth within the colluvium; generally the horizons C1 and C2 are more plastic and contain more clay fraction than the weathered bedrock and the bedrock. The geotechnical profiles also show the homogeneity of index properties and clay content. There is little difference between horizons C2 and WB, which have a smaller clay fraction and lower plasticity than C1 (which is weathered), and the bedrock B.

Alluvial soils

Figure 4 shows the lithological and geotechnical profiles for recent alluvial deposits (a1, Fig. 4a) and pre-Würm terraced deposits (a3, Fig. 4b). The recent alluvial deposits (clayey silt with sand) are covered by silty clay and clayey silt derived from the weathering of the alluvial soils and from the downslope transportation of soils of the surrounding hills. The geotechnical soil profile does not show significant geotechnical differences with depth within the weathered alluvial soils. The oldest alluvial deposits consist of a sand and gravel layer (old alluvium) and are covered by reddish yellow clays, up to 3 m thick, derived from the weathering of older alluvial sediments.

Calcareous and Fe–Mn nodules are often found within the soil mass. The upper clay is stiff to very stiff and generally is fissured; it has a very high shrink–swell potential. The soils are internally sheared with shiny slickenside surfaces; they are overconsolidated with a block or prism-shaped structure. Geotechnical properties vary with depth in relation to the degree of weathering and the pedogenic processes (e.g. migration–accumulation of clay or calcium carbonate). Mechanical and chemical weathering (cycles of drying and wetting) cause deterioration of the shear strength of these soils with time.

Geotechnical characterization of the studied soils

The results of the geotechnical tests vary considerable for the different soils. Nine soil samples were selected to represent the upper colluvial soil and the weathered bedrock; two soil samples were collected on the slip surface of shallow landslides (samples 12 and 15). The remaining samples represent the weathered alluvial soils; only one sample is related to the unweathered alluvial deposits (sample 3). The geotechnical properties are summarized in Table 1 and Figure 5.

The tested soils can be divided into the following groups.

(1) Recent alluvial deposits of the Main Level of the Po Plain (a1) (sample 1): silty clay, with some percentage of sand (14%), classified as CH according to the Unified Soil Classification System.

(2) Recent alluvial deposits of the Staffora River (right tributary of the Po River) (a1) (samples 2 and 3): silty clay and clayey silt, classified as CH and CL. The samples were collected on the same soil profile (Fig. 4b) at different depth and correspond to the alluvial soil (sample 3) and its weathering products (sample 2); sample 2 is characterized by slickensides. The plasticity and the clay fraction decrease with depth as demonstrated by the VB, which decreases from 5 g per 100 g of soil (sample 2) to 2.6 g per 100 g of soil (sample 3).

(3) Alluvial deposits belonging to the second order of terraces (a2) (samples 4–7): clayey silts (C = 33–41%) containing calcareous and Fe–Mn nodules and classified as CH (PI = 30). Sample 5 corresponds to a layer of a reddish very high plasticity clay (PI = 81, C = 75%, VB = 10.8 g per 100 g of soil) found at 6 m depth, which can be probably referred to a palaeo-soil.

(4) Alluvial deposits belonging to the third order of terraces (a3) (samples 8–11 and 16): the soils are very heterogeneous. On the basis of grain-size distribution, the soils were classified as silty clay, clayey silt and clay. The plot of Atterberg limits on a Casagrande plasticity chart shows that the soils can be classified as CH and OH–MH (sample 10).

(5) Colluvial soils derived by the weathering and downslope transportation of the Zebedassi Limestone (c1) (samples 12, 17–22 and 24): clayey silt classified as CH (C = 37–53%, PI = 31–49). The VB ranges from 4.6 to 8.8 g per 100 g of soils. Sample 19 corresponds to a horizon characterized by a low content of clay (C = 29%) and low plasticity (PI = 27). Nevertheless it has a high VB of 5.1 g per 100 g of soil; this is probably related to the presence of expansive lattice-type minerals such as smectite. Sample 12 was collected on the slip surface of a shallow landslide.

(6) Weathered Zebedassi Limestone (WB) (sample 23): clayey silt, classified as CH with no significant differences from the colluvial soils developed on it.

(7) Weathered bedrock belonging to Monte Piano Marls and Antognola Marls (WB) (samples 13 and 14): clayey silt of low plasticity (CL), with very low VB (1.7–3.2 g per 100 g of soil).

(8) Weathered bedrock belonging to Chaotic Complex (WB) (sample 15): clay of high plasticity (CH) collected on the slip surface of a shallow landslside.

The calcium carbonate content is high in the colluvial soils and in the weathered bedrock (in some case $CaCO_3$ is

Table 1. *Geotechnical properties of the soils subjected to direct shear strength tests using the Kanji method*

Sample	*Type*	*Geology*	*Description*	*Depth* (m)	Wl (%)	PI	C (%)	VB (g per 100 g)	ACB	Sa ($m^2 g^{-1}$)	$\phi' r$ (°)	$CaCO_3$ (%)
1	a1	Alluvial deposits	Silty clay	2.3	60	41	49	6.5	13.3	136	10.0	9
2	a1	Alluvial deposits	Silty clay with slickensides	2.4	68	47	58	5.0	8.7	127	9.0	
3	a1	Alluvial deposits	Clayey silt	4.5	33	18	34	2.6	7.7	54	17.0	
4	a2	Alluvial deposits	Silt and clay	2.5	55	32	33	4.9	14.9	103	14.5	3
5	a2	Alluvial deposits	Silt and clay with calcareous nodules	6.7	104	81	75	10.8	14.3	225	4.7	5
6	a2	Alluvial deposits	Silty clay	0.8	54	30	41	4.9	11.8	101	16.0	5
7	a2	Alluvial deposits	Clayey silt	2.2	54	31	33	5.8	17.5	121	14.0	
8	a3	Alluvial deposits	Clay with calcareous nodules	1.3	72	52	50	8.2	16.3	173	8.0	8
9	a3	Alluvial deposits	Clay with calcareous nodules	2.1	57	38	45	7.1	16.0	148	6.0	
10	a3	Alluvial deposits	Clayey silt with calcareous nodules	1.5	52	22	24	3.8	15.6	82	21.0	7
11	a3	Alluvial deposits	Clay with calcareous nodules	1.9	63	40	53	7.3	13.7	147	11.0	5
16	a3	Alluvial deposits	Clayey silt	0.9	47	27	38	4.1	10.9	86	15.0	7
12	c(c1)	Zebedassi Limestone	Clay	1.2	90	70	61	9.6	15.7	200	5.5	
13	c(wb)	Antognola Marls	Clayey silt	0.5	31	11	34	1.7	5.1	36	17.0	
14	c(wb)	Monte Piano Marls	Clayey silt	0.5	45	24	31	3.2	10.4	67	16.0	45
15	c(wb)	Chaotic Complex	Clay	0.5	94	68	52	9.9	18.8	207	4.0	
17	c(c1)	Zebedassi Limestone	Silty clay	0.2	68	46	48	7.8	16.1	169	8.2	19
18	c(c1)	Zebedassi Limestone	Silty clay	1.1	70	49	53	8.8	16.6	119	7.6	31
19	c(c1)	Zebedassi Limestone	Silt	1.4	59	27	29	5.1	17.7	107	9.1	31
20	c(c1)	Zebedassi Limestone	Clayey silt	0.5	56	32	47	4.6	9.9	97	11.3	37
21	c(c1)	Zebedassi Limestone	Clayey silt	0.2	69	46	49	5.7	11.6	201	7.1	8
22	c(c1)	Zebedassi Limestone	Clayey silt	0.8	64	43	45	7.0	15.7	147	6.4	11
23	c(wb)	Zebedassi Limestone	Clayey silt	1.7	66	45	39	5.9	15.1	124	8.5	9
24	c(c1)	Zebedassi Limestone	Clayey silt	0.4	60	31	37	6.1	16.4	129	8.9	36

Wl, liquid limit; Wp, plastic limit; PI, plasticity index; C, clay fraction; VB, value of blue; ACB, clay fraction index of activity; Sa, active surface of clay fraction; $\phi' r$, residual shear angle.

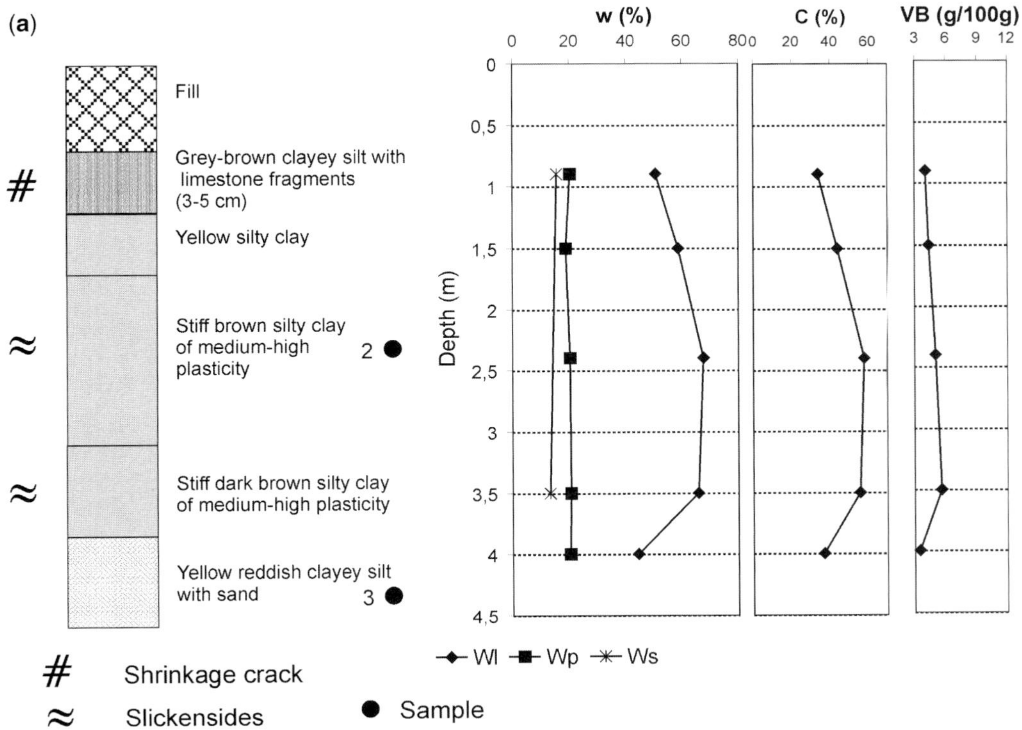

Fig. 4. Lithological and geotechnical profiles of some alluvial soils. Abbreviations as in Figure 3, and C, clay content; SP, swelling pressure.

up to 30%); alluvial soils are characterized by low calcium carbonate content (<10%). According to Lautrin's (1989) classification, based on the blue activity index (ACB), the studied soils (except samples 3 and 13) belong to the class of soils containing an active and very active clay fraction, as ACB is respectively as high as 8 and 18 (sample 15). The expansive clay minerals (smectite, swelling chlorite) are present in high percentages in colluvial soils; depending on the degree of weathering the minerals of alluvial soils consist of vermiculite and quartz.

The drained residual strength envelopes obtained from the shear tests on alluvial soils are shown in Figure 6a. The residual friction angle ranges from 4.7° to 21°, reflecting the extreme variability of the geotechnical properties in relation to the different pedological processes (Fig. 7). The highest values ($\phi'r = 17°$ and 21°) are related to the unweathered alluvial deposits (sample 3) and to sample 10, which is characterized by a low amount of clay fraction (C = 24%). The lowest value ($\phi'r = 4.7°$) is associated with sample 5, which is characterized by high plasticity (PI = 81), high clay fraction (C = 75%) and high VB (10.8 g per 100 g of soil).

The residual friction angle of colluvial soils and the weathered bedrock generally ranges between 5.5° and 11.3° (Fig. 8). Very low values are associated with samples 15 and 12 ($\phi'r = 4°$ and 5.5°, respectively), which were collected on slip surfaces. The weathered bedrock of Monte Piano Marls and of Antognola Marls is characterized by high values of the residual friction angle (16–17°) (Fig. 6b). The failure envelopes are linear to a good approximation.

Comparison with published correlations

Residual friction angles measured in the laboratory for the studied soils were compared with selected published correlations based on liquid limit (Wl), plasticity index (PI) and clay fraction (C) (Tables 2 & 3). Correlations between $\phi'r$ and PI were published by Voight (1973), Kanji (1974), Seycek (1978) and (Fleischer, Bucher and Vaughan *et al.*, all cited by Lupini *et al.* 1981). Fleisher (see Lupini *et al.* 1981) presented a relationship between residual friction angle and the PI for various stiff clays, based on ring shear tests with varying strain rates. This correlation generally overestimates the residual friction angle for colluvial soils and underestimates this parameter for alluvial soils; in some cases (PI > 50), it is not applicable. Voight (1973) also found a relationship between residual friction angle and PI, which overestimates the $\phi'r$ of alluvial soils and

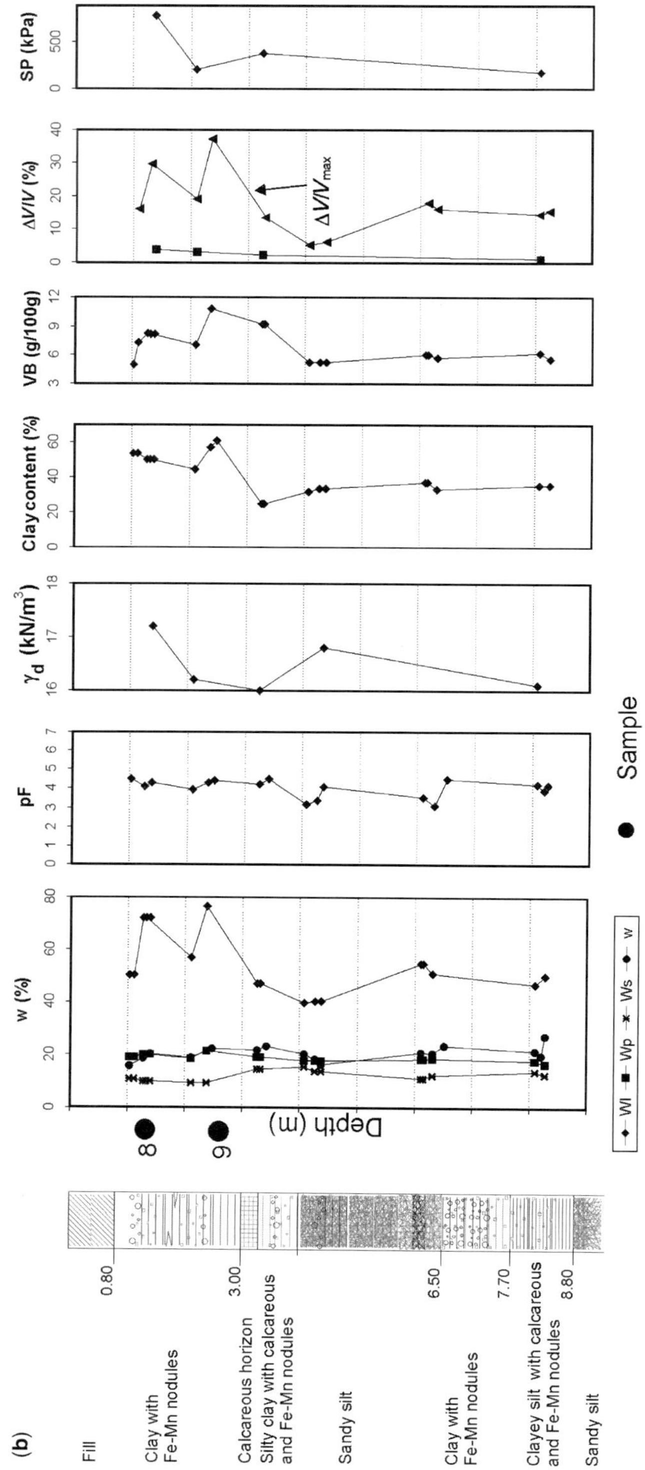

Fig. 4. *Continued.*

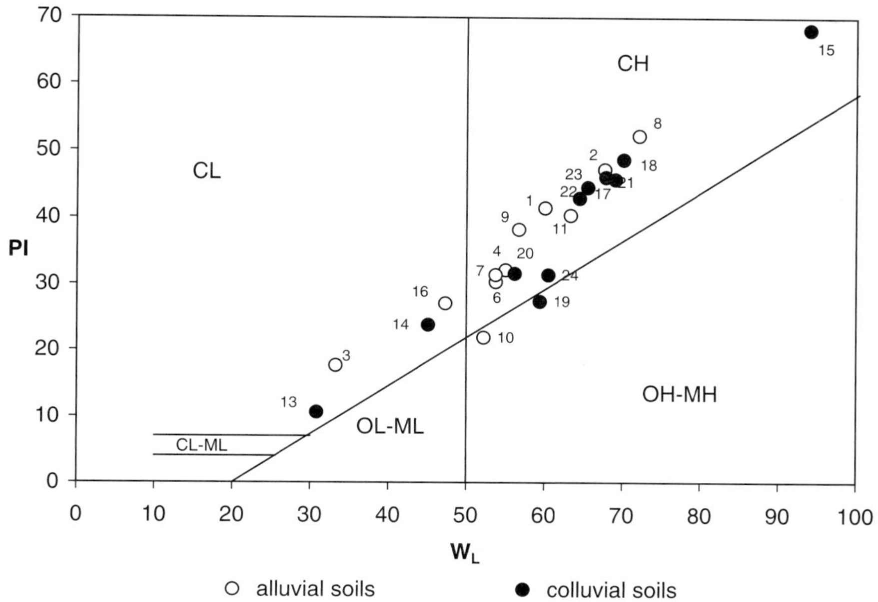

Fig. 5. Casagrande plasticity chart. Test numbers refer to Table 1.

underestimates the residual friction angle of colluvial soils. Kanji (1974) found a good correlation for the residual friction angle, obtained with reversal shear box tests on the interfaces between soils and a polished rock surface, in terms of the plasticity index for values of PI of 5–350. This correlation, obtained from the same test procedures as used in this study, is fairly good for colluvial soils but for alluvial soils it underestimates the $\phi' r$. Bucher (see Lupini *et al.* 1981) used two strain-controlled ring shear devices of different design and reversal shear box tests to study the influence of stress history, stress level, shear rate and temperature on residual friction angle. He found a relationship between PI and residual friction angle similar to that of Vaughan *et al.* (see Lupini *et al.* 1981), who postulated a discontinuous relationship between residual friction angle and PI, obtained with tests in the ring shear apparatus on natural clays of medium activity (0.7–1). Both correlations show considerable scatter. Seycek (1978) summarized the results of reversal shear box tests and results from other researchers, and reported that there was a better correlation between residual friction angle and PI than with any other parameter. This correlation provides the best agreement with the laboratory tests for the studied soils but shows a large scatter of data. For the soils analysed the $\phi' r$ values fall generally within the ranges already established in the literature by the above-mentioned researchers, except for samples 5 and 15, which are characterized by high plasticity (PI >65).

Correlations between $\phi' r$ and clay fraction were published by Skempton (1964), Borowicka (1965), Binnie *et al.* (1967) and Blondeau & Josseaume (1976); these studies indicated a reduction in residual friction angle with increasing clay content. Skempton (1964) reviewed the effect of residual strength on slope stability, and postulated a general correlation between residual strength measured in the shear box and multiple reversals of shear and clay fraction. Binnie *et al.* (1967) found the same relationship for natural soils of high plasticity (Mangla and Jari clays) containing field shear surfaces, tested in the shear box. Borowicka (1967) reported reversal shear box tests on various clay soils produced artificially by mixing in the laboratory. He found that brittleness increased and residual shear angle decreased with increasing clay fraction. Blondeau & Josseaume (1976) investigated the residual shear strength of clay of varying plasticity and found a relationship between $\phi' r$ and the clay fraction. Most of the previous residual strength correlations based on clay size fraction, except that of Borowicka, overestimate the residual friction angle. Samples 5, 15 and 12, and samples 19, 20 and 9 generally fall outside the ranges established in the literature by correlations using the clay fraction. The first group of soils (5, 15 and 12) is characterized by a high clay content and a high plasticity index; the soils of the second group have a low clay content; nevertheless, the low values of $\phi' r$ are probably associated with the mineralogical composition (high smectite content) as demonstrated by the high VB. From Tables 2 and 3 it appears that the band of correlations found by Blondeau & Josseaume fits the studied soils fairly well. The rest of the correlations generally fall above the dataset, except the correlation by Borowicka, which falls below most of the dataset.

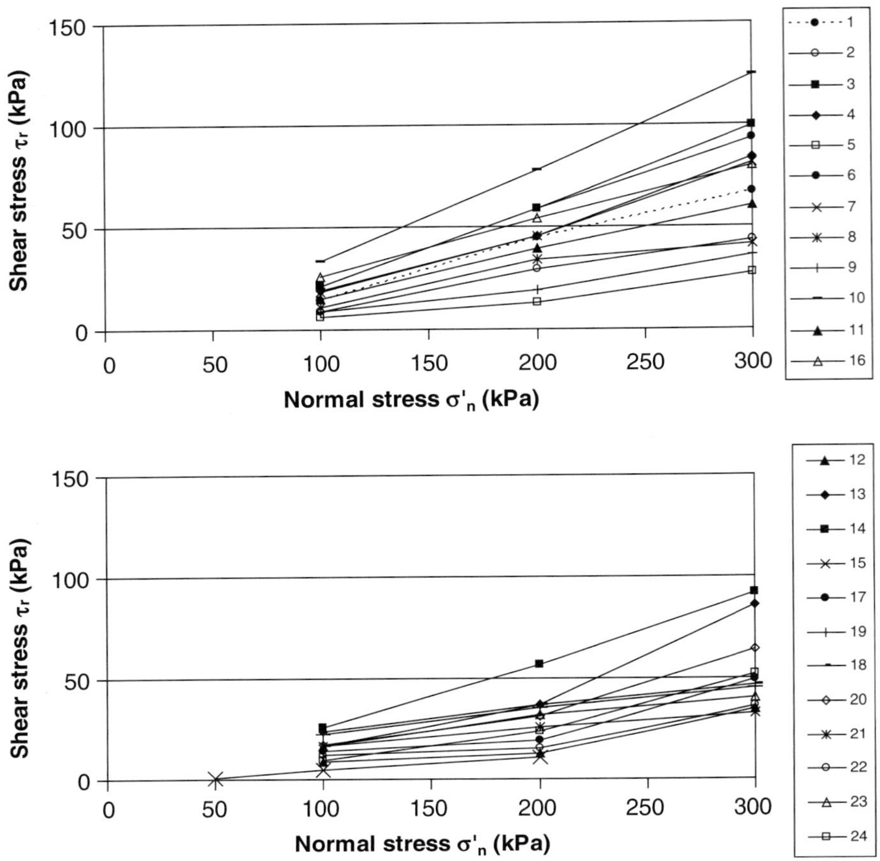

Fig. 6. Residual shear strength envelopes. (**a**) Alluvial soils; (**b**) colluvial soils.

Stark & Eid (1994) showed that there is a linear relationship between the residual friction angle and liquid limit up to liquid limit values slightly less than 100%, when the relationship reaches asymptotic values. They also showed that the suggested relationship is a function of soil clay fraction as well as effective normal stress (from 100 to 700 kPa). Most of the studied soils are outside the range proposed by Stark & Eid.

Collotta *et al.* (1989) proposed a correlation that relates $\phi'r$ to a parameter that combines C, Wl and PI in the following manner:

$$\phi'r = (C)^2 \times Wl \times PI \times 10^{-5} \quad (4)$$

where C is the clay fraction. The proposed correlation is less steep and contains less scatter than other correlations. It applies to soil with more than 80% passing the No. 40 ASTM sieve. These correlations overestimate the residual friction angle of the studied soils.

To investigate the possibility of a better correlation between $\phi'r$ and Atterberg limits, Wesley (2003) plotted the residual friction angle against the distance above or below the A-line of the Casagrande plasticity chart. This distance is expressed as ΔPI, given by

$$\Delta PI = PI - 0.73 \times (Wl - 20). \quad (5)$$

Only those soils with Wl above 50 were included in the plot. There is also in this case a considerable scatter; however, the studied soils lie above the A-line (except soils 10 and 19) and are expected to have very low $\phi'r$ values.

In general, it appears that the studied soils compared better with the plasticity index-based correlations than the clay fraction-based correlations. Existing correlations based on only clay-size fraction and the plasticity index fit the colluvial soils better, but they tend to overestimate the residual friction angle both for colluvial soils and for alluvial soils. The comparison between the published correlations and

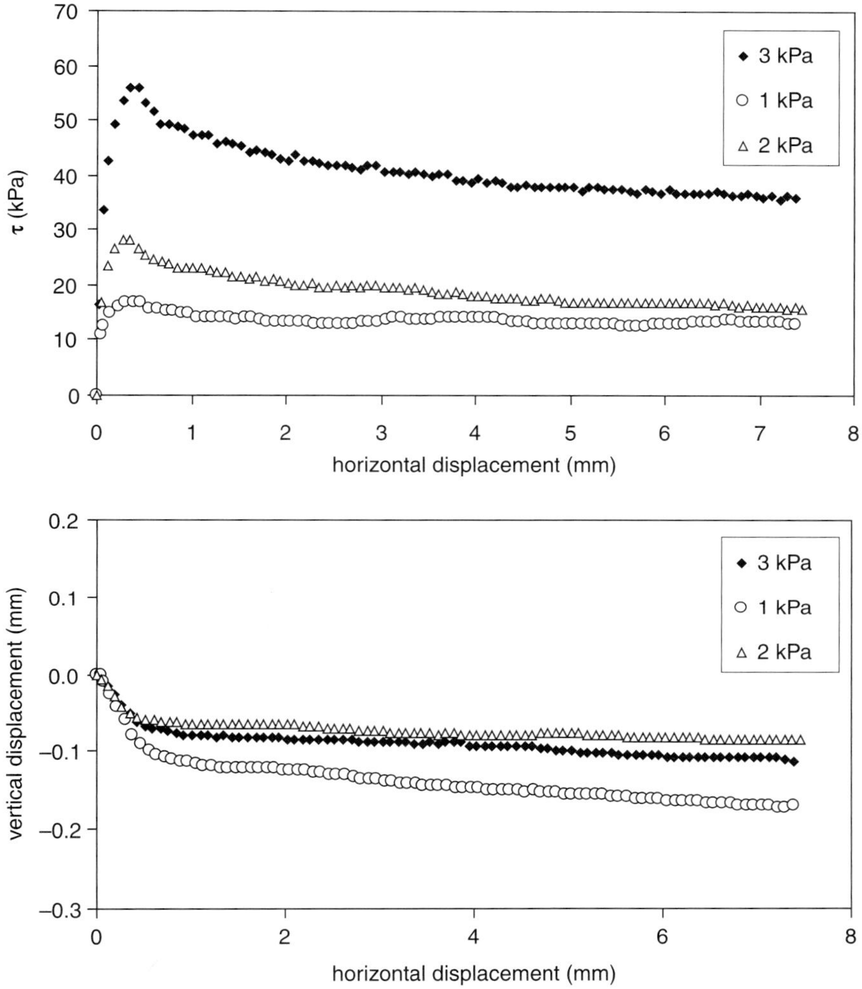

Fig. 7. Results of direct shear tests, soil 16. (**a**) Shear stress v. horizontal displacement; (**b**) vertical displacement v. horizontal displacement.

the data for samples 5, 12 and 15, characterized by high clay content and high plasticity, is not very good. All these correlations are characterized by a large amount of scatter in the data. Therefore, they are useful to evaluate trends but not for practical applications. For example, in the correlation of Wesley when the ΔPI is ± 5, the variation in $\phi' r$ is of the order of 5° to nearly 25°.

The results obtained could be related to various factors: (1) the effect of variable sample preparation on the measurement of index properties (the degree of induration or aggregation of overconsolidated clay has an important effect on the measurement of liquid limit and clay fraction; Mesri & Cepeda-Diaz 1986); (2) the effect of sample preparation procedure in residual strength testing (remoulded or undisturbed specimens, natural or artificial soils) and the type of equipment used (e.g. ring shear tests, reversal shear box tests); (3) the different geological characteristics of the soils analysed; (4) the influence of effective normal stress on drained residual friction angle (e.g. Stark & Eid's correlation used effective normal stress range of 100–700 kPa, whereas the data reported here are for an effective normal stress range of 100–300 kPa).

Proposal for new correlation method

In Figure 9a–d the residual friction angle $\phi' r$ is plotted against clay fraction C, liquid limit Wl, plasticity index PI

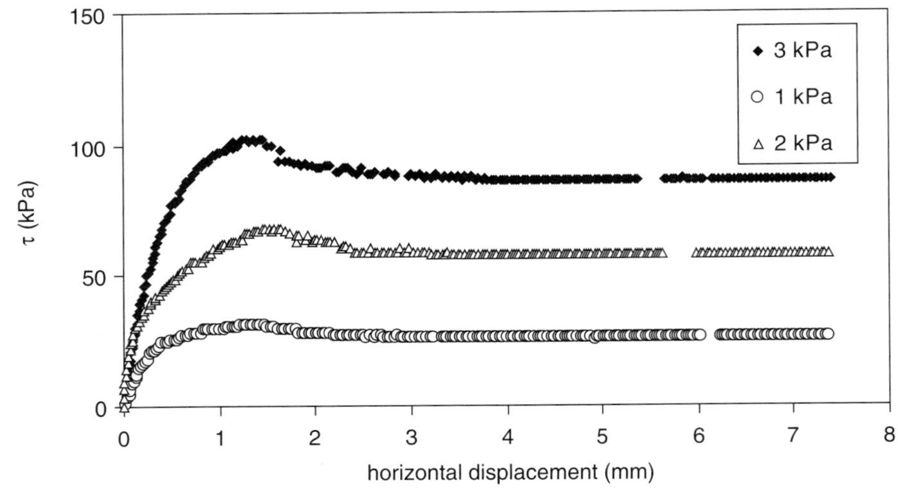

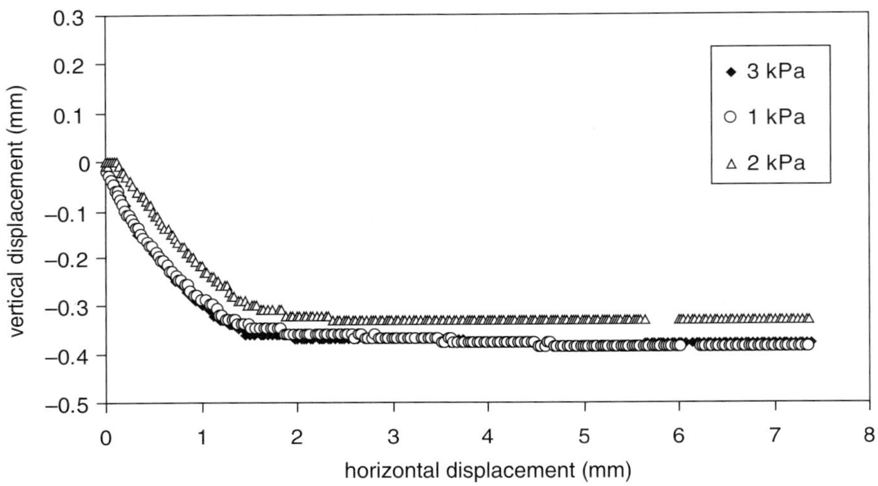

Fig. 8. Results of direct shear tests, soil 22. (**a**) Shear stress v. horizontal displacement; (**b**) vertical displacement v. horizontal displacement.

and calcium carbonate content. The correlation of $\phi'r$ with clay content is negative ($\phi'r$ decreases with increasing clay fraction), but the data show some scatter. A better correlation was found between $\phi'r$ and the liquid limit and plasticity index for colluvial soils. The magnitude of the residual friction angle decreases with increasing liquid limit and plasticity index. The variation of the effective residual friction angle shows a positive correlation with carbonate content. Although the gradient of the correlation line is low this trend can be recognized for colluvial soils. Samples with less than *c.* 25% calcite content have low residual strength; samples with greater than 25% calcite content have elevated residual strength, which increases with calcite content. These results are in good agreement with those obtained by other researchers (Hawkins & McDonald 1993). Because of their heterogeneity, alluvial soils do not show significant correlations with the index properties (Wl, PI and C).

The mineralogy of the clay particles and the influence of pore water chemistry, which control the residual shearing resistance, may not be reflected in a systematic way by PI, C or Wl (Lupini *et al.* 1981). Kenney (1967) reported results of reversal shear tests using 2.5 mm thick specimens placed between carborundum discs. He studied the influence of mineralogy, normal effective stress, rate of strain and pore water chemistry on residual shear angle and concluded that mineralogy was the most important factor and that there was no satisfactory relationship between residual shear angle and plasticity index.

Table 2. *Residual shear friction angle of the studied soils derived from empirical correlations for alluvial soils*

Reference	*Method*	*Sample*											
		1	*2*	*3*	*4*	*5*	*6*	*7*	*8*	*9*	*10*	*11*	*16*
Skempton (1964)	$\phi'r$\C	9\20	8.5\18	14\26.5	*15\25*	*9.5\20*	13\24	*15\27*	*10.5\21*	*11.5\23*	20\31	10\20	13\25
Borowicka (1965)	$\phi'r$\C	**7**	**7**	**12.5**	15	5.0\10	**9**	14	**7**	*7*	19.5	**7**	**11**
Binnie *et al.* (1967)	$\phi'r$\C	*14*	*12.5*	*19*	*20*	–	*17*	*19*	*15*	*16*	*23*	*14*	*17*
Blondeau & Josseaume (1976)	$\phi'r$\C	6.5\14.5	6\12	9\22	10\20	4\7	7\19	10\23	7\15	*7.5\15.5*	13.5\27	6.5\13.5	8\18
Collotta *et al.* (1989)	$\phi'r$\CALIP	**7.5\9**	**7.5\8.5**	*>26*	*19\23*	–	14.5\18	*19.5\23.5*	7.5\8.5	*10\12.5*	*>26*	**7.5\9**	*19.5\23.5*
Fleischer (cited by Lupini *et al.* 1981)	$\phi'r$\PI	–	–	17	**10**	–	**10**	**10**	*9*	*9.5*	**13**	**9**	**11**
Voight (1973)	$\phi'r$\PI	*13.5*	*12.5*	*23.5*	*17*	*7*	*18.5*	*18*	*12*	*16*	*22*	14	*18*
Kanji (1974)	$\phi'r$\PI	**8.8**	**8.4**	**12.9**	**9.9**	*6.6*	**10.2**	**10.0**	8.0	*9.2*	**11.8**	**9.0**	**10.7**
Bucher (cited by Lupini *et al.* 1981)	$\phi'r$\PI	7.5\15	7.5\12.5	16\34	7\21	–	10\26	9\25	*8.5\13*	*8.5\18.5*	13\34	8.5\17	11\30
Seycek (1978)	$\phi'r$\PI	5\18.5	4.5\17	8\28	5\20	5\12	4.5\23	4.5\22.5	4.5\17	4.5\20	6\26	4.5\19	5\22
Vaughan *et al.* (cited by Lupini *et al.* 1981)	$\phi'r$\PI	8\13	8\13	*22\31*	*8\13*	–	**8\13**	**8\13**	8\13	*8\13*	*22 \30*	8\13	*8\13*
Wesley (2003)	$\phi'r$\ΔPI	**5\10**	5\10	–	7\17	4\7	6\21	6\21	4.5\8.5	5\9	8\28	4\12	–
Laboratory test	$\phi'r$ (°)	10	9	17	14.5	4.7	16	14	8	6	21	11	15

Italic values, the residual friction angle is overestimated; bold values, the residual friction angle is underestimated.

Table 3. *Residual shear friction angle of the studied soils derived from empirical correlations for colluvial soils*

Reference	Method	Sample											
		12	*13*	*14*	*15*	*17*	*18*	*19*	*20*	*21*	*22*	*23*	*24*
Skempton (1964)	ϕ'_r\C	*8.5–17*	17\27	17\27	*9.5–20*	*12\22*	*11\21*	*17\27*	10\22	10\21	*11\22*	*12\24*	*13\25*
Borowicka (1965)	ϕ'_r\C	*7*	**15**	**15**	*8*	**7**	**7**	*15*	**8**	*8*	*8*	*10*	*10*
Binnle *et al.* (1967)	ϕ'_r\C	–	18	*20*	*14*	*16*	*15*	*20*	*15*	*15*	*16*	*17*	17
Blondeau & Josseaume (1976)	ϕ'_r\C	5.5–10.5	9\21	9\21	*7–14*	6\15	5\14	*10\24*	8\19	6\15	*7\16*	7\16	8\20
Collotta *et al.* (1989)	ϕ'_r\CALIP	–	*25/30*	*22/25*	*7.5–8.5*	7.5–9	7.5–8.5	*22–27*	11/14.0	*7.5–8.5*	*8.5–10.5*	*10.0–12*	*14.0–17*
Fleischer (cited by Lupini *et al.* 1981)	ϕ'_r\PI	–	*27*	**13**	–	8	*8*	*11*	**10**	*9*	*9*	*9*	*10*
Voight (1973)	ϕ'_r\PI	*8*	*28*	*21*	*9*	*13*	*13*	*17*	*17*	*13*	*13*	*13*	*17*
Kanji (1974)	ϕ'_r\PI	*7.0*	16.2	**11.3**	*7.1*	8.4	8.2	*10.6*	**10.0**	*8.5*	*8.7*	8.6	*10.0*
Bucher (cited by Lupini *et al.* 1981)	ϕ'_r\PI	–	*21\33*	12\32	–	7\13	7\13	*10\29*	9\23	8\14	*8\15*	8\14	*9\24*
Seycek (1978)	ϕ'_r\PI	4.5–13	17\30	5\24	4.5–13.5	5\17	5\17	5\23	5\21	5\18	5\18	5\18	5\22
Vaughan *et al.* (cited by Lupini *et al.* 1981)	ϕ'_r\PI	–	*23\33*	*21\30*	–	8\13	*8\13*	8\13	8\13	8\13	*8\13*	8\13	8\13
Wesley (2003)	ϕ'_r\ΔPI	5\7	–	–	5\8	5\10	5\9	9\26	6\20	5\10	5\11	5\11	8\25
Laboratory test	ϕ'_r (°)	5.5	17	16	4	8.2	7.6	9.1	11.3	7.1	6.4	8.5	8.9

Italic values, the residual friction angle is overestimated; bold values, the residual friction angle is underestimated.

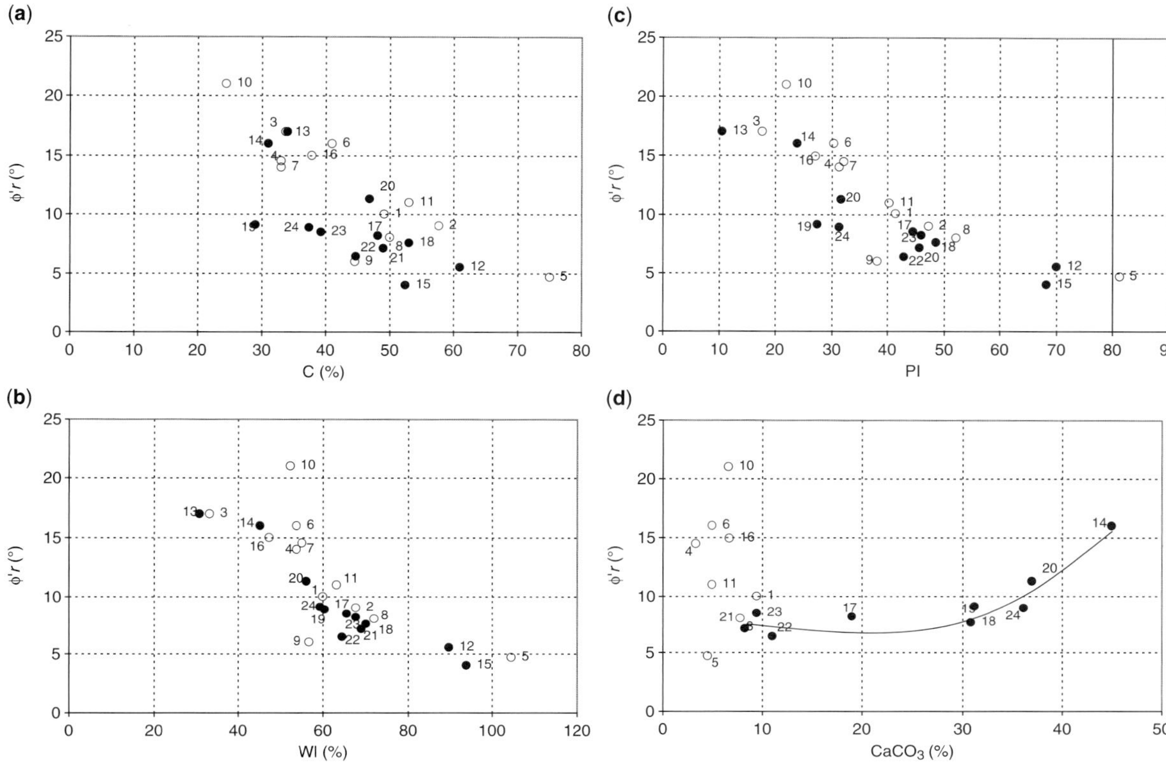

Fig. 9. Correlations between the residual shear angle and clay content (**a**), Wl (**b**), PI (**c**) and carbonate content (**d**) for alluvial soil (open circles) and colluvial soils (filled circles).

The fact that soils with similar clay content give different residual shear strength results indicates the influence of the minerals present within the clay fraction (Fig. 9a). The correlations in Figure 9 confirm that the residual friction angle is influenced by both the clay fraction and the mineralogical composition of the clay. Both factors are reflected by the VB. Thus the VB is correlated with a range of soil properties: the clay fraction, the liquid limit, the plasticity index and the calcium carbonate content (Fig. 10). The total active specific surface (Sa) and the index of activity of the clay fraction (ACB) are indicative of the mineral type present and they were also compared with the clay fraction (Fig. 11). The VB, ACB and Sa do not correlate with the clay fraction; soils with similar clay content give different VB, ACB and Sa; this confirms the influence of the type of clay mineral within the clay fraction. It is also evident from Figure 10 that there is a definite trend of increasing methylene blue adsorption with increasing PI and Wl. There is also a good correlation between VB and the carbonate content for colluvial soils, where the influence of the carbonate content is evident for values up to 25%.

Data from colluvial and alluvial soils were used to develop a correlation between residual friction angle and the VB (Fig. 12). Separate correlations were developed for alluvial and colluvial soils. Figure 12 reveals that the measured VB could reasonably be correlated with the residual friction angle value, generally within a range of about 1–2° units. The data fall close to the average trend and the proposed correlation contains less scatter than others proposed in the literature.

The best correlation between the residual friction angle and VB is related to alluvial soils for VB up to 4 g per 100 g of soil. Some problems are related to the soils with low VB (<4 g per 100 g of soil), such as samples 3 (unweathered alluvial deposits) and 10 (weathered alluvial soils a3). For these soils the data scatter is higher ($\phi' r_{predicted} - \phi' r_{measured} = 3$–$4°$). Sample 3 is characterized by low VB, related to the limited clay fraction (C = 34%) and the presence of inactive minerals (ACB = 7.7), and thus is supposed to have an higher residual friction angle. The clay fraction of sample 10 contains active clay mineral (ACB = 16), but this fraction represents only 24% of the soil, and this results in a higher residual friction angle than that predicted by the correlation.

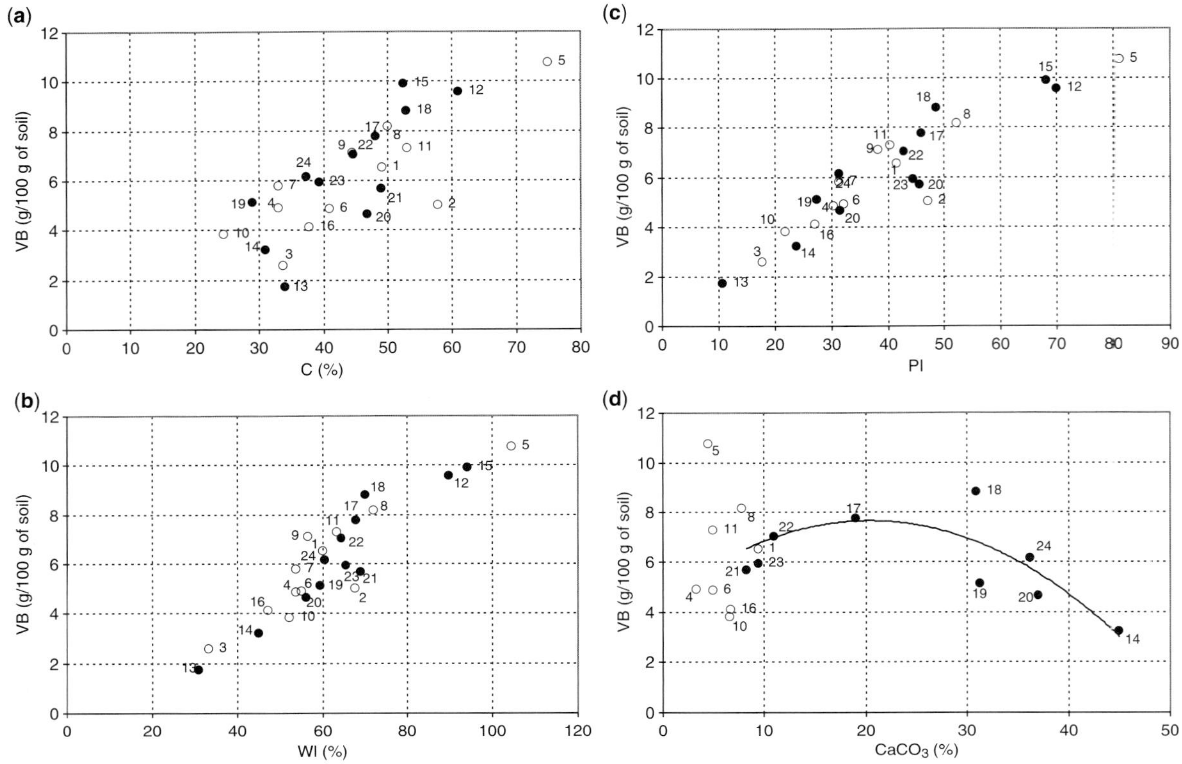

Fig. 10. Correlations between the VB and clay content (**a**), Wl (**b**), PI (**c**) and carbonate content (**d**) for alluvial soil (open circles) and colluvial soils (filled circles).

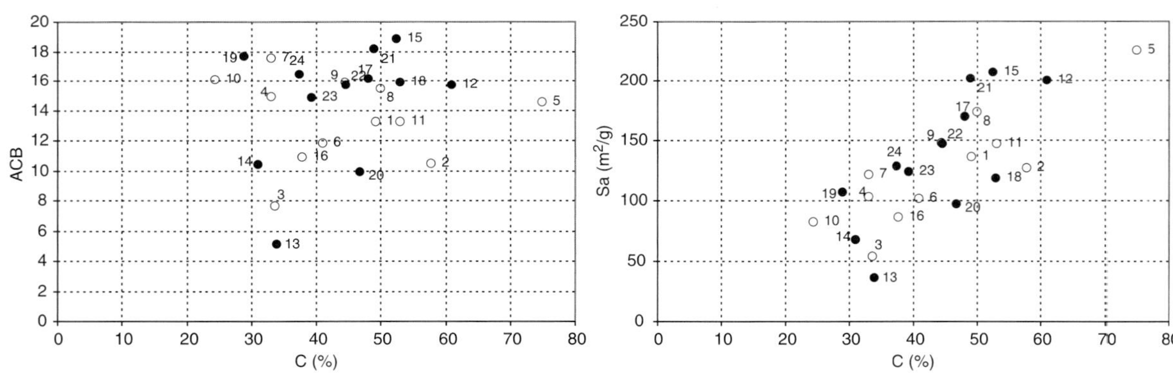

Fig. 11. Correlations between the index of activity of the clay fraction (ACB), the total active specific surface (Sa) and the clay fraction.

Discussion and conclusion

Shallow landslides affecting colluvial soils on Tertiary bedrock and weathered clays on alluvial soils are a recurrent problem in many areas of the Oltrepo Pavese, and they are mostly related to the reactivation of pre-existing landslides. Therefore knowledge of the residual shear strength is important for slope stability analysis and landslide hazard assessment.

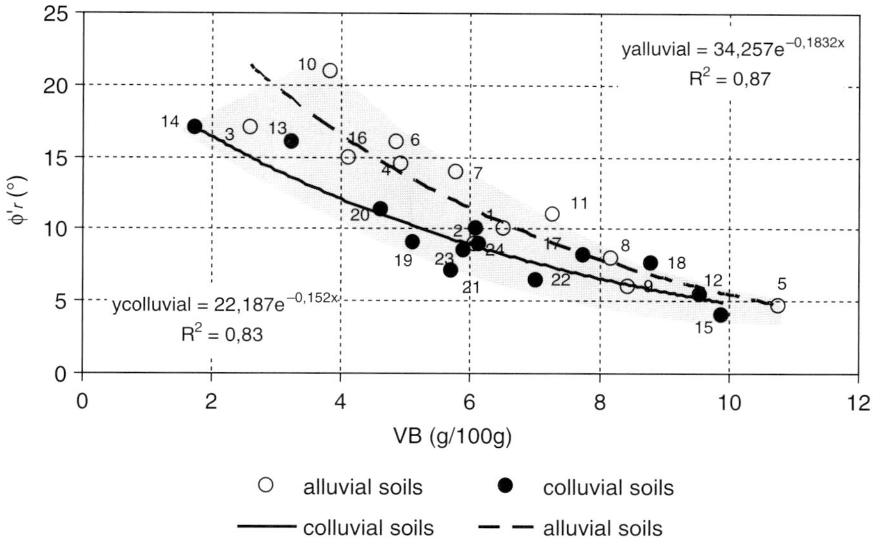

Fig. 12. Correlation between the VB and the residual shear friction angle for alluvial soil (open circles) and colluvial soils (filled circles).

Site investigation revealed that the soil profiles of colluvial soils can be divided into three zones based on the amount of weathering that has occurred: colluvium, which is composed of totally weathered bedrock and of the material derived from downslope transportation, partially weathered bedrock and bedrock. Geotechnical profiles show homogeneity of index properties and clay content. The failure surfaces are located along the contact between the weathered material and the underlying weathered bedrock. Samples 10 and 15, collected on slip surfaces, are characterized by very low residual friction angle (5.5–4°). The correlations in the literature generally overestimate the value of $\phi'r$ and in some cases the soil parameters fall outside the range proposed by earlier researchers. Good agreement has been obtained between the residual friction angle resulting from the VB–$\phi'r$ relationship and the values measured by laboratory tests. Colluvial soils of Zebedassi Limestone (samples 17–24), characterized by shallow complex landslides, show homogeneous geotechnical characteristics. $\phi'r$ ranges between 6.4 and 11.3°. For these colluvial soils the variation of the residual friction angle shows a positive correlation with carbonate content: samples with less than *c.* 25% calcite content have low residual strength, whereas those with greater than 25% calcite content have high residual strength, which increases with increasing calcite content. The colluvial soils of Antognola Marls and Monte Piano Marls (samples 13 and 14) are characterized by lower densities of shallow landslides (earth slides). For these soils $\phi'r$ has values of 16° and 17°, which are in agreement with the values determined from the correlations in the literature. However, the data scatter is high in the VB–$\phi'r$ correlation for low VB.

The alluvial deposits are formed of sand and gravel layers and are covered by clays, up to 3 m thick, derived from the weathering of the alluvial sediments. Mechanical and chemical weathering (cycles of drying and wetting) cause deterioration with time of the shear strength of the alluvial soils, which are internally sheared with shiny slickensided surfaces. The localized shear displacements have reduced the shear resistance to the residual condition. The results of the geotechnical tests vary considerably for the different soils. The residual friction angle ranges from 4.7° to 21°. The correlations in the literature are characterized by a large amount of scatter in the data and tend to overestimate or underestimate the residual friction angle. Although $\phi'r$ shows large variation in alluvial soils the correlation between $\phi'r$ and VB is good, especially for VB > 4 g per 100 g of soil.

For colluvial soils the magnitude of the residual strength decreases with increasing liquid limit and plasticity index, whereas there is not a good correlation with the clay fraction. The fact that soils with similar clay content give different residual shear strength results indicates the influence of the type of minerals present within the clay fraction. Because of their geotechnical heterogeneity alluvial soils do not show significant correlations with the index properties.

Comparison of the residual shear strength data with published correlations is not straightforward. This may be due to various factors: (1) the results may have been affected by sample preparation and testing procedures (if a correlation is to be used for estimating drained residual strength based on Atterberg limits and/or clay fraction, then these properties should be determined following the same sample preparation procedures as were used in the development of the correlation); (2) the studied soils have different geological characteristics (e.g. different clay mineralogy or different soil type) from the soils used to develop the correlations (e.g. residual soils for the method of Wesley 2003); (3) the index properties do not reflect in a systematic way the mineralogy of the clay particles, which control the residual shearing resistance. This strongly suggests that the correlations for residual strength should be used with caution; they cannot be universally applicable to all soils and appear to be valid for a limited range of soil tested. Because of the scatter of data the correlations are useful to evaluate trends but not for practical applications.

The methylene blue stain test makes it possible to characterize both alluvial and colluvial soils in an exhaustive manner by taking into account both the quantity and type of clay minerals present. For the same clay content, VB varies as a function of active clay mineral content (smectite). A correlation was found between VB and $\phi' r$ for the tested soils. The correlation is good for alluvial soils and colluvial soils derived by weathering of Zebedassi Limestones. The dye adsorption technique is an easy, cheap, rapid (it requires about 15–60 min) and efficient means of estimating the residual friction angle of a clay soil. The correlation established in this study contains less scatter than others and indicates that a measured VB will relate to a residual friction angle with an accuracy of 1–2°. The correlation is good for VB up to 4 g per 100 g of soils; for lower values the data scatter is higher ($\phi' r_{\text{predicted}} - \phi' r_{\text{measured}} = 3\text{–}4°$). Thus, the practical use of the correlation is limited for the colluvial soils derived by weathering of Monte Piano Marls and Antognola Marls, which are characterized by high residual friction angle and low VB. This confirms that the method proposed is strongly dependent on the geological and geotechnical characteristics of soils examined; it would be necessary to test its efficacy in other geological contexts.

The method cannot replace the laboratory test, but it could help in a rapid assessment of $\phi' r$ for design purposes and hazard assessment. The correlation is useful as a guide for designers when remedial works are urgent. It gives reliable qualitative indications when the input data cannot be obtained from an extensive laboratory test programme.

References

AFNOR 1998. *Mesure de la quantité et de l'activité de la fraction argileuse: determination de la valeur de bleu de méthylène d'un sol par l'essai a la tache.* Norme française homologuée, **NF P94-068**.

ASTM 1994. *Annual Book of ASTM Standards.* ASTM, Philadelphia, PA.

Beatrizzotti, G., Bellinzona, G., Beltrami, G., Boni, A., Braga, G., Marchetti, G. & Mosna, S. 1969. *Geological Map of Italy Fg. 71—Voghera*, 2nd edn. Servizio Geologico d'Italia, Rome.

Binnie, M., Clark, J. F. F. & Skempton, A. W. 1967. The effect of discontinuities in clay bedrock on the design of dams in the Mangle project. *In*: *Transactions of 9th International Congress on Large Dams*, Vol. 1, 165–183.

Blondeau, F. & Josseaume, H. 1976. Mesure de la résistance au cisaillement résiduelle en laboratoire. *Bulletin de Liaison des Laboratoires des Ponts et Chaussées*, **65**, 90–106.

Borowicka, H. 1965. The influence of the colloidal content on the shear strength of clay. *In*: *Proceedings of 6th International Conference on Soil Mechanics and Foundation Engineering*, Vol. 1, 175–178.

Calvello, M., Lasco, M., Vassallo, R. & Di Maio, C. 2005. Compressibility and residual shear strength of smectitic clays: influence of pore aqueous solutions and organic solvents. *Rivista Italiana di Geotecnica*, **1**, 34–46.

Cancelli, A. 1977. Residual shear strength and stability analysis of a landslide in fissured overconsolidated clays. *Bulletin of the International Association of Engineering Geology*, **16**, 193–197.

Chiappone, A. & Scavia, C. 1999. La prova al blu di metilene nella caratterizzazione geotecnica: un'applicazione allo studio nella stabilità dei versanti. *Rivista Italiana di Geotecnica*, **4**, 36–51.

Chiappone, A., Marello, S., Scavia, C. & Setti, M. 2004. Clay mineral characterization through the methylene blue test: comparison with other experimental techniques and applications of the method. *Canadian Geotechnical Journal*, **41**, 1168–1178.

Collotta, T., Cantoni, R., Pavesi, U., Ruberl, E. & Moretti, P. C. 1989. A correlation between residual friction angle gradation and the properties of cohesive soils. *Géotechnique*, **39**, 343–346.

Cruden, D. & Varnes, D. J. 1996. Landslide types and processes. *In*: Turner, A. K. & Schuster, R. L. (eds) *Landslides, Investigation and Mitigation.* Transportation Research Board Special Report 247. National Academy Press, WA, 36–75.

Di Maio, C. 1996. The influence of pore fluid composition on the residual shear strength of some natural clayey soils. *In*: Senneset, K. (ed.) *Proceedings of 7th International Symposium on Landslides, Trondheim*, Vol. 2. Balkema, Rotterdam, 1189–1194.

Erguler, Z. A. & Ulusay, R. 2003. A simple test and predictive models for assessing swell potential of Ankara (Turkey) Clay. *Engineering Geology*, **67**, 331–352.

Fityus, S. G., Smith, D. W. & Jennar, A. M. 2000. Surface area using methylene blue adsorption as a measure of soil expansivity. *GeoEng 2000*, **2**, 601.

Hawkins, A. B. & McDonald, C. 1993. The influence of granular calcareous particles on the residual shear strength of Fuller's earth clay. *Quarterly Journal of Engineering Geology*, **26**, 321–325.

Jamiolkowski, M. & Pasqualini, E. 1976. *Sulla scelta dei parametri geotecnici che intervengono nelle verifiche di stabilità dei pendii naturali ed artificiali.* Atti Istituto di Scienze delle Costruzioni, Politecnico di Torino, **319**.

Kanji, M. A. 1974. The relationship between drained friction angles and Atterberg limits of natural soil. *Géotechnique*, **24**, 671–674.

Kaya, A. & Kwong, J. K. P. 2007. Evaluation of common practice empirical procedures for residual friction angles of soils: Hawaiian amorphous material rich colluvial soil case study. *Engineering Geology*, **92**, 49–58.

Kenney, T. C. 1967. Influence of mineralogical composition on the residual strength of natural soils. *In*: *Proceedings of Oslo*

Geotechnical Conference Shear Strength of Natural Soils and Rocks, Oslo, Vol. 1, 123–129.

LAUTRIN, D. 1989. Utilisation pratiques des paramètres dérivés de l'essai au bleu de méthylène dans les projets de génie civil. *Bulletin de Liaison des Laboratoires des Ponts et Chaussées*, **160**, 29–41.

LEROUEIL, S. 2001. Natural slopes and cuts: movement and failure mechanisms. *Géotechnique*, **51**, 197–243.

LO PRESTI, D. & FROIO, F. 2004. Residual shear strength on soft rocks and fine-grained soils. *Rivista Italiana di Geotecnica*, **3**, 48–74.

LUPINI, J. F., SKINNER, A. E. & VAUGHAN, P. R. 1981. The drained residual strength of cohesive soil. *Géotechnique*, **31**, 181–213.

MAGNAN, J. P. & YOUSSEFIAN, G. 1989. Essai au bleu de méthylène et classification gèotechnique des sols. *Bulletin de Liaison des Laboratoires des Ponts et Chaussées*, **159**, 93–104.

MEISINA, C. 2000. Predicting swelling/shrinkage potential using the blue methylene method: some examples in Italian clayey soils. *In*: *Proceedings of International Conference on Geotechnical and Geological Engineering Geoeng2000, Melbourne, 19–24 November 2000*, 1–6.

MEISINA, C. 2006. Engineering geological units for the characterisation of weathered clayey soils responsible for shallow landslides. *Natural Hazards and Earth System Sciences*, **6**, 1–14.

MESRI, G. & CEPEDA-DIAZ, A. F. 1986. Residual strength of clays and shales. *Géotechnique*, **36**, 269–274.

MESRI, G. & SHAHIEN, M. 2003. Residual strength mobilized in first time slope failures. *Journal of Geotechnical and Geoenvironmental Engineering*, **129**, 12–31.

ROUIGUIA, A. 1996. L'influence de quelques facteurs sur la resistance residuelle des argiles. *Revue Française de Géotechnique*, **74**, 55–69.

SCHAEFFNER, M. 1989. Introduction de la valeur de bleu de méthylène d'un sol dans la classification des sols de la recommandation pour les terrassements routiers. *Bulletin de Liaison des Laboratoires des Ponts et Chaussées*, **163**, 9–16.

SEYCEK, J. 1978. Residual shear strength of soils. *Bulletin of Engineering Geology and the Environment*, **17**, 73–75.

SKEMPTON, A. W. 1964. Long-term stability of clay slopes. *Géotechnique*, **14**, 77–101.

SKEMPTON, A. W. 1985. Residual strength of clays in landslides, folded strata and the laboratory. *Géotechnique*, **35**, 3–18.

STARK, T. D. & EID, H. T. 1994. Drained residual shear strength of cohesive soils. *Journal of Geotechnical Engineering, ASCE*, **120**, 856–871.

TOURENQ, C. & TRAN NGOC LAN 1989. Mise en évidence des argiles par l'essai au bleu de méthylène. application aux sols, roches et granulats. *Bulletin de Liaison des Laboratoires des Ponts et Chaussées*, **159**, 79–92.

VOIGHT, B. 1973. Correlation between Atterberg plasticity limits and residual shear strength of natural soil. *Géotechnique*, **27**, 125–136.

WAN, Y. & KWONG, K. 2002. Shear strength of soils containing amorphous clay-size materials in a slow-moving landslide. *Engineering Geology*, **65**, 293–303.

WESLEY, L. D. 2003. Residual strength of clays and correlations using Atterberg limits. *Géotechnique*, **53**, 669–672.

YIN, J.-H. 2002. Stress–strain–strength characteristics of a marine soil with different clay contents. *ASTM Geotechnical Testing Journal*, **25**, 459–462.

Index

Page numbers in *italic* denote figures. Page numbers in **bold** denote tables.